The Writing Teacher's Sourcebook

The Writing Teacher's Sourcebook

Third Edition

GARY TATE

EDWARD P. J. CORBETT

NANCY MYERS

New York Oxford
OXFORD UNIVERSITY PRESS
1994

Oxford University Press

Oxford New York Toronto
Delhi Bombay Calcutta Madras Karachi
Kuala Lumpur Singapore Hong Kong Tokyo
Nairobi Dar es Salaam Cape Town
Melbourne Auckland Madrid

and associated companies in

Berlin Ibadan

Published by Oxford University Press, Inc.,
200 Madison Avenue, New York, New York 10016

Library of Congress Cataloging-in-Publication Data
The writing teacher's sourcebook
[edited by] Gary Tate, Edward P. J. Corbett,
Nancy Myers.—3rd ed.
p. cm. Includes bibliographical references.
ISBN 0-19-508306-7
1. English language—Rhetoric—Study and teaching.
I. Tate, Gary.
II. Corbett, Edward P. J.
III. Myers, Nancy.
PE1404.W74 1994
808'.042'0711—dc20 93-37865

1 3 5 7 9 8 6 4 2

Printed in the United States of America
on acid-free paper

Preface

In the Preface of the first edition (1981) and the second edition (1988) of *The Writing Teacher's Sourcebook,* Gary Tate and Edward P. J. Corbett remarked that whereas they had to search long and hard to find enough good articles to fill their earlier anthologies, *Teaching Freshman Composition* (1967) and *Teaching High School Composition* (1970), their new problem was making a selection from the astounding abundance of good articles now available to them. They still face the problem of making a judicious selection from the plethora of excellent articles available to them, but now they face a new problem. They had decided that for the third edition of *The Writing Teacher's Sourcebook* they would concentrate on pedagogical articles—that is, on articles that dealt primarily with the *teaching* of written composition. Their new problem, as they soon discovered, was that there was a scarcity of pedagogical articles.

As a result of the enhanced professionalism of writing teachers that has taken place in the last twenty years or so and the remarkable growth in the number of graduate courses in rhetoric and composition, most of the articles being published in the many new composition journals that had been established dealt with the history and the theory of rhetoric and composition. And the new collections of essays, such as Richard L. Graves, ed., *Rhetoric and Composition: A Sourcebook for Teachers and Writers,* 3rd ed. (1990), and Erika Lindeman and Gary Tate, eds., *An Introduction to Composition Studies* (1991), also featured historical and theoretical articles. There definitely was a need for a collection of articles that emphasized pedagogy.

So Tate and Corbett engaged a third editor for the new edition, Nancy Myers, a young teacher who could provide them with a perspective that, as antiquated teachers, they did not command. As a relatively new teacher of composition, Nancy Myers was well aware of the anxieties of inexperienced teachers who were concerned about what they should be doing in their writing classes next Monday morning. What the three of us discovered when we went searching for essays that could help young teachers decide what they could do in next Monday's classes was that there were very few *purely* pedagogical articles available. When we did find a purely pedagogical article that we judged to be genuinely useful to writing teachers, we latched on to it, but in order to fill out our anthology with useful articles, we had to modify our original notion of what constituted a helpful pedagogical article. We decided that even if an article dealt predominantly with the history or theory of rhetoric or composition, that article could be useful if it suggested, either explicitly or implicitly, some fruitful practice or approach that writing teachers could adopt in their classrooms.

With that modified criterion, we were able to retain fifteen of the thirty-four essays (40.5%) that appeared in the second edition of *The Writing Teacher's Sourcebook,* but we were also able to find an additional twenty-two pedagogically helpful essays to fill out our roster of selections. Many of these essays were written by some of our most esteemed veteran teachers of writing. Six of the articles in this edition were written by teachers who had won the prestigious Richard Braddock Award for the best article in composition published in *College Composition and Communication* in a particular year, and at least three other articles were written by teachers who won awards offered by other journals or organizations. But several of the articles in this third edition were written by teachers who are relatively new to the profession. We are very happy about this latter fact.

Whereas the thirty-four essays of the second edition were arranged in ten sections, the thirty-seven essays of this edition are arranged in nine sections. Six of these sections in the third edition bear headings only slightly different in wording from the headings in the second edition. For instance:

Second edition	*Third edition*
Perspectives	Some Overviews
Composing Process	Composing and Revising
Teaching	Teachers
Audience	Audiences

But there are three new sections in this edition: Classrooms, Assigning and Responding, and Computers.

For the second edition, we decided that instead of running the risk of overwhelming the users of the book with an extensive list of additional articles on rhetoric and composition, we would list no more than six pertinent articles at the end of each of the ten sections of the book, and at the back of the book, we would provide An Annotated List of Some Important Books on Writing or the Teaching of Writing. Well, we were soon informed by teachers of graduate courses in the teaching of composition and by some student users of the book that students needed to be overwhelmed with extensive lists of additional books and articles on rhetoric and composition. Many teachers asked their students to select a particular facet of rhetoric or composition, to do some in-depth research on that facet, and then write a paper as the major project of the seminar. Neophyte teachers desperately needed bibliographic resources for such projects. So for this edition, we have provided at the back of the book a section labeled Additional Sources, which is subdivided into three sections: Journals—a list of thirteen journals that publish articles of interest to college writing teachers; Suggested Readings—a list of more than thirty books that focus, as do the articles in the body of this book, on pedagogical theory and practice; and Bibliographic Resources—divided into two subsections: Serial Bibliographies and Specialized Bibliographies. Students may be overwhelmed by this mass of resources, but they certainly will gets lots of help from these listings.

And we hope that teachers of writing will get lots of help also from the limited number of articles that we have been able to bring together under one cover. Veteran teachers of writing are likely to mutter, "I wish a book like this had been available when I started my teaching career." We can only mutter in response, "Better late than never."

Fort Worth G. T.
Columbus E. P. J. C.
September 1993 N. M.

Contents

PERSPECTIVES

Richard Fulkerson. Four Philosophies of Composition 3

James A. Berlin. Contemporary Composition: The Major Pedagogical Theories 9

Maxine Hairston. Diversity, Ideology, and Teaching Writing 22

Donald Lazere. Teaching the Political Conflicts: A Rhetorical Schema 35

TEACHERS

Rosemary Deen. Notes to Stella 55

Peter Elbow. Embracing Contraries in the Teaching Process 65

Lad Tobin. Reading Students, Reading Ourselves: Revising the Teacher's Role in the Writing Class 77

Susan C. Jarratt. Teaching Across and Within Differences 92

Donald M. Murray. The Listening Eye: Reflections on the Writing Conference 96

CLASSROOMS

Terry Dean. Multicultural Classrooms, Monocultural Teachers 105

Rae Rosenthal. Male and Female Discourse: A Bilingual Approach to English 101 119

Harvey S. Wiener. Collaborative Learning in the Classroom: A Guide to Evaluation 132

Hephzibah Roskelly. The Risky Business of Group Work 141

COMPOSING AND REVISING

Sondra Perl. Understanding Composing 149

Nancy Sommers. Between the Drafts 155

James A. Reither. Writing and Knowing: Toward Redefining the Writing Process 162

ASSIGNING AND RESPONDING

Christopher C. Burnham. Crumbling Metaphors: Integrating Heart and Brain Through Structured Journals 173

Richard L. Larson. The "Research Paper" in the Writing Course: A Non-Form of Writing 180

Jeanne Fahnestock and Marie Secor. Teaching Argument: A Theory of Types 186

Catherine E. Lamb. Beyond Argument in Feminist Composition 195

Brooke K. Horvath. The Components of Written Response: A Practical Synthesis of Current Views 207

Jerry Farber. Learning How to Teach: A Progress Report 224

AUDIENCES

Douglas B. Park. The Meanings of "Audience" 233

Lisa Ede and Andrea Lunsford. Audience Addressed/Audience Invoked: The Role of Audience in Composition Theory and Pedagogy 243

Peter Elbow. Closing My Eyes as I Speak: An Argument for Ignoring Audience 258

STYLES

Robert J. Connors. Static Abstractions and Composition 279

Winston Weathers. Teaching Style: A Possible Anatomy 294

Elizabeth D. Rankin. Revitalizing Style: Toward a New Theory and Pedagogy 300

Richard Ohmann. Use Definite, Specific, Concrete Language 310

BASIC WRITING

Mina P. Shaughnessy. Diving In: An Introduction to Basic Writing 321

Min-zhan Lu. Redefining the Legacy of Mina Shaughnessy: A Critique of the Politics of Linguistic Innocence 327

David Bartholomae. The Study of Error 338

Mike Rose. Remedial Writing Courses: A Critique and a Proposal 353

Elza C. Tiner. Elements of Classical and Medieval Rhetoric in the Teaching of Basic Composition 371

COMPUTERS

Gail E. Hawisher and Cynthia L. Selfe. The Rhetoric of Technology and the Electronic Writing Class 381

Dawn Rodrigues and Raymond Rodriques. How Word Processing Is Changing Our Teaching: New Technologies, New Approaches, New Challenges 391

Wayne M. Butler and James L. Kinneavy. The Electronic Discourse Community: god, Meet Donald Duck 400

ADDITIONAL SOURCES

Journals 415

Suggested Readings 415

Bibliographic Resources 417

PERSPECTIVES

Richard Fulkerson

Four Philosophies of Composition

My research interest in philosophies of composition and their curricular and pedagogical implications had two immediate causes.[1] The first was my reading of Charles Silberman's *Crisis in the Classroom* (1970). Although the book is primarily about elementary and high schools and although I am not persuaded of the soundness of Silberman's proposals, I found the book valuable in two ways. First, it highlighted the existence of serious problems in American education. Second and more important, Silberman said the problems were the result more of mindlessness than of maliciousness. The problem was not, he said, that evil or incompetent people were in charge but that educators exhibited a consistent mindlessness about relating means to desired ends. The second precipitating event was my rereading of M. H. Abrams's *The Mirror and the Lamp* (1953). Abrams analyzes four overriding theories of literature and literary criticism, each emphasizing one of the four elements in an artistic transaction.[2] Since the elements in an artistic transaction are the same as those in any communication, it seemed that Abrams's four theories might also be relevant to composition.

Any theory making the reader primary and judging literature by its effect, Abrams labels *pragmatic*. When the universe shared by artist and auditor becomes the primary element and measure of success, then, Abrams says, we have a *mimetic* theory, such as that of Pope and the Neo-Classical period. Emphasis on the personal views of the artist, such as in the Romantic period, Abrams labels the *expressive* position. And finally, theories emphasizing only the internal relationships within the artifact, Abrams calls *objective* criticism.

Abrams's analysis made me wonder whether a parallel set of four philosophies of composition might exist, each one stressing a different element in the communicative transaction. If so, each would provide—as do the philosophies Abrams outlined—both a description of the composition process and a method of evaluating the composed product. Furthermore, the existence of four such philosophies might help to explain both the widely recognized variations in English teachers' evaluations and (perhaps) what Kitzhaber in *Themes, Theories, and Therapy* referred to as the "bewildering variety" of freshman composition courses.

For application to composition, I prefer to make two shifts in Abrams's terminology.

From *College Composition and Communication* 30 (December 1979): 343-48.

1. The research upon which this article is based was supported by a grant from the East Texas State University Office of Organized Research in the summer of 1976. The author wishes to thank Professor H. M. Lafferty, Chairman, and the Committee on Organized Research for their support.

2. M. H. Abrams, *The Mirror and the Lamp: Romantic Theory and the Critical Tradition* (New York: Norton, 1953), pp. 2-29.

I will keep the term *expressive* for philosophies of composition emphasizing the writer, and the term *mimetic* for philosophies emphasizing correspondence with "reality." But philosophies emphasizing the effect on a reader I will call *rhetorical,* and philosophies emphasizing traits internal to the work I will call *formalist.*

My thesis is that this four-part perspective helps give a coherent view of what goes on in composition classes. All four philosophies exist in practice. They give rise to vastly different ways of judging student writing, vastly different courses to lead students to produce such writing, vastly different textbooks and journal articles. Moreover, the perspective helps to clarify, though not to resolve, a number of the major controversies in the field, including the "back-to-the-basics" cry and the propriety of dialectal variations in student writing.

Let me clarify each of these four philosophies and simultaneously attempt to classify a number of major theorists according to the value philosophies implicit in their writings.

Adherents of formalist theories judge student work primarily by whether it shows certain internal forms. Some teachers, for example, judge a paper a failure if it contains one comma splice or five spelling errors. Those are judgments based purely on form. Indeed, the most common type of formalist value theory is a grammatical one: good writing is "correct" writing at the sentence level. In the classroom, one studies errors of form—in order to avoid them. But forms other than grammatical can also be the teacher's key values. I have heard of metaphorical formalists, sentence-length formalists, and topic-sentence formalists, to name a few. And Janet Emig in *The Composing Processes of Twelfth Graders* (1971) concluded that "most of the criteria by which students' school-sponsored writing is evaluated concern the accidents rather than the essences of discourse—that is, spelling, punctuation, penmanship, and length" (p. 93), four formalist criteria.

Few major writers accept formalist values, but two seem to me to do so. Francis Christensen elevated the form he designated as the cumulative sentence to primacy in value. He wanted "sentence acrobats." Similarly in his provocative and provoking *The Philosophy of Composition* (1977), E. D. Hirsch builds an elaborate argument for what he calls "relative readability." Such a phrase has its origins in a reader-centered value theory, but Hirsch elevates such sentences to goods in themselves rather than goods because they communicate to the reader. He says that "relative readability is an intrinsic and truly universal norm of writing" (p. 89) and that "there are ways of making quite objective comparisons between passages . . . on the criterion of communicative efficiency" (p. 61). In other words, we can take a pair of passages and determine which is the better embodiment of "semantic intent"—without reference to a reader, or to the writer using them, or to the reality they reflect. This is a formalist philosophy of composition. And one assumes that Hirsch is designing a program to teach students at the University of Virginia how to produce relatively readable sentences. One's value theory shapes his or her pedagogy.

Although formalists are hard to find these days, adherents of the other three positions and of courses built around them are not.

Expressionism as a philosophy about what writing is good for and what makes for good writing became quite common in the late sixties and early seventies, perhaps gaining its chief emphasis with the famous Dartmouth Conference in 1967. Expressionists cover a wide range, from totally accepting and non-directive teachers, some of whom insist that one neither can nor should evaluate writing, to much more directive, experiential teachers who design classroom activities to maximize student self-discovery. The names most commonly associated with the expressive value-position are John Dixon (in England), Ken Macrorie, and Lou Kelly—the latter two of whom have written textbooks about their courses.[3] There were, in fact, quite a few expressive textbooks, including James Miller and Stephen Judy's *Writing in Reality* (1978) and a fascinating book by Dick Friedrich and David Kuester called *It's Mine and I'll Write It That Way* (1972). Half of this book consists of suggestions for classroom activities, while the alternating chapters are a journal kept by one of the authors about a class he taught using those materials. Expressivists value writing that is about personal subjects, and such journal-keeping is an absolute essential. Another keynote for expressivists is the desire to have writing contain an interesting, credible, honest, and personal voice; hence the title of Donald C. Stewart's text, *The Authentic Voice* (1972).

Expressive views even show up in some surprising places. Ross Winterowd remarks early in *The Contemporary Writer* (1975) that "most people in the real world outside of school do not need to write very much" (p. 4). Consequently, he tells the student reader, "There's a very good chance that learning to do self-expressive writing will constitute the greatest benefit that you gain from *The Contemporary Writer*" (p. 8). Using Jung's comments on the function of dreams, Winterowd says that the purpose of such writing is "to restore our psychological balance" by reestablishing "the total psychic equilibrium" (p. 8). Taking "psychic equilibrium" as the major goal of writing leads to quite a different evaluative position from taking, say, "changes in audience opinion" as the prime goal.

The most common presentation of the third philosophy of composition, the mimetic, says that a clear connection exists between good writing and good thinking. The major problem with student writing is that it is not solidly thought out. Hence, we should either teach students how to think or help them learn enough about various topics to have something worth saying, or we should do both. Thus the first mimetic approach emphasizes logic and reasoning, sometimes formal logic as in Monroe Beardsley's *Writing with Reason* (1976), sometimes less formal logic as in Ray Kytle's *Clear Thinking for Composition* (2nd ed. 1973). Sometimes the mimetic-logical emphasis is on propaganda analysis—the detecting of hidden assumptions, emotional appeals, and fallacies in reasoning. All discourse does contain unstated assumptions; the problem with such assumptions arises only if they are also unacceptable assumptions, and usually, they are unacceptable because they violate reality as we know it. That is, an unstated or hidden assumption may be a writing weakness if writing is viewed from a mimetic perspective. Similarly, a fallacy is fallacious precisely because it contradicts what we accept as truth. Hastiness is not the problem with a hasty generalization; the problem is that reality as we know it rarely squares with such generalizations. Thus the teaching of sound reasoning as a basis for good writing is an essentially mimetic practice.

The other major mimetic approach says that students do not write well on significant matters because they do not know enough. One resulting methodology is to emphasize research during the prewriting stages; another is to emphasize heuristic systems. Still

---

3. Ken Macrorie, *Telling Writing* (New York: Hayden, 1970), and Lou Kelly, *From Dialogue to Discourse: An Open Approach* (Glenview, IL: Scott, Foresman, 1972).

another is the use of a topically arranged anthology of readings. If a student reads four essays taking both sides on a controversial issue—say, capital punishment—then he or she supposedly will know enough to be able to write about that topic; that is, the writing will be closer to the "real situation" and thus better—from a mimetic perspective.

In almost any issue of *College Composition and Communication,* several writers espouse the fourth philosophy, the rhetorical one. Such a philosophy says, in essence, good writing is writing adapted to achieve the desired effect on the desired audience. If the same verbal construct is directed to a different audience, then it may have to be evaluated differently.

Leading adherents of this view are E. P. J. Corbett, Richard Larson, and most other theorists who emphasize an adaptation of classical rhetoric. This group also includes the textbook writers who emphasize a writer's commitment to his or her reader as shaping discourse: Robert Gorrell and Charlton Laird in the *Modern English Handbook* (6th ed. 1976), Michael Adelstein and Jean Pival in *The Writing Commitment* (1976), and James McCrimmon in *Writing with a Purpose* (6th ed. 1976).

Given this four-part perspective, I was much intrigued by one of Peter Elbow's articles.[4] I had already read his *Writing without Teachers* (1973) and had had some trouble classifying him. But in this article, Elbow explained that his theories of free writing, collaborative criticism, and audience adaptation are really classical theories masquerading as modern theories. That is, he said that although most teachers judge student writing either on the basis of its truth or its formal correctness, his courses are built on judging student writing by its effect on an audience. Aristotle in modern dress.

Classifying Elbow as a rhetoricist illustrates an important point. As I said, teaching procedures have to harmonize with evaluative theories. More precisely, one's philosophy about what writing is for leads to a theory of what constitutes good writing. That philosophy, in turn, leads to a concept of pedagogical goals, and the goals lead, in turn, to classroom procedures. But the relationship between goals and procedures is complex, because handled in certain ways, procedures that might usually be associated with one value position can be used to reach quite another end. Elbow's techniques *seem* to put him with the expressivists, but at least in 1968, Elbow saw himself, accurately I think, as a rhetoricist.

Similarly, I mentioned previously the expressivist penchant for student journals, but I also use journals, and I use them most often in the service of a quite different goal: critical thinking about reading assignments. To the extent that I use journals to teach writing, I am using them in the service of a mimetic set of values.

A reasonable response to this model might be, "Yes, those are four positions; they exist, but they are not mutually exclusive. Nor are they a problem. One can either hold to all of them simultaneously or pick and choose among them, depending on which one seems appropriate for a given piece of writing." James Kinneavy's very impressive *A Theory of Discourse* (1971), in fact, proposes that there are four types of writing growing from the four elements in a communicative act: reference discourse, expressive discourse, persuasive discourse, and literary discourse, each to be judged on its own terms. Moreover, the National Assessment of Educational Progress uses three types of writing growing out of three of the communicative elements: expressive (self-centered), expository (world-centered), and persuasive (reader-centered).

4. Peter Elbow, "A Method for Teaching Writing," *College English,* 30 (Nov., 1968), 123ff.

That a separate type of writing arises from an emphasis on each communicative element is certainly an attractive position, one that might provide an elegant basis for designing comprehensive composition courses. But such a view leads to both theoretical and practical problems, not the least of which is that it gives us no direction in selecting which writing types merit greater emphasis in our courses. Furthermore, when used as an approach to evaluating writing, such a classification runs into all the problems inherent in determining intent in a text. Most teachers have seen student writing that was impossible to classify as one of Kinneavy's four types of discourse and that would be evaluated quite differently depending upon which of the four philosophies one applied.

My research has convinced me that in many cases composition teachers either fail to have a consistent value theory or fail to let that philosophy shape pedagogy. In Silberman's terms, they are guilty of mindlessness. A fairly common writing assignment, for example, directs the student to "state and explain clearly your opinion about X" (censorship, abortion, the Dallas Cowboys). There is nothing wrong with such an assignment. But if a student does state his or her opinion and if the opinion happens to be based on gross ignorance or to contain major contradictions, the teacher must, to be consistent, ignore such matters. The topic as stated asks for opinion; it does not ask for good opinion, judged by whatever philosophy. In short, the assignment implies an expressive value-theory. It does not say, "Express your opinion to persuade a reader" (which would imply a rhetorical theory), or "Express your opinion so that it makes sense" (which would imply a mimetic theory), or even "Express your opinion correctly" (implying a formalist theory). To give the bald assignment and then judge it from any of the perspectives not implied is to be guilty of value-mode confusion.

Modal confusion is not easy to locate, since one almost has to be inside a classroom to see it. A few instances, however, have been reported in the literature. Walker Gibson has told the story of visiting a high school class in which the teacher was emphasizing the idea of a *persona* within a piece of literature. She then shifted to composition and had a boy read aloud a paper he had written. She next asked the class, "Now what's the trouble here? What voice did you hear in that paper? Is it Jimmy's voice? Is that Jimmy speaking or is it some artificial, insincere voice?" To this, the student responded, "I don't *have* to sound like *me*." Gibson notes that he agreed with the student; the teacher had failed to relate her appreciation of the author-speaker distinction in literature to the student's writing.[5] For me, she had shown in her evaluation that she was at least momentarily committed to an expressive philosophy: "Good writing is sincere writing; it sounds like the real author." There is nothing wrong with an expressive philosophy, but there is something seriously wrong with classroom methodology which implies one variety of value judgment when another will actually be employed. That is modal confusion, mindlessness.

The worst instance of modal confusion I have come across was reported by Lawrence Langer in a *Chronicle of Higher Education* article entitled "The Human Use of Language: Insensitive Ears Can't Hear Honest Prose" (January 24, 1977). He tells the story of a forty-year-old student who in childhood had been in a Nazi concentration camp in which her parents had been killed. She had never been able to talk about the experience except to other former inmates, not even to her husband and children. Her first assignment in freshman composition was to write a paper on something that was of great

5. Quoted in McCrimmon, *Writing with a Purpose*, 6th ed., p. 181.

importance to her. She resolved to handle her childhood trauma on paper in an essay called "People I Have Forgotten." Langer quotes the entire paper of eight paragraphs and about three hundred words, calling it "not a confrontation, only a prelude." It is a moving and painful piece with a one-sentence opening paragraph, "Can you forget your own Father and Mother? If so—how or why?" The paper was returned with a large D-minus on the last page, emphatically circled. The only comment was "Your theme is not clear—you should have developed your first paragraph. You talk around your subject."

From the perspective of my four-part model, there was a conflict of evaluative modes at work here. The assignment seemed to call for writing that would be judged expressively, but the teacher's brief comment was *not* written from an expressivist point of view. It may imply a formalist perspective (good writing requires directness and development). Or it may rest on an unstated rhetorical perspective (for a reader's benefit this paper needs more directness and development). It is scarcely adequate in either case, and in either case, this sort of judgment was not what the student had been led to expect. There was once more a mindless failure to relate the outcome valued to the means adopted. My hope is that the four-part paradigm I have adopted from Abrams may reduce such mindlessness in the future.

James A. Berlin

Contemporary Composition: The Major Pedagogical Theories

A number of articles attempting to make sense of the various approaches to teaching composition have recently appeared. While all are worth considering, some promote a common assumption that I am convinced is erroneous.[1] Since all pedagogical approaches, it is argued, share a concern for the elements of the composing process—that is, for writer, reality, reader, and language—their only area of disagreement must involve the element or elements that ought to be given the most attention. From this point of view, the composing process is always and everywhere the same because writer, reality, reader, and language are always and everywhere the same. Differences in teaching theories, then, are mere cavils about which of these features to emphasize in the classroom.

I would like to say at the start that I have no quarrel with the elements that these investigators isolate as forming the composing process, and I plan to use them myself. While it is established practice today to speak of the composing process as a recursive activity involving prewriting, writing, and rewriting, it is not difficult to see the writer-reality-audience-language relationship as underlying, at a deeper structural level, each of these three stages. In fact, as I will later show, this deeper structure determines the shape that instruction in prewriting, writing, and rewriting assumes—or does not assume, as is sometimes the case.

I do, however, strongly disagree with the contention that the differences in approaches to teaching writing can be explained by attending to the degree of emphasis given to universally defined elements of a universally defined composing process. The differences in these teaching approaches should instead be located in diverging definitions of the composing process itself—that is, in the way the elements that make up the process—writer, reality, audience, and language—are envisioned. Pedagogical theories in writing courses are grounded in rhetorical theories, and rhetorical theories do not differ in the simple undue emphasis of writer or audience or reality or language or some combination of these. Rhetorical theories differ from each other in the way writer, reality, audience, and language are conceived—both as separate units and in the way the units relate to each other. In the case of distinct pedagogical approaches, these four

From *College English* 44 (December 1982): 765-77.

1. I have in mind Richard Fulkerson, "Four Philosophies of Composition," *College Composition and Communication*, 30 (1979), 343-48; David V. Harrington, et. al., "A Critical Survey of Resources for Teaching Rhetorical Invention," *College English*, 40 (1979), 641-61; William F. Woods, "Composition Textbooks and Pedagogical Theory 1960-80," *CE*, 43 (1981), 393-409.

elements are likewise defined and related so as to describe a different composing process, which is to say a different world with different rules about what can be known, how it can be known, and how it can be communicated. To teach writing is to argue for a version of reality, and the best way of knowing and communicating it—to deal, as Paul Kameen has pointed out, in the metarhetorical realm of epistemology and linguistics.[2] And all composition teachers are ineluctably operating in this realm, whether or not they consciously choose to do so.

Considering pedagogical theories along these lines has led me to see groupings sometimes similar, sometimes at variance, with the schemes of others. The terms chosen for these categories are intended to prevent confusion and to be self-explanatory. The four dominant groups I will discuss are the Neo-Aristotelians or Classicists, the Positivists or Current-Traditionalists, the Neo-Platonists or Expressionists, and the New Rhetoricians. As I have said, I will be concerned in each case with the way that writer, reality, audience, and language have been defined and related so as to form a distinct world construct with distinct rules for discovering and communicating knowledge. I will then show how this epistemic complex makes for specific directives about invention, arrangement, and style (or prewriting, writing, and rewriting). Finally, as the names for the groups suggest, I will briefly trace the historical precedents of each, pointing to their roots in order to better understand their modern manifestations.

My reasons for presenting this analysis are not altogether disinterested. I am convinced that the pedagogical approach of the New Rhetoricians is the most intelligent and most practical alternative available, serving in every way the best interests of our students. I am also concerned, however, that writing teachers become more aware of the full significance of their pedagogical strategies. Not doing so can have disastrous consequences, ranging from momentarily confusing students to sending them away with faulty and even harmful information. The dismay students display about writing is, I am convinced, at least occasionally the result of teachers unconsciously offering contradictory advice about composing—guidance grounded in assumptions that simply do not square with each other. More important, as I have already indicated and as I plan to explain in detail later on, in teaching writing we are tacitly teaching a version of reality and the student's place and mode of operation in it. Yet many teachers (and I suspect most) look upon their vocations as the imparting of a largely mechanical skill, important only because it serves students in getting them through school and in advancing them in their professions. This essay will argue that writing teachers are perforce given a responsibility that far exceeds this merely instrumental task.[3]

2. "Rewording the Rhetoric of Composition," *PRE/TEXT,* 1 (1980), 39. I am indebted to Professor Kameen's classification of pedagogical theories for the suggestiveness of his method; my conclusions, however, are substantially different.

3. There is still another reason for pursuing the method I recommend, one that explains why rhetorical principles are now at the center of discussions in so many different disciplines. When taken together, writer, reality, audience, and language identify an epistemic field—the basic conditions that determine what knowledge will be knowable, what not knowable, and how the knowable will be communicated. This epistemic field is the point of departure for numerous studies, although the language used to describe it varies from thinker to thinker. Examples are readily available. In *Science and the Modern World* (New York: Macmillan, 1926). A. N. Whitehead sees this field as a product of the "fundamental assumptions which adherents of all variant systems within the epoch unconsciously presuppose" (p. 71). Susanne Langer, in *Philosophy in a New Key* (Cambridge, Mass.: Harvard University Press, 1979), calls it the "tacit, fundamental way of seeing things" (p. 6). Michael Polanyi uses the terms "tacit knowledge" in *Personal Knowledge* (Chicago: Univer-

I begin with revivals of Aristotelian rhetoric not because they are a dominant force today—far from it. My main purpose in starting with them is to show that many who say that they are followers of Aristotle are in truth opposed to his system in every sense. There is also the consideration that Aristotle has provided the technical language most often used in discussing rhetoric—so much so that it is all but impossible to talk intelligently about the subject without knowing him.

In the Aristotelian scheme of things, the material world exists independently of the observer and is knowable through sense impressions. Since sense impressions in themselves reveal nothing, however, to arrive at true knowledge it is necessary for the mind to perform an operation upon sense data. This operation is a function of reason and amounts to the appropriate use of syllogistic reasoning, the system of logic that Aristotle himself developed and refined. Providing the method for analyzing the material of any discipline, this logic offers, as Marjorie Grene explains, "a set of general rules for scientists (as Aristotle understood science) working each in his appropriate material. The rules are rules of validity, not psychological rules" (*A Portrait of Aristotle* [London: Faber and Faber, 1963], p. 69). Truth exists in conformance with the rules of logic, and logic is so thoroughly deductive that even induction is regarded as an imperfect form of the syllogism. The strictures imposed by logic, moreover, naturally arise out of the very structure of the mind and of the universe. In other words, there is a happy correspondence between the mind and the universe, so that, to cite Grene once again, "As the world is, finally, so is the mind that knows it" (p. 234).

Reality for Aristotle can thus be known and communicated, with language serving as the unproblematic medium of discourse. There is an uncomplicated correspondence between the sign and the thing, and—once again emphasizing the rational—the process whereby sign and thing are united is considered a mental act: words are not a part of the external world, but both word and thing are a part of thought.[4]

Rhetoric is of course central to Aristotle's system. Like dialectic—the method of discovering and communicating truth in learned discourse—rhetoric deals with the realm of the probable, with truth as discovered in the areas of law, politics, and what might be called public virtue. Unlike scientific discoveries, truth in these realms can never be stated with absolute certainty. Still, approximations to truth are possible. The business of rhetoric then is to enable the speaker—Aristotle's rhetoric is preeminently oral—to find the means necessary to persuade the audience of the truth. Thus rhetoric is primarily concerned with the provision of inventional devices whereby the speaker may discover his or her argument, with these devices naturally falling into three categories: the ratio-

---

sity of Chicago Press, 1962). Michel Foucault, in *The Order of Things* (1971; rpt. New York: Vintage Books, 1973), speaks of the "episteme," and Thomas Kuhn, in *Structure of Scientific Revolutions* (Chicago: University of Chicago Press, 1970), discusses at length the "paradigm" that underlies a scientific discipline. The historian Hayden White, in *Metahistory: The Historical Imagination in Nineteenth-Century Europe* (Baltimore: Johns Hopkins University Press, 1973), has translated the elements of the composing process into terms appropriate to the writing of history, seeing the historical field as being made up of the historian, the historical record, the historical accounts, and an audience. One compelling reason for studying composition theory is that it so readily reveals its epistemic field, thus indicating, for example, a great deal about the way a particular historical period defines itself—a fact convincingly demonstrated in Murray Cohen's *Sensible Words: Linguistic Practice in England 1640-1785* (Baltimore: Johns Hopkins University Press, 1977), a detailed study of English grammars.

4. See Gerald L. Bruns, *Modern Poetry and the Idea of Language* (New Haven, Ct.: Yale University Press, 1974), p. 34.

nal, the emotional, and the ethical. Since truth is rational, the first is paramount and is derived from the rules of logic, albeit applied in the relaxed form of the enthymeme and example. Realizing that individuals are not always ruled by reason, however, Aristotle provides advice on appealing to the emotions of the audience and on presenting one's own character in the most favorable light, each considered with special regard for the audience and the occasion of the speech.

Aristotle's emphasis on invention leads to the neglect of commentary on arrangement and style. The treatment of arrangement is at best sketchy, but it does display Aristotle's reliance on the logical in its commitment to rational development. The section on style is more extensive and deserves special mention because it highlights Aristotle's rationalistic view of language, a view no longer considered defensible. As R. H. Robins explains:

> The word for Aristotle is thus the minimal meaningful unit. He further distinguishes the meaning of a word as an isolate from the meaning of a sentence; a word by itself "stands for" or "indicates" . . . something, but a sentence affirms or denies a predicate of its subject, or says that its subject exists or does not exist. One cannot now defend this doctrine of meaning. It is based on the formal logic that Aristotle codified and, we might say, sterilized for generations. The notion that words have meaning just by standing for or indicating something, whether in the world at large or in the human mind (both views are stated or suggested by Aristotle), leads to difficulties that have worried philosophers in many ages, and seriously distorts linguistic and grammatical studies.[5]

It should be noted, however, that despite this unfavorable estimate, Robins goes on to praise Aristotle as in some ways anticipating later developments in linguistics.

Examples of Aristotelian rhetoric in the textbooks of today are few indeed. Edward P. J. Corbett's *Classical Rhetoric for the Modern Student* (1971) and Richard Hughes and Albert Duhamel's *Principles of Rhetoric* (1967) revive the tradition. Most textbooks that claim to be Aristotelian are operating within the paradigm of what has come to be known as Current-Traditional Rhetoric, a category that might also be called the Positivist.

The Positivist or Current-Traditional group clearly dominates thinking about writing instruction today. The evidence is the staggering number of textbooks that yearly espouse its principles. The origins of Current-Traditional Rhetoric, as Albert Kitzhaber showed in his dissertation (University of Washington, 1953) on "Rhetoric in American Colleges," can be found in the late nineteenth-century rhetoric texts of A. S. Hill, Barrett Wendell, and John F. Genung. But its epistemological stance can be found in eighteenth-century Scottish Common Sense Realism as expressed in the philosophy of Thomas Reid and James Beattie, and in the rhetorical treatises of George Campbell, Hugh Blair, and to a lesser extent, Richard Whately.

For Common Sense Realism, the certain existence of the material world is indisputable. All knowledge is founded on the simple correspondence between sense impressions and the faculties of the mind. This so far sounds like the Aristotelian world view, but is in fact a conscious departure from it. Common Sense Realism denies the value of

---

5. *Ancient and Mediaeval Grammatical Theory in Europe* (London: G. Bell and Sons, 1951), pp. 20-21.

the deductive method—syllogistic reasoning—in arriving at knowledge. Truth is instead discovered through induction alone. It is the individual sense impression that provides the basis on which all knowledge can be built. Thus the new scientific logic of Locke replaces the old deductive logic of Aristotle as the method for understanding experience. The world is still rational, but its system is to be discovered through the experimental method, not through logical categories grounded in a mental faculty. The state of affairs characterizing the emergence of the new epistemology is succinctly summarized by Wilbur Samuel Howell:

> The old science, as the disciples of Aristotle conceived of it at the end of the seventeenth century, had considered its function to be that of subjecting traditional truths to syllogistic examination, and of accepting as new truth only what could be proved to be consistent with the old. Under that kind of arrangement, traditional logic had taught the methods of deductive analysis, had perfected itself in the machinery of testing propositions for consistency, and had served at the same time as the instrument by which truths could be arranged so as to become intelligible and convincing to other learned men. . . . The new science, as envisioned by its founder, Francis Bacon, considered its function to be that of subjecting physical and human facts to observation and experiment, and of accepting as new truth only what could be shown to conform to the realities behind it.[6]

The rhetoric based on the new logic can be seen most clearly in George Campbell's *Philosophy of Rhetoric* (1776) and Hugh Blair's *Lectures on Rhetoric and Belles Lettres* (1783). The old distinction between dialectic as the discipline of learned discourse and rhetoric as the discipline of popular discourse is destroyed. Rhetoric becomes the study of all forms of communication: scientific, philosophical, historical, political, legal, and even poetic. An equally significant departure in this new rhetoric is that it contains no inventional system. Truth is to be discovered outside the rhetorical enterprise—through the method, usually the scientific method, of the appropriate discipline, or, as in poetry and oratory, through genius.

The aim of rhetoric is to teach how to adapt the discourse to its hearers—and here the uncomplicated correspondence of the faculties and the world is emphasized. When the individual is freed from the biases of language, society, or history, the senses provide the mental faculties with a clear and distinct image of the world. The world readily surrenders its meaning to anyone who observes it properly, and no operation of the mind—logical or otherwise—is needed to arrive at truth. To communicate, the speaker or writer—both now included—need only provide the language which corresponds either to the objects in the external world or to the ideas in his or her own mind—both are essentially the same—in such a way that it reproduces the objects and the experience of them in the minds of the hearers (Cohen, pp. 38-42). As Campbell explains, "Thus language and thought, like body and soul, are made to correspond, and the qualities of the one exactly to co-operate with those of the other."[7] The emphasis in this rhetoric is

---

6. *Eighteenth-Century British Logic and Rhetoric* (Princeton, N.J.: Princeton University Press, 1971), pp. 5-6.

7. *The Philosophy of Rhetoric*, ed. Lloyd F. Bitzer (Carbondale: Southern Illinois University Press, 1963), p. 215.

on adapting what has been discovered outside the rhetorical enterprise to the minds of the hearers. The study of rhetoric thus focuses on developing skills in arrangement and style.

Given this epistemological field in a rhetoric that takes all communication as its province, discourse tends to be organized according to the faculties to which it appeals. A scheme that is at once relevant to current composition theory and typical in its emulation of Campbell, Blair, and Whately can be found in John Francis Genung's *The Practical Elements of Rhetoric* (1886).[8] For Genung the branches of discourse fall into four categories. The most "fundamental" mode appeals to understanding and is concerned with transmitting truth, examples of which are "history, biography, fiction, essays, treatises, criticism." The second and third groups are description and narration, appealing again to the understanding, but leading the reader to "feel the thought as well as think it." For Genung "the purest outcome" of this kind of writing is poetry. The fourth kind of discourse, "the most complex literary type," is oratory. This kind is concerned with persuasion and makes its special appeal to the will, but in so doing involves all the faculties. Genung goes on to create a further distinction that contributed to the departmentalization of English and Speech and the division of English into literature and composition. Persuasion is restricted to considerations of experts in the spoken language and poetry to discussions of literature teachers, now first appearing. College writing courses, on the other hand, are to focus on discourse that appeals to the understanding—exposition, narration, description, and argumentation (distinct now from persuasion). It is significant, moreover, that college rhetoric is to be concerned solely with the communication of truth that is certain and empirically verifiable—in other words, not probabilistic.

Genung, along with his contemporaries A. S. Hill and Barrett Wendell, sets the pattern for most modern composition textbooks, and their works show striking similarities to the vast majority of texts published today.[9] It is discouraging that generations after Freud and Einstein, college students are encouraged to embrace a view of reality based on a mechanistic physics and a naive faculty psychology—and all in the name of a convenient pedagogy.

The next theory of composition instruction to be considered arose as a reaction to current-traditional rhetoric. Its clearest statements are located in the work of Ken Macrorie, William Coles, Jr., James E. Miller and Stephen Judy, and the so-called "Pre-Writing School" of D. Gordon Rohman, Albert O. Wlecke, Clinton S. Burhans, and Donald Stewart (see Harrington, et al., pp. 645-647). Frequent assertions of this view, however, have appeared in American public schools in the twentieth century under the veil of including "creative expression" in the English curriculum.[10] The roots of this view of rhetoric in America can be traced to Emerson and the Transcendentalists, and its ultimate source is to be found in Plato.

---

8. For a more detailed discussion of Genung see my "John Genung and Contemporary Composition Theory: The Triumph of the Eighteenth Century," *Rhetoric Society Quarterly*, 11 (1981), 74-84.

9. For an analysis of modern composition textbooks, see James A. Berlin and Robert P. Inkster, "Current-Traditional Rhetoric: Paradigm and Practice," *Freshman English News*, 8 (1980), 1-4, 13-14.

10. Kenneth J. Kantor, "Creative Expression in the English Curriculum: A Historical Perspective," *Research in the Teaching of English*, 9 (1975), 5-29.

In the Platonic scheme, truth is not based on sensory experience since the material world is always in flux and thus unreliable. Truth is instead discovered through an internal apprehension, a private vision of a world that transcends the physical. As Robert Cushman explains in *Therepeia* (Chapel Hill: University of North Carolina Press, 1958), "The central theme of Platonism regarding knowledge is that truth is not brought to man, but man to the truth" (p. 213). A striking corollary of this view is that ultimate truth can be discovered by the individual, but cannot be communicated. Truth can be learned but not taught. The purpose of rhetoric then becomes not the transmission of truth, but the correction of error, the removal of that which obstructs the personal apprehension of the truth. And the method is dialectic, the interaction of two interlocutors of good will intent on arriving at knowledge. Because the respondents are encouraged to break out of their ordinary perceptual set, to become free of the material world and of past error, the dialectic is often disruptive, requiring the abandonment of long held conventions and opinions. Preparing the soul to discover truth is often painful.

Plato's epistemology leads to a unique view of language. Because ultimate truths cannot be communicated, language can only deal with the realm of error, the world of flux, and act, as Gerald L. Bruns explains, as "a preliminary exercise which must engage the soul before the encounter with 'the knowable and truly real being' is possible" (p. 16). Truth is finally inexpressible, is beyond the resources of language. Yet Plato allows for the possibility that language may be used to communicate essential realities. In the *Republic* he speaks of using analogy to express ultimate truth, and in the *Phaedrus,* even as rhetoric is called into question, he employs an analogical method in his discussion of the soul and love. Language, it would appear, can be of some use in trying to communicate the absolute, or at least to approximate the experience of it.

The major tenets of this Platonic rhetoric form the center of what are commonly called "Expressionist" textbooks. Truth is conceived as the result of a private vision that must be constantly consulted in writing. These textbooks thus emphasize writing as a "personal" activity, as an expression of one's unique voice. In *Writing and Reality* (New York: Harper and Row, 1978), James Miller and Stephen Judy argue that "all good writing is *personal,* whether it be an abstract essay or a private letter," and that an important justification for writing is "to sound the depths, to explore, and to discover." The reason is simple: "Form in language grows from content—something the writer has to say—and that something, in turn, comes directly from the self" (pp. 12, 15). Ken Macrorie constantly emphasizes "Telling Truths," by which he means a writer must be "true to the feeling of his experience." His thrust throughout is on speaking in "an authentic voice" (also in Donald Stewart's *The Authentic Voice: A Pre-Writing Approach to Student Writing,* based on the work of Rohman and Wlecke), indicating by this the writer's private sense of things.[11] This placement of the self at the center of communication is also, of course, everywhere present in Coles' *The Plural I* (New York: Holt, Rinehart, and Winston, 1978).

One obvious objection to my reading of these expressionist theories is that their conception of truth can in no way be seen as comparable to Plato's transcendent world of ideas. While this cannot be questioned, it should also be noted that no member of this school is a relativist intent on denying the possibility of any certain truth whatever. All

11. *Telling Writing* (Rochelle Park, N.J.: Hayden Book Company, 1978), p. 13.

believe in the existence of verifiable truths and find them, as does Plato, in private experience, divorced from the impersonal data of sense experience. All also urge the interaction between writer and reader, a feature that leads to another point of similarity with Platonic rhetoric—the dialectic.

Most expressionist theories rely on classroom procedures that encourage the writer to interact in dialogue with the members of the class. The purpose is to get rid of what is untrue to the private vision of the writer, what is, in a word, inauthentic. Coles, for example, conceives of writing as an unteachable act, a kind of behavior that can be learned but not taught. (See especially the preface to *The Plural I.*) His response to this denial of his pedagogical role is to provide a classroom environment in which the student learns to write—although he or she is not taught to write—through dialectic. *The Plural I,* in fact, reveals Coles and his students engaging in a dialogue designed to lead both teacher and class—Coles admits that he always learns in his courses—to the discovery of what can be known but not communicated. This view of truth as it applies to writing is the basis of Coles' classroom activity. Dialogue can remove error, but it is up to the individual to discover ultimate knowledge. The same emphasis on dialectic can also be found in the texts of Macrorie and of Miller and Judy. Despite their insistence on the self as the source of all content, for example, Miller and Judy include "making connections with others in dialogue and discussion" (p. 5), and Macrorie makes the discussion of student papers the central activity of his classroom.

This emphasis on dialectic, it should be noted, is not an attempt to adjust the message to the audience, since doing so would clearly constitute a violation of the self. Instead the writer is trying to use others to get rid of what is false to the self, what is insincere and untrue to the individual's own sense of things, as evidenced by the use of language—the theory of which constitutes the final point of concurrence between modern Expressionist and Platonic rhetorics.

Most Expressionist textbooks emphasize the use of metaphor either directly or by implication. Coles, for example, sees the major task of the writer to be avoiding the imitation of conventional expressions because they limit what the writer can say. The fresh, personal vision demands an original use of language. Rohman and Wlecke, as well as the textbook by Donald Stewart based on their research, are more explicit. They specifically recommend the cultivation of the ability to make analogies (along with meditation and journal writing) as an inventional device. Macrorie makes metaphor one of the prime features of "good writing" (p. 21) and in one form or another takes it up again and again in *Telling Writing*. The reason for this emphasis is not hard to discover. In communicating, language does not have as its referent the object in the external world or an idea of this object in the mind. Instead, to present truth language must rely on original metaphors in order to capture what is unique in each personal vision. The private apprehension of the real relies on the metaphoric appeal from the known to the unknown, from the public and accessible world of the senses to the inner and privileged immaterial realm, in order to be made available to others. As in Plato, the analogical method offers the only avenue to expressing the true.

The clearest pedagogical expression of the New Rhetoric—or what might be called Epistemic Rhetoric—is found in Ann E. Berthoff's *Forming/Thinking/Writing: The Composing Imagination* (Rochelle Park, N.J.: Hayden, 1978) and Richard L. Young,

Alton L. Becker, and Kenneth L. Pike's *Rhetoric: Discovery and Change* (New York: Harcourt Brace Jovanovich, 1970). These books have behind them the rhetorics of such figures as I. A. Richards and Kenneth Burke and the philosophical statements of Susan Langer, Ernst Cassirer, and John Dewey. Closely related to the work of Berthoff and Young, Becker, and Pike are the cognitive-developmental approaches of such figures as James Moffett, Linda Flower, Andrea Lunsford, and Barry Kroll. While their roots are different—located in the realm of cognitive psychology and empirical linguistics—their methods are strikingly similar. In this discussion, however, I intend to call exclusively upon the textbooks of Berthoff and of Young, Becker, and Pike to make my case, acknowledging at the start that there are others that could serve as well. Despite differences, their approaches most comprehensively display a view of rhetoric as epistemic, as a means of arriving at truth.

Classical Rhetoric considers truth to be located in the rational operation of the mind, Positivist Rhetoric in the correct perception of sense impressions, and Neo-Platonic Rhetoric within the individual, attainable only through an internal apprehension. In each case knowledge is a commodity situated in a permanent location, a repository to which the individual goes to be enlightened.

For the New Rhetoric, knowledge is not simply a static entity available for retrieval. Truth is dynamic and dialectical, the result of a process involving the interaction of opposing elements. It is a relation that is created, not pre-existent and waiting to be discovered. The basic elements of the dialectic are the elements that make up the communication process—writer (speaker), audience, reality, language. Communication is always basic to the epistemology underlying the New Rhetoric because truth is always truth for someone standing in relation to others in a linguistically circumscribed situation. The elements of the communication process thus do not simply provide a convenient way of talking about rhetoric. They form the elements that go into the very shaping of knowledge.

It is this dialectical notion of rhetoric—and of rhetoric as the determiner of reality—that underlies the textbooks of Berthoff and of Young, Becker, and Pike. In demonstrating this thesis I will consider the elements of the dialectic alone or in pairs, simply because they are more easily handled this way in discussion. It should not be forgotten, however, that in operation they are always simultaneously in a relationship of one to all, constantly modifying their values in response to each other.

The New Rhetoric denies that truth is discoverable in sense impression since this data must always be interpreted—structured and organized—in order to have meaning. The perceiver is of course the interpreter, but she is likewise unable by herself to provide truth since meaning cannot be made apart from the data of experience. Thus Berthoff cites Kant's "Percepts without concepts are empty; concepts without percepts are blind" (p. 13). Later she explains: "The brain puts things together, composing the percepts by which we can make sense of the world. We don't just 'have' a visual experience and then by thinking 'have' a mental experience; the mutual dependence of seeing and knowing is what a modern psychologist has in mind when he speaks of 'the intelligent eye' " (p. 44). Young, Becker, and Pike state the same notion:

> Constantly changing, bafflingly complex, the external world is not a neat, well-ordered place replete with meaning, but an enigma requiring interpretation. This inter-

> pretation is the result of a transaction between events in the external world and the mind of the individual—between the world "out there" and the individual's previous experience, knowledge, values, attitudes, and desires. Thus the mirrored world is not just the sum total of eardrum rattles, retinal excitations, and so on; it is a creation that reflects the peculiarities of the perceiver as well as the peculiarities of what is perceived. (p. 25)

Language is at the center of this dialectical interplay between the individual and the world. For Neo-Aristotelians, Positivists, and Neo-Platonists, truth exists prior to language so that the difficulty of the writer or speaker is to find the appropriate words to communicate knowledge. For the New Rhetoric truth is impossible without language since it is language that embodies and generates truth. Young, Becker, and Pike explain:

> Language provides a way of unitizing experience: a set of symbols that label recurring chunks of experience. . . . Language depends on our seeing certain experiences as constant or repeatable. And seeing the world as repeatable depends, in part at least, on language. A language is, in a sense, a theory of the universe, a way of selecting and grouping experience in a fairly consistent and predictable way. (p. 27)

Berthoff agrees: "The relationship between thought and language is dialectical: ideas are conceived by language; language is generated by thought" (p. 47). Rather than truth being prior to language, language is prior to truth and determines what shapes truth can take. Language does not correspond to the "real world." It creates the "real world" by organizing it, by determining what will be perceived and not perceived, by indicating what has meaning and what is meaningless.

The audience of course enters into this play of language. Current-Traditional Rhetoric demands that the audience be as "objective" as the writer; both shed personal and social concerns in the interests of the unobstructed perception of empirical reality. For Neo-Platonic Rhetoric the audience is a check to the false note of the inauthentic and helps to detect error, but it is not involved in the actual discovery of truth—a purely personal matter. Neo-Aristotelians take the audience seriously as a force to be considered in shaping the message. Still, for all its discussion of the emotional and ethical appeals, Classical Rhetoric emphasizes rational structures, and the concern for the audience is only a concession to the imperfection of human nature. In the New Rhetoric the message arises out of the interaction of the writer, language, reality, and the audience. Truths are operative only within a given universe of discourse, and this universe is shaped by all of these elements, including the audience. As Young, Becker, and Pike explain:

> The writer must first understand the nature of his own interpretation and how it differs from the interpretations of others. Since each man segments experience into discrete, repeatable units, the writer can begin by asking how his way of segmenting and ordering experience differs from his reader's. How do units of time, space, the visible world, social organization, and so on differ? . . .
>
> Human differences are the raw material of writing—differences in experiences and ways of segmenting them, differences in values, purposes, and goals. They are our reason for wishing to communicate. Through communication we create community,

> the basic value underlying rhetoric. To do so, we must overcome the barriers to communication that are, paradoxically, the motive for communication. (p. 30)

Ann E. Berthoff also includes this idea in her emphasis on meaning as a function of relationship.

> *Meanings are relationships.* Seeing means ''seeing relationships,'' whether we're talking about seeing as *perception* or seeing as *understanding.* ''I see what you mean'' means ''I understand how you put that together so that it makes sense.'' The way we make sense of the world is to see something *with respect to, in terms of, in relation to* something else. We can't make sense of one thing by itself; it must be seen as being *like* another thing; or *next to, across from, coming after* another thing; or as a repetition of another thing. *Something* makes sense—is meaningful—only if it is taken with *something else.* (p. 44)

The dialectical view of reality, language, and the audience redefines the writer. In Current-Traditional Rhetoric the writer must efface himself; stated differently, the writer must focus on experience in a way that makes possible the discovery of certain kinds of information—the empirical and rational—and the neglect of others—psychological and social concerns. In Neo-Platonic Rhetoric the writer is at the center of the rhetorical act, but is finally isolated, cut off from community, and left to the lonely business of discovering truth alone. Neo-Aristotelian Rhetoric exalts the writer, but circumscribes her effort by its emphasis on the rational—the enthymeme and example. The New Rhetoric sees the writer as a creator of meaning, a shaper of reality, rather than a passive receptor of the immutably given. ''When you write,'' explains Berthoff, ''you don't follow somebody else's scheme; you design your own. As a writer, you learn to make words behave the way you want them to. . . . Learning to write is not a matter of learning the rules that govern the use of the semicolon or the names of sentence structures, nor is it a matter of manipulating words; it is a matter of making meanings, and that is the work of the active mind'' (p. 11). Young, Becker, and Pike concur: ''We have sought to develop a rhetoric that implies that we are all citizens of an extraordinarily diverse and disturbed world, that the 'truths' we live by are tentative and subject to change, that we must be discoverers of new truths as well as preservers and transmitter of old, and that enlightened cooperation is the preeminent ethical goal of communication'' (p. 9).

This version of the composing process leads to a view of what can be taught in the writing class that rivals Aristotelian rhetoric in its comprehensiveness. Current-Traditional and Neo-Platonic Rhetoric deny the place of invention in rhetoric because for both truth is considered external and self-evident, accessible to anyone who seeks it in the proper spirit. Like Neo-Aristotelian Rhetoric, the New Rhetoric sees truth as probabilistic, and it provides students with techniques—heuristics—for discovering it, or what might more accurately be called creating it. This does not mean, however, that arrangement and style are regarded as unimportant, as in Neo-Platonic Rhetoric. In fact, the attention paid to these matters in the New Rhetoric rivals that paid in Current-Traditional Rhetoric, but not because they are the only teachable part of the process. Structure and language are a part of the formation of meaning, are at the center of the discovery of truth, not simply the dress of thought. From the point of view of pedagogy,

New Rhetoric thus treats in depth all the offices of classical rhetoric that apply to written language—invention, arrangement, and style—and does so by calling upon the best that has been thought and said about them by contemporary observers.

In talking and writing about the matters that form the substance of this essay, at my back I always hear the nagging (albeit legitimate) query of the overworked writing teacher: But what does all this have to do with the teaching of freshman composition? My answer is that it is more relevant than most of us are prepared to admit. In teaching writing, we are not simply offering training in a useful technical skill that is meant as a simple complement to the more important studies of other areas. We are teaching a way of experiencing the world, a way of ordering and making sense of it. As I have shown, subtly informing our statements about invention, arrangement, and even style are assumptions about the nature of reality. If the textbooks that sell the most copies tell us anything, they make abundantly clear that most writing teachers accept the assumptions of Current-Traditional Rhetoric, the view that arose contemporaneously with the positivistic position of modern science. Yet most of those who use these texts would readily admit that the scientific world view has demonstrated its inability to solve the problems that most concern us, problems that are often themselves the result of scientific "breakthroughs." And even many scientists concur with them in this view—Oppenheimer and Einstein, for example. In our writing classrooms, however, we continue to offer a view of composing that insists on a version of reality that is sure to place students at a disadvantage in addressing the problems that will confront them in both their professional and private experience.

Neo-Platonic, Neo-Aristotelian, and what I have called New Rhetoric are reactions to the inadequacy of Current-Traditional Rhetoric to teach students a notion of the composing process that will enable them to become effective persons as they become effective writers. While my sympathies are obviously with the last of these reactions, the three can be considered as one in their efforts to establish new directions for a modern rhetoric. Viewed in this way, the difference between them and Current-Traditional Rhetoric is analogous to the difference Richard Rorty has found in what he calls, in *Philosophy and the Mirror of Nature* (Princeton, N.J.: Princeton University Press, 1979), hermeneutic and epistemological philosophy. The hermeneutic approach to rhetoric bases the discipline on establishing an open dialogue in the hopes of reaching agreement about the truth of the matter at hand. Current-Traditional Rhetoric views the rhetorical situation as an arena where the truth is incontrovertibly established by a speaker or writer more enlightened than her audience. For the hermeneuticist truth is never fixed finally on unshakable grounds. Instead it emerges only after false starts and failures, and it can only represent a tentative point of rest in a continuing conversation. Whatever truth is arrived at, moreover, is always the product of individuals calling on the full range of their humanity, with esthetic and moral considerations given at least as much importance as any others. For Current-Traditional Rhetoric truth is empirically based and can only be achieved through subverting a part of the human response to experience. Truth then stands forever, a tribute to its method, triumphant over what most of us consider important in life, successful through subserving writer, audience, and language to the myth of an objective reality.

One conclusion should now be incontestable. The numerous recommendations of the ''process''-centered approaches to writing instruction as superior to the ''product''-centered approaches are not very useful. Everyone teaches the process of writing, but everyone does not teach the *same* process. The test of one's competence as a composition instructor, it seems to me, resides in being able to recognize and justify the version of the process being taught, complete with all of its significance for the student.

Maxine Hairston

Diversity, Ideology, and Teaching Writing

Where We Have Come From

In 1985, when I was chair of CCCC, as my chair's address I gave what might be called my own State of the Profession Report. On the whole it was a positive report. I rejoiced in the progress we had made in the previous fifteen years in establishing our work as a discipline and I pointed out that we were creating a new paradigm for the teaching of writing, one that focused on process and on writing as a way of learning. I asserted that we teach writing for its own sake, as a primary intellectual activity that is at the heart of a college education. I insisted that writing courses must not be viewed as service courses. Writing courses, especially required freshman courses, should not be *for* anything or *about* anything other than writing itself, and how one uses it to learn and think and communicate.

I also warned in my Chair's address that if we hoped to flourish as a profession, we would have to establish our psychological and intellectual independence from the literary critics who are at the center of power in most English departments; that we could not develop our potential and become fully autonomous scholars and teachers as long as we allowed our sense of self worth to depend on the approval of those who define English departments as departments of literary criticism.

We've continued to make important strides since 1985. We have more graduate programs in rhetoric and composition, more tenure track positions in composition created each year, more and larger conferences, and so many new journals that one can scarcely keep up with them. In those years, I've stayed optimistic about the profession and gratified by the role I've played in its growth.

Where We Seem to Be Heading

Now, however, I see a new model emerging for freshman writing programs, a model that disturbs me greatly. It's a model that puts dogma before diversity, politics before craft, ideology before critical thinking, and the social goals of the teacher before the educational needs of the student. It's a regressive model that undermines the progress we've made in teaching writing, one that threatens to silence student voices and jeopardize the process-oriented, low-risk, student-centered classroom we've worked so hard to establish as the norm. It's a model that doesn't take freshman English seriously in its

From *College Composition and Communication* 43 (May 1992): 179-93.

own right but conceives of it as a tool, something to be used. The new model envisions required writing courses as vehicles for social reform rather than as student-centered workshops designed to build students' confidence and competence as writers. It is a vision that echoes that old patronizing rationalization we've heard so many times before: students don't have anything to write about so we have to give them topics. Those topics used to be literary; now they're political.

I don't suggest that all or even most freshman writing courses are turning this way. I have to believe that most writing teachers have too much common sense and are too concerned with their students' growth as writers to buy into this new philosophy. Nevertheless, everywhere I turn I find composition faculty, both leaders in the profession and new voices, asserting that they have not only the right, but the duty, to put ideology and radical politics at the center of their teaching.

Here are four revealing quotations from recent publications. For instance, here is James Laditka in the *Journal of Advanced Composition:*

> All teaching supposes ideology; there simply is no value free pedagogy. For these reasons, my paradigm of composition is changing to one of critical literacy, a literacy of political consciousness and social action. (361)

Here is Charles Paine in a lead article in *College English:*

> Teachers need to recognize that methodology alone will not ensure radical visions of the world. An appropriate course content is necessary as well. . . . [E]quality and democracy are not transcendent values that inevitably emerge when one learns to seek the truth through critical thinking. Rather, if those are the desired values, the teacher must recognize that he or she must influence (perhaps manipulate is the more accurate word) students' values through charisma or power—he or she must accept the role as manipulator. Therefore it is of course reasonable to try to inculcate into our students the conviction that the dominant order is repressive. (563-564)

Here is Patricia Bizzell:

> We must help our students . . . to engage in a rhetorical process that can collectively generate . . . knowledge and beliefs to displace the repressive ideologies an unjust social order would prescribe. . . . I suggest that we must be forthright in avowing the ideologies that motivate our teaching and research. For instance, [in an experimental composition course he teaches at Purdue] James Berlin might stop trying to be value-neutral and anti-authoritarian in the classroom. Berlin tells his students he is a Marxist but disavows any intention of persuading them to his point of view. Instead, he might openly state that this course aims to promote values of sexual equality and left-oriented labor relations and that this course will challenge students' values insofar as they conflict with these aims. Berlin and his colleagues might openly exert their authority as teachers to try to persuade students to agree with their values instead of pretending that they are merely investigating the nature of sexism and capitalism and leaving students to draw their own conclusions. (670)

Here is C. H. Knoblauch:

> We are, ultimately, compelled to choose, to make, express, and act upon our commitments, to denounce the world, as Freire says, and above all oppression and whatever

> arguments have been called upon to validate it. Moreover our speech may well have to be boldly denunciative at times if it is to affect its hearers in the midst of their intellectual and political comfort. . . . We are obliged to announce ourselves so that, through the very process of self-assertion, we grow more conscious of our axioms. . . . The quality of our lives as teachers depends on our willingness to discover through struggle ever more fruitful means of doing our work. The quality of our students' lives depends on [it]. (''Rhetorical'' 139)

These quotations do not represent just a few instances that I ferreted out to suit my thesis; you will find similar sentiments if you leaf through only a few of the recent issues of *College English, Rhetoric Review, College Composition and Communication, Journal of Advanced Composition, Focuses,* and others. Some names that you might look for in addition to the ones I've quoted are James Berlin, John Trimbur, Lester Faigley, Richard Ohmann, and Linda Brodkey. At least forty percent of the essays in *The Right to Literacy,* the proceedings of a 1988 conference sponsored by the Modern Language Association in Columbus, Ohio, echo such sentiments, and a glance at the program for the 1991 CCCC convention would confirm how popular such ideas were among the speakers. For that same convention, the publisher HarperCollins sponsored a contest to award grants to graduate students to attend; the topic they were asked to write on was ''Describe the kind of freshman writing course you would design.'' Nearly all of the contestants described a politically-focused course. All ten essays in the 1991 MLA publication *Contending with Words* recommend turning writing courses in this direction.

Distressingly often, those who advocate such courses show open contempt for their students' values, preferences, or interests. For example, in an article in *College English,* Ronald Strickland says, ''The teacher can best facilitate the production of knowledge by adapting a confrontational stance toward the student. . . . Above all, the teacher should avoid the pretense of detachment, objectivity, and autonomy.'' He admits that his position ''conflicts with the expectations of some students [and] these students make it difficult for me to pursue my political/intellectual agenda'' (293).

David Bleich dismisses his students' resistance with equal ease:

> There is reason to think that students want to write about what they say they don't want to write about. They want a chance to write about racism, classism, and homophobia even though it makes them uncomfortable. But what I think makes them most uncomfortable is to surrender the paradigm of individualism and to see that paradigm in its sexist dimensions.

He cites his students' religion as one of the chief obstacles to their enlightenment:

> Religious views collaborate with the ideology of individualism and with sexism to censor the full capability of what people can say and write. . . . By ''religious values'' I mean belief in the savability of the individual human soul. The ideal of the nuclear family, as opposed to the extended or communal family, permits the overvaluation of the individual child and the individual soul. (167)

And here is Dale Bauer in an article from *College English:*

> I would argue that political commitment—especially feminist commitment—is a legitimate classroom strategy and rhetorical imperative. The feminist agenda offers a goal

> toward our students' conversions to emancipatory critical action. . . . In teaching identification and teaching feminism, I overcome a vehement insistence on pluralistic relativism or on individualism.

Bauer acknowledges that her students resist her political agenda. She says,

> There is an often overwhelming insistence on individualism and isolation . . . [They] labor at developing a critical distance to avoid participating in "the dialectic of resistance and identification."

Bauer quotes one of her students as saying in an evaluation,

> "The teacher consistently channels class discussions around feminism and does not spend time discussing the comments that oppose her beliefs. In fact, she usually twists them around to support her beliefs."

Bauer dismisses such objections, however, claiming she has to accept her authority as rhetor because "anything less ends up being an expressivist model, one which reinforces . . . the dominant patriarchal culture" (389).

Often these advocates are contemptuous of other teachers' approaches to teaching or the goals those teachers set for their students. For example, Lester Faigley assails the advice given about writing a job application letter in a standard business writing text:

> In the terms of [the Marxist philosopher] Althusser, [the applicant who writes such a letter] has voluntarily assented his subjectivity within the dominant ideology and thus has reaffirmed relations of power. By presenting himself as a commodity rather than as a person, he has not only made an initial gesture of subservience like a dog presenting its neck, but he has also signaled his willingness to continue to be subservient. (251)

In discussing Linda Flower's cognitive, problem-solving approach to teaching writing, James Berlin calls it, "the rationalization of economic activity. The pursuit of self-evident and unquestioned goals in the composing process parallels the pursuit of self-evident and unquestioned profit-making goals in the corporate market place." (What a facile non-logical leap!) He continues in the same article to deride Donald Murray's and Peter Elbow's approaches to writing because of their focus on the individual, saying

> Expressionist rhetoric is inherently and debilitatingly divisive of political protest. . . . Beyond that, expressionist rhetoric is easily co-opted by the very capitalist forces it opposes. After all, this rhetoric can be used to reinforce the entrepreneurial virtues capitalism values most: individualism, private initiative, the confidence for risk taking, the right to be contentious with authority (especially the state). (491)

How We Got Here

But how did all this happen? Why has the cultural left suddenly claimed writing courses as their political territory?

There's no simple answer, of course. Major issues about social change and national priorities are involved, and I cannot digress into those concerns in this essay. But my first response is, "You see what happens when we allow writing programs to be run by

English departments?'' I'm convinced that the push to change freshman composition into a political platform for the teacher has come about primarily because the course is housed in English departments.

As the linguistics scholar John Searle pointed out in a detailed and informative article in *The New York Review of Books,* the recent surge of the cultural left on major American campuses has centered almost entirely in English departments. He says,

> The most congenial home left for Marxism, now that it has been largely discredited as a theory of economics and politics, is in departments of literary criticism. And [because] many professors of literature no longer care about literature in ways that seemed satisfactory to earlier generations . . . they teach it as a means of achieving left-wing political goals or as an occasion for exercises in deconstruction, etc. (38)

I theorize that the critical literary theories of deconstruction, post-structuralism (both declining by now), and Marxist critical theory have trickled down to the lower floors of English departments where freshman English dwells. Just as they have been losing their impact with faculty above stairs, they have taken fresh root with those dwelling below.

Deconstructionists claim that the privileged texts of the canon are only reflections of power relations and the dominant class structures of their eras. Thus the job of the literary critic is to dissect Shakespeare or Milton or Eliot or Joyce to show how language reflects and supports the ''cultural hegemony'' of the time. They also claim that all meaning is indeterminate and socially constructed; there is no objective reality nor truth that can be agreed on.

Marxist criticism echoes these sentiments. For example, Ronald Strickland writes in *College English:*

> Marxist critics have demonstrated that conventional literary studies have been more complicitous . . . than any other academic discipline in the reproduction of the dominant ideology. . . . Traditional English studies helps to maintain liberal humanism through its emphasis on authorial genius. . . . [Thus] there is a political imperative to resist the privileging of individualism in this practice, for, as Terry Eagleton has demonstrated, it amounts to a form of coercion in the interests of conservative, elitist politics. (293)

All these claims strike me as silly, simplistic, and quite undemonstrable. Nevertheless, if one endorses these intellectual positions—and sympathizes with the politics behind them—it's easy to go to the next step and equate conventional writing instruction with conventional literary studies. Then one can say that because standard English is the dialect of the dominant class, writing instruction that tries to help students master that dialect merely reinforces the status quo and serves the interest of the dominant class. An instructor who wants to teach students to write clearly becomes part of a capitalistic plot to control the workforce. What nonsense! It seems to me that one could argue with more force that the instructor who fails to help students master the standard dialect conspires against the working class.

How easy for theorists who, by the nature of the discipline they have chosen, already have a facile command of the prestige dialect to denigrate teaching that dialect to students. Have they asked those students what *they* want to learn? And how easy for these same theorists to set up straw men arguments that attack a mechanistic, structuralist, literature-based model of composition and call it "conservative, regressive, deterministic, and elitist" (Knoblauch, "Literacy" 76) when they know such models have long been discredited in the professional literature.

But I think this is what happens when composition theorists remain psychologically tied to the English departments that are their base. Partly out of genuine interest, I'm sure, but also out of a need to belong to and be approved by the power structure, they immerse themselves in currently fashionable critical theories, read the authors that are chic—Foucault, Bahktin, Giroux, Eagleton, and Cixous, for example—then look for ways those theories can be incorporated into their own specialty, teaching writing.

This, according to Searle's article, means that they subscribe to a view of the role of the humanities in universities that is

> based on two primary assumptions. 1. They believe that Western civilization in general, and the United States in particular, are in large part oppressive, patriarchal, hegemonic, and in need of replacement or at least transformation. 2. The primary function of teaching the humanities is political; they [the cultural left] do not really believe the humanities are valuable in their own right except as a means of achieving social transformation. (38)

Searle goes on to point out that this debate about what is "hegemonic," "patriarchal," or "exclusionary" has been focused almost entirely in English departments.

I find it hard to believe that most English professors seriously hold these opinions or that they are ready to jettison their lifelong commitment to the humanities, but evidently significant numbers do. News releases and many professional articles suggest that these attitudes have permeated the Modern Language Association, and the associate chair of the English Department at the University of Texas recently said in a colloquium of the College of Liberal Arts that the "mission of English departments is always to oppose the dominant culture."

For those who agree, how natural to turn to the freshman writing courses. With a huge captive enrollment of largely unsophisticated students, what a fertile field to cultivate to bring about political and social change. Rhetoric scholars who go along will also get new respect now that they have joined the ideological fray and formed alliances with literature faculty who have been transforming their own courses.

Composition faculty who support such change can bring fresh respectability and attention to those often despised introductory English courses now that they can be used for "higher purposes." They may even find some regular faculty who will volunteer to teach freshman writing when they can use it for a political forum. Five years ago the regular faculty in our department at Texas tried to get rid of freshman English altogether by having it taught entirely in extension or at the local community college; this past year, many of those who had previously advocated abandoning the course were in the forefront of the battle to turn it into a course about racism and sexism. Now the course was suddenly worth their time.

The opportunity to make freshman English a vehicle for such social crusades is particularly rich: in many universities, graduate students in English teach virtually all of the sections, graduate students who are already steeped in post-structuralism and deconstruction theory, in the works of Foucault, Raymond Williams, Terry Eagleton, and Stanley Fish, and in feminist theory. Too often they haven't been well trained in how to teach writing and are at a loss about what they should be doing with their students. How easy then to focus the course on their own interests, which are often highly political. Unfortunately, when they try to teach an introductory composition course by concentrating on issues rather than on craft and critical thinking, large numbers of their students end up feeling confused, angry—and cheated.

I also believe that two major social forces outside the liberal arts are contributing to creating the environment that has given rise to this new model.

The first is the tremendous increase in diversity of our student population, especially in states like California and Texas and in all our major cities. With changing demographics, we face an ethnic and social mix of students in our classes that previews for us what our institutions are going to be like in the year 2000. These students bring with them a kaleidoscope of experiences, values, dialects, and cultural backgrounds that we want to respond to positively and productively, using every resource we can to help them adapt to the academic world and become active participants in it. The code words for our attempts to build the kind of inclusive curriculum that we need have become "multiculturalism" and "cultural diversity." They're good terms, of course. Any informed and concerned educator endorses them in the abstract. The crucial question, however, is how one finds concrete ways to put them into practice, and also how one guards against their becoming what Richard Weaver called "god terms" that can be twisted to mean anything an ideologue wants them to mean.

As writing teachers, I think all of us are looking for ways to promote genuine diversity in our classes and yet keep two elements that are essential for any state-of-the-art composition course.

First, students' own writing must be the center of the course. Students need to write to find out how much they know and to gain confidence in their ability to express themselves effectively. They do not need to be assigned essays to read so they will have something to write about—they bring their subjects with them. The writing of others, except for that of their fellow students, should be supplementary, used to illustrate or reinforce.

Second, as writing teachers we should stay within our area of professional expertise: helping students to learn to write in order to learn, to explore, to communicate, to gain control over their lives. That's a large responsibility, and all that most of us can manage. We have no business getting into areas where we may have passion and conviction but no scholarly base from which to operate. When classes focus on complex issues such as racial discrimination, economic injustices, and inequities of class and gender, they should be taught by qualified faculty who have the depth of information and historical competence that such critical social issues warrant. Our society's deep and tangled cultural conflicts can neither be explained nor resolved by simplistic ideological formulas.

But one can run a culturally diverse writing course without sacrificing any of its integrity as a writing course. Any writing course, required or not, can be wonderfully diverse, an exciting experience in which people of different cultures and experience learn about difference first-hand. More about that shortly.

Forces from Outside

The second major force I see at work is directly political. There's no question in my mind that this new radical stance of many composition faculty is in some ways a corollary of the angry response many intellectuals have to the excesses of right-wing, conservative forces that have dominated American politics for the past decade. Faculty in the liberal arts tend to be liberals who are concerned about social problems and dislike the trends we've seen in cutting funds for human services and for education. We're sick over the condition of our country: one child in five living in poverty; one person in eight hungry; 33 million people with no health insurance; a scandalous infant mortality rate; hundreds of thousands homeless. Yet we see our government spend billions on a dubious war. No need to go on—we all know the terrible inequities and contradictions of our society.

As educators of good will, we shouldn't even have to mention our anger about racism and sexism in our society—that's a given, as is our commitment to work to overcome it. I, for one, refuse to be put on the defensive on such matters of personal conscience or to be silenced by the fear that someone will pin a label on me if I don't share his or her vision of the world or agree on how to improve it. *Ad hominem* arguments don't impress me.

But it's entirely understandable that academics who are traditional liberals sympathize at first with those who preach reform, even when they sound more radical than we'd like. On the surface we share common ground: we'd all like to bring about a fairer, more compassionate society. But I fear that we are in real danger of being co-opted by the radical left, coerced into acquiescing to methods that we abhor because, in the abstract, we have some mutual goals. Some faculty may also fear being labeled "right-wing" if they oppose programs that are represented as being "liberating." But we shouldn't be duped. Authoritarian methods are still authoritarian methods, no matter in what cause they're invoked. And the current battle is *not* one between liberals and conservatives. Those who attempt to make it so—columnists like George Will—either do not understand the agenda of the cultural left, or they make the association in order to discredit liberal goals. Make no mistake—those on the cultural left are not in the least liberal; in fact, they despise liberals as compromising humanists. They're happy, however, to stir up traditional liberal guilt and use it for their purposes.

What's Wrong with Their Goals?

Why do I object so strongly to the agenda that these self-styled radical teachers want to establish for composition courses and freshman English in particular?

First, I vigorously object to the contention that they have a right—even a *duty*—to use their classrooms as platforms for their own political views. Such claims violate all academic traditions about the university being a forum for the free exchange of ideas, a place where students can examine different points of view in an atmosphere of honest and open discussion, and, in the process, learn to think critically. It is a teacher's obligation to encourage diversity and exploration, but diversity and ideology will not flourish together. By definition, they're incompatible.

By the logic of the cultural left, any teacher should be free to use his or her classroom to promote any ideology. Why not facism? Racial superiority? Religious fundamentalism? Anti-abortion beliefs? Can't any professor claim the right to indoctrinate students simply because he or she is right? The argument is no different from that of any true believers who are convinced that they own the truth and thus have the right to force it on others. My colleague John Ruszkiewicz compares them to Milton's "the new forcers of conscience.'' We don't have to look far to see how frightening such arguments really are. They represent precisely the kind of thinking that leads to ''re-education camps'' in totalitarian governments, to putting art in the service of propaganda, and to making education always the instrument of the state.

Those who want to bring their ideology into the classroom argue that since any classroom is necessarily political, the teacher might as well make it openly political and ideological. He or she should be direct and honest about his or her political beliefs; then the students will know where they stand and everyone can talk freely. Is any experienced teacher really so naive as to believe that? Such claims are no more than self-serving rationalizations that allow a professor total freedom to indulge personal prejudices and avoid any responsibility to be fair. By the same reasoning, couldn't one claim that since we know it is impossible to find absolute, objective truths, we might just as well abandon the search for truth and settle for opinion, superstition and conjecture? Would that advance our students' education? Couldn't one also say that since one can never be completely fair with one's children, one might as well quit trying and freely indulge one's biases and favoritism? It's astonishing that people who purport to be scholars can make such specious arguments.

The real political truth about classrooms is that the teacher has all the power; she sets the agenda, she controls the discussion, and she gives the grades. She also knows more and can argue more skillfully. Such a situation is ripe for intellectual intimidation, especially in required freshman composition classes, and although I think it is unprofessional for teachers to bring their ideology into any classroom, it is those freshman courses that I am especially concerned about.

The Threat to Freshman Courses

I believe that the movement to make freshman English into courses in which students must write about specific social issues threatens all the gains we have made in teaching writing in the last fifteen years. I also think that rather than promoting diversity and a genuine multicultural environment, such courses actually work against those goals. Here are my reasons.

First, we know that students develop best as writers when they can write about something they care about and want to know more about. Only then will they be motivated to invest real effort in their work; only then can we hope they will avoid the canned, clichéd prose that neither they nor we take seriously. Few students, however, will do their best when they are compelled to write on a topic they perceive as politically charged and about which they feel uninformed, no matter how thought-provoking and important the instructor assumes that topic to be. If freshmen choose to write about issues involving race, class, and gender, that's fine. They should have every encouragement. I believe all topics in a writing class should be serious ones that push students to think and to say something substantial. But the topic should be their choice, a careful and thoughtful choice, to be sure, but not what someone else thinks is good for them.

Second, we know that young writers develop best as writers when teachers are able to create a low-risk environment that encourages students to take chances. We also know that novice writers can virtually freeze in the writing classroom when they see it as an extremely high-risk situation. Apprehensive about their grades in this new college situation, they nervously test their teachers to see what is expected of them, and they venture opinions only timidly. It is always hard to get students to write seriously and honestly, but when they find themselves in a classroom where they suspect there is a correct way to think, they are likely to take refuge in generalities and responses that please the teacher. Such fake discourse is a kind of silence, the silence we have so often deplored when it is forced on the disadvantaged. But when we stifle creative impulse and make students opt for survival over honesty, we have done the same thing. In too many instances, the first lesson they will learn as college students is that hypocrisy pays—so don't try to think for yourself.

My third objection to injecting prescribed political content into a required freshman course is that such action severely limits freedom of expression for both students and instructors. In my view, the freshman course on racism and sexism proposed at the University of Texas at Austin in the spring of 1990 would have enforced conformity in both directions. Students would have had no choice of what to write about, and the instructors who were graduate students would have had no choice about what to teach. Even if they felt unqualified to teach the material—and many did—or believed that the prescribed curriculum would work against their students' learning to write—and many did—they had to conform to a syllabus that contradicted their professional judgment and, often, their personal feelings. That course has since been revised and the freshman course in place since the fall of 1991 offers choices to both students and teachers.

New Possibilities for Freshman Courses

I believe we can make freshman English—or any other writing course—a truly multicultural course that gives students the opportunity to develop their critical and creative abilities and do it in an intellectually and ethically responsible context that preserves the heart of what we have learned about teaching writing in the past two decades.

First, I resist the effort to put any specific multicultural content at the center of a writing course, particularly a freshman course, and particularly a required course. Multicultural issues are too complex and diverse to be dealt with fully and responsibly in an English course, much less a course in which the focus should be on writing, not reading. Too often attempts to focus on such issues encourage stereotyping and superficial thinking. For instance, what English teacher wouldn't feel presumptuous and foolish trying to introduce Asian culture into a course when he or she can quickly think of at least ten different Asian cultures, all of which differ from each other drastically in important ways? What about Hispanic culture? Can the teacher who knows something of Mexico generalize about traditions of other Hispanic cultures? Can anyone teach the "black experience"? Do black men and women whose forebears come from Haiti and Nigeria and Jamaica share the experiences and heritage of African-Americans? Is Southern culture a valid topic for study? Many people think so. What about Jewish culture? But I don't need to labor the point. I only want to highlight the concerns any of us should have when the push for so-called multicultural courses threatens the integrity of our discipline and the quality of our teaching.

I believe, however, that we can create a culturally inclusive curriculum in our writing classes by focusing on the experiences of our students. *They* are our greatest multicultural resource, one that is authentic, rich, and truly diverse. Every student brings to class a picture of the world in his or her mind that is constructed out of his or her cultural background and unique and complex experience. As writing teachers, we can help students articulate and understand that experience, but we also have the important job of helping every writer to understand that each of us sees the world through our own particular lens, one shaped by unique experiences. In order to communicate with others, we must learn to see through their lenses as well as try to explain to them what we see through ours. In an interactive classroom where students collaborate with other writers, this process of decentering so one can understand the "other" can foster genuine multicultural growth.

Imagine, for example, the breadth of experience and range of difference students would be exposed to in a class made up of students I have had in recent years.

One student would be from Malawi. The ivory bracelet he wears was put on his arm at birth and cannot be removed; he writes about his tribal legends. Another student is a young Vietnamese man who came to America when he was eight; he writes about the fear he felt his first day in an American school because there were no walls to keep out bullets. Another is a young Greek woman whose parents brought her to America to escape poverty; she writes about her first conscious brush with sexism in the Greek Orthodox church. One student is the son of illegal aliens who followed the harvests in Texas; he writes with passion about the need for young Hispanics to get their education. A young black man writes about college basketball, a culture about which he is highly knowledgeable. A young man from the Texas panhandle writes about the traditions of cowboy boots and the ethical dimensions of barbed wire fences. Another young black man writes about the conflicts he feels between what he is learning in astronomy, a subject that fascinates him, and the teachings of his church.

It's worth noting here that religion plays an important role in the lives of many of our students—and many of us, I'm sure—but it's a dimension almost never mentioned by those who talk about cultural diversity and difference. In most classrooms in which there is an obvious political agenda, students—even graduate students—are very reluctant to reveal their religious beliefs, sensing they may get a hostile reception. And with reason—remember the quotation from David Bleich. But a teacher who believes in diversity must pay attention to and respect students with deep religious convictions, not force them too into silence.

Real diversity emerges from the students themselves and flourishes in a collaborative classroom in which they work together to develop their ideas and test them out on each other. They can discuss and examine their experiences, their assumptions, their values, and their questions. They can tell their stories to each other in a nurturant writing community. As they are increasingly exposed to the unique views and experiences of others, they will begin to appreciate differences and understand the rich tapestry of cultures that their individual stories make up. But they will also see unified motifs and common human concerns in that tapestry.

In this kind of classroom not all writing should be personal, expressive writing. Students need a broader range of discourse as their introduction to writing in college. The teacher can easily design the kinds of writing assignments that involve argument and exposition and suggest options that encourage cross-cultural awareness. For instance, some suggested themes for development might be these: family or community rituals; power relationships at all levels; the student's role in his or her family or group; their roles as men and women; the myths they live by; cultural tensions within groups. There are dozens more rich possibilites that could be worked out with the cooperation of colleagues in other departments and within the class itself.

The strength of all the themes I've mentioned is that they're both individual and communal, giving students the opportunity to write something unique to them as individuals yet something that will resonate with others in their writing community. The beauty of such an approach is that it's *organic*. It grows out of resources available in each classroom, and it allows students to make choices, then discover more about others and themselves through those choices. This approach makes the teacher a midwife, an agent for change rather than a transmitter of fixed knowledge. It promotes a student-centered classroom in which the teacher doesn't assume, as our would-be forcers of conscience do, that he or she owns the truth. Rather the students bring their own truths, and the teacher's role is to nurture change and growth as students encounter individual differences. Gradually their truths will change, but so will ours because in such a classroom one continually learns from one's students.

This is the kind of freshman English class from which students can emerge with confidence in their ability to think, to generate ideas, and to present themselves effectively to the university and the community. It is a class built on the scholarship, research, and experience that has enabled us to achieve so much growth in our profession in the last fifteen years. It is the kind of classroom we can be proud of as a discipline. I don't think we necessarily have to take freshman English out of English departments in order to establish this model, but we do have to assert our authority as writing professionals within our departments and fiercely resist letting freshman English be used for anyone else's goals. We must hold on to the gains we have made and teach writing in the ways we know best. Above all, we must teach it for the *students'* benefit, not in the service of politics or anything else.

Freshman English is a course particularly vulnerable to takeover because English departments in so many universities and colleges refuse to take it seriously and thus don't pay much attention to what happens in it. They can wake up, however, to find that some political zealots take the course very seriously indeed and will gladly put it to their own uses. The scores of us who have been studying, writing, speaking, and publishing for two decades to make freshman English the solid intellectual enterprise that it now is must speak out to protect it from this kind of exploitation. It is time to resist, time to speak up, time to reclaim freshman composition from those who want to politicize it.

What is at stake is control of a vital element in our students' education by a radical few. We can't afford to let that control stand.

Works Cited

Bauer, Dale. "The Other 'F' Word: Feminist in the Classroom." *College English* 52 (1990): 385-396.

Berlin, James A. "Rhetoric and Ideology in the Writing Class." *College English* 50 (1988): 477-494.

Bizzell, Patricia. "Beyond Anti-Foundationalism to Rhetorical Authority: Problems in Defining 'Cultural Literacy.' " *College English* 52 (1990): 661-675.

Bleich, David. "Literacy and Citizenship: Resisting Social Issues." Lunsford, Moglen, and Slevin 163-169.

Faigley, Lester. "The Study of Writing and the Study of Language." *Rhetoric Review* 7 (Spring 1989): 240-256.

Harkin, Patricia, and John Schilb. *Contending with Words: Composition and Rhetoric in a Postmodern Age*. New York: MLA, 1991.

Knoblauch, C. H. "Literacy and the Politics of Education." Lunsford, Moglen, and Slevin 74-80.

———. "Rhetorical Constructions: Dialogue and Commitment." *College English* 50 (1988): 125-140.

Laditka, James N. "Semiology, Ideology, Praxis: Responsible Authority in the Composition Classroom." *Journal of Advanced Composition* 10.2 (Fall 1990): 357-373.

Lunsford, Andrea A, Helen Moglen, and James Slevin, eds. *The Right to Literacy*. New York: MLA and NCTE, 1990.

Paine, Charles. "Relativism, Radical Pedagogy, and the Ideology of Paralysis." *College English* 51 (1989): 557-570.

Searle, John. "The Storm over the University." Rev. of *Tenured Radicals,* by Roger Kimball; *The Politics of Liberal Education,*, ed. by Darryl L. Gless and Barbara Hernstein Smith; and *The Voice of Liberal Learning: Michael Oakeshott on Education,* ed. by Timothy Fuller. *The New York Review of Books* 6 Dec. 1990: 34-42.

Strickland, Ronald. "Confrontational Pedagogy and Traditional Literary Studies." *College English* 52 (1990): 291-300.

Weaver, Richard M. *The Ethics of Rhetoric*. Chicago: Henry Regnery, 1953.

Donald Lazere

Teaching the Political Conflicts: A Rhetorical Schema

During the 1990-91 academic year, reports erupted into the national press about attempts at the University of Texas at Austin and the University of Massachusetts at Amherst to address controversial political issues such as racism and sexism in freshman writing courses. One reason these attempts provoked disputes, within their own English departments as well as publicly, is that little basis has been established within the discipline of composition delineating either a theoretical framework or ethical guidelines for dealing with political controversies in writing courses. While I do not have the local knowledge necessary to judge the particular conception and implementation of the courses at Austin and Amherst, I want briefly to address the larger theoretical issues and then go on to outline my own model for incorporating critical thinking about politics in writing courses, an approach which has evolved over some twenty-five years of college teaching and a decade of presenting workshops based on this model.

Overviews

Under the opposition of Maxine Hairston and other critics to courses like those at Austin and Amherst lie not only quarrels with politicizing writing instruction but broader theoretical assumptions and emphases that have dominated the profession for most of the last three decades—assumptions that I believe have imposed crippling restrictions on our field. The major emphasis in theory, courses, and textbooks has been on basic writing and the generation and exposition of one's own ideas, to the neglect of more advanced levels of writing that involve critical thinking in evaluating others' ideas (particularly in the public discourse of politics and mass media)—i.e., semantics, logic and argumentative rhetoric, and their application to writing critical, argumentative, and research papers and other writing from sources. The consequence of these dominant attitudes has been a failure of responsibility in the English profession to emphasize those aspects of composition that bear quite legitimately on the development of critical civic literacy—a failure that has contributed by default to the present, universally deplored state of political illiteracy, apathy, and semantic pollution by public doublespeak in America.

Obviously, I endorse the theoretical conceptions of courses like those at Austin and

From *College Composition and Communication* 43 (May 1992): 194-213.

Amherst. But I also share the concern of critics that such courses can all too easily be turned into an indoctrination to the instructor's particular ideology or, at best, into classes in political science. This concern has certainly been warranted by the tendency of some leftist teachers and theorists to assume that all students and colleagues agree—or *should* agree—with their views, rather than formulating their approach in a manner that takes respectful account of opposing views. My own political leanings are toward democratic socialism, and I believe that college English courses have a responsibility to expose students to socialist viewpoints because those views are virtually excluded from all other realms of the American cognitive, rhetorical, semantic, and literary universe of discourse.[1] I am firmly opposed, however, to instructors imposing socialist (or feminist, or Third-World, or gay) ideology on students as the one true faith—just as much as I am opposed to the present, generally unquestioned (and even unconscious) imposition of capitalist, white-male, heterosexual ideology that pervades American education and every other aspect of our culture.

I assert, then, that our primary aim should be to broaden the ideological scope of students' critical thinking, reading, and writing capacities so as to empower them to make their own autonomous judgments on opposing ideological positions in general and on specific issues. And it is just this aim that justifies introducing political issues in writing courses, within a *rhetorical* framework quite different from anything students are apt to encounter in political science or other social science courses. My concept of that framework consists of a version of what Gerald Graff advocates in literary theory as "teaching the conflicts"—introducing as explicit subject matter the issues of political partisanship and bias, as examples of the subjective, socially constructed elements in perceptions of reality and of the way ideology consciously or unconsciously pervades teaching, learning, and other influential realms of public discourse, including news reporting, mass culture, and of course political rhetoric itself. (By addressing these issues, *through a distinctively rhetorical approach,* writing courses can also become a vital part of the reorientation of English toward cultural studies.)

Part of my theoretical intention here is to indicate ways in which partisan political positions—like my own favoring socialist views on economic matters or those on sexism and racism emphasized in the courses at Austin and Amherst—can be introduced within a rhetorical schema that is acceptable to teachers and students of any reasoned political persuasion. In this way, I believe the left agenda of prompting students to question the subjectivity underlying socially constructed modes of thinking can be reconciled with the conservative agenda of objectivity and nonpartisanship. This approach obliges teachers to raise in class the question of their own partisan biases and how they can most honestly be dealt with in pedagogy and grading; I have found that students are immensely relieved at being able to discuss this taboo subject openly, to come to an open accord with the instructor about what guidelines are most fair, and to evaluate the instructor's fairness at the end of the course accordingly.

The most economical way of concretizing my theoretical position and tacitly answering likely objections to it is to provide an outline of the schema's central components in the form of four units of study that can be integrated into a writing course such as the one I teach as my department's second term of freshma n English, a course devoted

1. I have developed this argument in "Literacy and Mass Media" and *American Media and Mass Culture*.

to argumentative and source-based writing and the research paper.[2] These units coincide with the preliminary stages of researching and writing a term paper on a topic of current public controversy, a paper which consists of a rhetorical analysis of sources expressing opposing ideological viewpoints on the topic. The units provide a pedagogical context for seven appendices, which form the substantive core of the schema. The appendices provide the kind of guides to locating and analyzing partisan sources that have only recently started to show up in writing textbooks (e.g. Mayfield 236-263), and I hope my version will prompt other writers to incorporate them in textbooks, as I am doing myself in a textbook in progress.

If, as is almost inevitable, I sometimes let my own partisanship bias my presentation, I think that, far from discrediting my general intention, this will only illustrate and validate it. An implicit message of my approach to teaching political conflict is that any effort to construct such a schema is itself bound to be captive, in some measure, to the partisan biases it sets out to analyze. The only possible way to transcend these biases is refinement through dialectical exchanges with those of differing ideologies. So much the better, then, if readers who find fault with my definitions, interrelations among ideological positions, or predictable lines of partisan rhetoric can suggest modifications that will bring the schema closer to the difficult ideal of acceptability to those of any reasoned ideology; I regularly modify and update it myself in response to student suggestions in class. I hope that my beginnings here will prompt ongoing professional debate on these points and thereby help bring such debate into the forefront of composition theory.

The content of this schema and the four course units it is keyed to do not dictate any particular pedagogical model. Nor do their emphases on specific subject matter related to politics contradict the current emphasis in the profession on the writing process; on the contrary, the sequence of topics and assignments shows how process instruction can be extended to critical thinking, reading, and writing about political subject matter, while the appendices constitute a heuristic for the process of working the analysis of partisan rhetoric into an outline for the term paper.

As a final introductory note, I need to respond to the concerns of Hairston and other critics that writing instructors are venturing out of their own field of expertise when they address political issues. The level at which these issues are analyzed in a course like the one I describe here is that at which they are addressed, not in scholarly studies, but in political speeches, news and entertainment media, op-ed columns, general-circulation journals of opinion, and other realms of public discourse to which everyone is exposed every day. The political vocabulary and information covered here are no more specialized than what every citizen in a democracy should be expected to know, even before taking a college argumentative and research writing course. Indeed, two of the main points that must be stressed throughout such a course are the difference in levels of rhetoric between public and scholarly treatments of political issues and the need for students to take more specialized courses to gain deeper knowledge of these issues.

---

2. I have incorporated this approach in *Composition for Critical Thinking*, a monograph-length description of a model for a two-term freshman English course; the same pedagogical approach tacitly informs "Mass Culture, Political Consciousness, and English Studies" and its development in *American Media and Mass Culture*, explicitly in the introduction and readings on critical pedagogy in the concluding section, "Alternatives and Cultural Activism."

Students *can* learn in writing classes, though, to develop a more complex and comprehensive rhetorical understanding of political events and ideologies than that provided by campaign propaganda and mass media—or, for that matter, by most social science courses, whose emphasis is empirical rather than rhetorical. Higher education in composition as well as literature has the unique, Emersonian mission of bringing to bear on current events the longer view, the synthesizing vision needed to counteract the hurriedness, atomization, and ideological hodgepodge that debase our public discourse as well as our overdepartmentalized curricula and overspecialized scholarship.

Four Units for Teaching Political Conflicts

Political Semantics. This first topic in the schema can be integrated into a standard argumentative and research writing course with a review, within a General Semantics perspective, of definition, denotation, and connotation. This review provides a context for discussion of racism and sexism through study of selections from the large body of writings analyzing the role of definition and connotative language in the social construction of racial and gender identity, and of other issues in which control of definitions functions as a form of social power. Analysis of readings with opposing viewpoints on race and gender can focus on the semantic intricacies involved in current disputes over "political correctness," limits on free speech, tolerance of intolerance, and "reverse discrimination."

The unit continues with study of the problems of subjectivity involved in defining political terms, including the way partisan biases color our perception of these terms' meaning, through ambiguous or selective definitions, unconcretized abstractions, connotative associations and slanting, etc. Students are assigned to look up in one or more dictionaries the following terms: "conservatism," "liberalism," "libertarianism," "radicalism," "right wing," "left wing," "fascism," "plutocracy," "capitalism," "socialism," "communism," "Marxism," "patriotism," "democracy," "totalitarianism," "freedom," and "free enterprise." Then they bring their different dictionaries to class and read aloud the multiple and varying definitions for each word. In this way, students learn that understanding these terms and using them accurately in spoken or written discourse are complicated, not only by each dictionary's giving several meanings for each word but by differences among various dictionaries (and from one edition to another of the same dictionary—a nice lesson in historical subjectivity). Furthermore, even the largest unabridged dictionary fails to cover the almost infinite number of senses in which "liberal," "conservative," "socialist," "communist," and "Marxist" are used throughout the world, or the equally immense diversity of political factions which identify themselves with each of these ideologies. In America alone, a conservative may be a Menckenian aristocrat, a Donald Trump-type corporate capitalist, a Moral Majority populist, a Ku Klux Klanner, a member of the Libertarian Party, etc. And yet our mass media chronically use *conservative* either without any definition at all or as a simplistic label, as though it had one and only one meaning. Many Democratic and Republican Party politicians consciously evade any consistent definition of their ideology in an unscrupulous attempt to woo the widest possible constituency; hence they almost inevitably must resort to doublespeak.

This dictionary exercise can point up another widespread semantic confusion in our public discourse, the false equation of political terms like "democracy," "freedom," "justice," "patriotism," and "dictatorship" with words referring to economic systems—"capitalism" or "free enterprise" and "socialism." One must again go beyond dictionary definitions to address the problematic relation between these political and economic terms, for partisans of varying ideologies posit differing connections between, say, freedom and democracy on one hand and capitalism and socialism on the other. In Appendix Six, I have attempted to present my own definitions of these relations as objectively as possible, but, as I make clear to students, scholars whose ideological convictions differ from mine might take issue and present a quite different set of definitions. Indeed, the larger rhetorical question (a vital one for both class exercises and theoretical inquiry by English scholars) is whether it is possible to arrive at definitions of these terms and relationships that can be agreed on by partisans of all differing ideologies.

Next, the writer seeking accuracy of definition needs to key these political terms to a spectrum of positions from far right to far left in the United States and the rest of the world (see Appendix Two). Rather than speaking of "the liberal *New York Times,*" one should explain and document the sense and degree of liberalism referred to. "Liberal" in relation to what other media? One might clarify the label by placing the *Times* to the left of *Time* but to the right of *The Nation*. The whole range of American news media—along with politicians and parties, individual journalists and scholars, and even figures in popular entertainment (like Clint Eastwood and Jane Fonda, Donald Duck and Doonesbury,[3] Madonna and Bruce Springsteen, "Dallas" and "Roseanne")—can be placed on this spectrum in such a precise way that their political identity can be agreed on to a large extent by those of every ideological persuasion. In distributing Appendix Two to students I make it clear that this is a very general overview that necessarily involves over-simplifications and some debatable placements, and that this schema needs regular updating due to shifts in the positions of countries, individual politicians, writers, and periodicals. Recent upheavals in the Communist world have compounded semantic complexities: left-wing has been equated historically with Communism as a political ideology, but if "left-wing" is defined as opposition to the status quo, does that make those trying to overturn the status quo in Communist countries leftists or rightists? (During the 1991 attempted coup in the Soviet Union, my students, in tracking American news media, found that they most often designated the Communist hardliners "right wing" and "conservative.")

Extending the right-to-left spectrum worldwide serves to call students' attention to the parochially limited span of ideology represented by the poles of the Republican and Democratic parties and of "conservatism" and "liberalism" that define the boundaries of most American political, journalistic, scholarly, and cultural discourse. Factions and positions that are considered liberal in the United States, for example, usually stay well within the limits of capitalist ideology, thus are considerably to the right of the labor, social-democratic, and communist parties with large constituencies in most other democratic countries today. (A politician or position labeled "moderate" in the United States

---

3. For analyses of implicit ideological viewpoints in comic books, see the two chapters from *How to Read Donald Duck,* by Ariel Dorfman and Armand Mattelart, and "From Menace to Messiah: The History and Historicity of Superman," by Tom Andrae, reprinted in *American Media and Mass Culture*.

is considered right-wing from today's European perspective, while many American "radicals" would be "moderates" in Europe. Similarly, many "ultra-conservatives" in American terminology appear "moderate" in comparison to fascistic countries.) Therefore, in order to expose themselves to a full range of ideological viewpoints, students need to seek out sources excluded from the mainstream of American discourse, though such sources may be hard to find in many communities. The most prominent of these ideologies in a worldwide perspective are democratic socialism and libertarian conservatism (both of which favor political freedom), communism and fascism (both of which are opposed to democracy and freedom but are nonetheless strong presences in today's world and therefore need to be studied and understood through their own spokespeople and not just through the distorting filters of second-hand accounts).

Throughout this unit, and the rest of the course, the instructor needs to indicate the parameters of this kind of rhetorical analysis and the need for students to expose themselves, in history, political science, or economics courses, to more systematic analysis of ideologies and the way they have actually been implemented throughout the world.

Psychological Blocks to Perceiving Bias. The psychological factors that lead student writers and readers into partisan or biased arguments are an essential aspect of critical thinking and argumentative rhetoric that is inadequately emphasized in most conventional approaches to composition. (The growing body of recent scholarship applying the psychology of critical thinking "dispositions" and of cognitive and moral development to composition is useful theoretical and pedagogical background here; see my "Critical Thinking in College English Studies.") This unit—keyed, along with the following one, to Appendices Five through Seven—focuses on the most common psychological blocks to critical thinking that students should watch for in their sources for their term paper, as well as in themselves while reading and writing on these sources, and in their teachers in this and other courses. These blocks include culturally conditioned assumptions (which frequently emerge as hidden premises in arguments), closed-mindedness, prejudice and stereotyping; authoritarianism, absolutism, and the inability to recognize ambiguity, irony, and relativity of point of view; ethnocentrism and parochialism; rationalization, wishful thinking, and sentimentality. (Despite its brevity and simplicity, Ray Kytle's *Clear Thinking for Composition* is the most useful textbook I know on these blocks.)

The topic of prejudice provides a further occasion for consideration of opposing viewpoints on racial, sexual, and class bias. To avert a one-sided approach to these charged issues, it may be best to introduce them through psychological studies like Allport's *The Nature of Prejudice,* Rokeach's *The Open and Closed Mind,* or the developmental principles of Perry and Kohlberg and their feminist critique in Gilligan. In regard to subjectivity in political ideology in general, the beginning point here can be the hypothesis that many students have lived all their lives in a parochial circle of people who all have pretty much the same set of beliefs, so that they are inclined to accept a culturally conditioned consensus of values as objective, uncontested truth. (Most of the students at my school are middle-class whites, so middle-class ethnocentrism is the focus of study; this and other units of the course might need to be adjusted to differing pools of students.) Students need to become aware that what they or their sources of information assume to be self-evident truths are often—though not always—only the opinion or interpretation of the truth that is held by their particular social class, political ideology, religion, racial or ethnic group, gender, nationality and geographical location, historical period, occupation, age group, etc. Furthermore, we are all inclined to tailor our "objective" beliefs to the shape of our self-interest; consequently, in controversies where our interests are involved, we are susceptible to wishful thinking, rationalization, selective vision, and other logical fallacies. (A basic example of semantic cleans and dirties is that "biased" is a word that always applies only to arguments favoring the other side; we instinctively label arguments that confirm our own biases as "impartial," "well-balanced," "judiciously supported with solid research.")

Modes of Biased and Deceptive Rhetoric. While authorities used as sources, such as scholars, professional researchers or journalists, public officials, and business or labor executives can—or should—be expected to have a more informed viewpoint than students on specialized subjects, students should be made aware that authorities are not immune from numerous causes of subjective bias. This unit, then, addresses modes of biased or self-interested arguments in sources, as well as of outright deception—another aspect of rhetoric inadequately addressed in many conventional composition and argumentation courses and textbooks that regard fallacious reasoning mainly in terms of impersonal, formal reasoning and unintentional fallacies. A more realistic approach to contemporary public discourse necessitates a systematic study of possible causes for bias (and the predictable rhetorical patterns they produce—see Appendix Seven) in conventional sources of information—including political partisanship, conflicts of interest, sponsored research and journalism, special pleading, and other forms of propaganda and pure lying that have come to be known as public doublespeak. The growing influence during the past twenty years of books and articles produced by scholars in research institutes like those in Appendix Three, whose sponsors—frequently foundations representing corporations or political lobbies with special interests in the subjects studied—calls for particular attention to the possible biases of such scholars.

Useful textbooks and teachers' guides for this unit include the works cited by Schrank, Harty, Lazere, Rank, Dieterich, and Lutz. The latter two were published by the NCTE Committee on Public Doublespeak, whose *Quarterly Review of Doublespeak* and annual Doublespeak and Orwell Awards are equally valuable classroom resources. These can be supplemented by comparative analysis of current issues of periodicals devoted to criticism of bias in media, such as *Extra!, Propaganda Review,* or *Lies of Our Times* on the left, versus *AIM Report, Repap Media Guide,* or *MediaWatch* on the right.

The next point to be made is that every ideology—political, religious, etc.—is predisposed toward its own distinct pattern of rhetoric that its conscious or unconscious partisans tend to follow on virtually any subject they are reading, writing, or speaking about. Critical readers need to learn to identify and understand the various ideologies apt to be found in current sources of information. Having done so, they can then to a large extent anticipate what underlying assumptions, lines of argument, rhetorical strategies, logical fallacies, and modes of semantic slanting to watch for in any partisan source. (See Appendices Five through Seven.)

This is not to say that partisan sources should be shunned. Indeed, a clear-cut, well-supported expression of a partisan position can be more valuable than a blandly nonpartisan one. Nor does partisanship in a source necessarily go along with biased or deceptive reasoning. One must judge a partisan argument on the basis of how fully and fairly it represents the opposing position and demonstrates why its own is more reasonable. Some partisan authors or journals are highly admirable on this score (and students should be encouraged in their papers to cite such examples, not just fallacious or deceptive ones). Others, unfortunately, predictably repeat the same one-sided, doctrinaire line year after year, whatever the subject, and they are to be read, if at all, with one's bias calculator close at hand.

Locating and Evaluating Partisan Sources. The foregoing discussion units culminate in an assignment, for the preliminary stages of the term paper, of an annotated bibliography and working outline, designed to locate and analyze articles and books with opposing partisan viewpoints on the chosen topic (Appendix One). These exercises can help prevent students from simply picking *American Spectator* or *In These Times,* a book published by Arlington House or Monthly Review Press, or a report from American Enterprise Institute or the Institute for Policy Studies off the library shelf to use as a source and quoting it as gospel, without a critical understanding of the sponsor's habitual viewpoint. Following this procedure enables students to replace categorical assertions in their papers with statements like these:

> Barbara Ehrenreich, writing in *Democratic Left,* a journal of the Democratic Socialists of America, presents a socialist analysis of the effects of Reaganomics on the gap between the rich and poor in the United States.
>
> Ed Rubenstein, in the conservative *National Review,* refutes statistics presented by leftists like Ehrenreich claiming that Reaganomic policies have widened the gap between the rich and poor.

Student writers can then go on to explain how the source's general ideological viewpoint applies to the particular issue in question, to analyze the rhetorical/semantic patterns accordingly, and to evaluate the source's arguments against opposing ones. In this way they can get beyond the parochial mentality of those who read and listen only to sources that confirm their preconceptions while deluding themselves that these sources impartially present a full range of information.

Lest this approach be misconstrued as an invitation to total relativity or scepticism, students are asked in the conclusion to their term papers not to make a final and absolute judgment on which side is right and wrong about the issue at hand, but to make a balanced summary of the strong and weak points made by each of the limited number of sources they have studied, and then to make—and support—their judgment about which sources have presented the best-reasoned case and the most thorough refutation of the other side's arguments. Grading for the paper and the course, then, becomes a matter of evaluating the quality of students' support for their judgments—regardless of what those judgments may be.

Conclusion

English faculties correctly resist attempts to make composition a "service course" providing only the technical skills needed for writing in other disciplines; however, composition can and should be a service course in the sense of fostering modes of critical thinking that are a prerequisite to studies in other disciplines—preeminently the social sciences—and to students' lifelong roles as citizens. To reiterate, this does not mean that composition should be turned into a social science. Nor does this conception of composition duplicate recent efforts to incorporate writing instruction within social science courses and other disciplines. Writing across the curriculum is a laudable enterprise, but it is not the same thing as what I am advancing, a program for studying political issues squarely within the discipline of English and through its distinctive humanistic concerns. By thus breaking through the arbitrary disciplinary constrictions that have diminished the scope of composition scholarship in recent decades, we can begin to restore the study of composition to its classical role as the center of education for citizenship.

Works Cited

Allport, Gordon. *The Nature of Prejudice.* Reading: Addison-Wesley, 1979.

Dieterich, Daniel, ed. *Teaching About Doublespeak.* Urbana, IL: NCTE. 1977.

Gilligan, Carol. *In a Different Voice: Psychological Theory and Women's Development.* Cambridge: Harvard UP, 1982.

Graff, Gerald. *Professing Literature: An Institutional History.* Chicago: U of Chicago P, 1987.

Hairston, Maxine. "Required Writing Courses Should Not Focus on Politically Charged Issues." *Chronicle of Higher Education* 23 January 1991: B2-3.

Harty, Sheila. *Hucksters in the Classroom: A Review of Industry Propaganda in Schools.* Washington: Center for the Study of Responsive Law, 1979.

Kohlberg, Lawrence. *Essays on Moral Development: Volume 1., The Philosophy of Moral Development.* San Francisco: Harper and Row, 1981.

Kytle, Ray. *Clear Thinking for Composition.* New York: Random House, 1986.

Lazere, Donald. *American Media and Mass Culture: Left Perspectives.* Berkeley: U of California P, 1987.

———. *Composition for Critical Thinking: A Course Description.* Rohnert Park, CA: Center for Critical Thinking and Moral Critique of Sonoma State U, 1986. ERIC, 1986. ED 273 959.

———. "Critical Thinking in College English Studies." ERIC Digest, 1987.

———. "Literacy and Mass Media: The Political Implications." *New Literary History* 18 (Winter 1987): 238-255. Rpt in *Reading in America: Literature and Social History.* Ed. Cathy Davidson. Johns Hopkins UP, 1989. 285-303.

———, guest editor. "Mass Culture, Political Consciousness, and English Studies." *College English* 38 (Nov. 1977).

Lutz, William A., ed. *After 1984: Doublespeak in a Post-Orwellian Age.* Urbana, IL: NCTE, 1989.

Mayfield, Marlys. *Thinking for Yourself: Developing Critical Thinking Skills Through Writing.* 2nd ed. Belmont, CA: Wadsworth, 1991.

Perry, William. *Forms of Intellectual and Ethical Development in the College Years.* New York: Holt, 1970.

Rank, Hugh. *Persuasion Analysis: A Companion to Composition.* Park Forest: Counter-Propaganda P, 1988.

———. *The Pitch.* Park Forest: Counter-Propaganda P, 1982.
Rokeach, Milton. *The Open and Closed Mind.* New York: Basic, 1960.
Schrank, Jeffrey. *Deception Detection.* Boston: Beacon P, 1979.
———. *Snap, Crackle, and Popular Taste: The Illusion of Free Choice in America.* New York: Dell, 1977.

APPENDIX ONE. *Assignment for an Annotated Bibliography and Working Outline*

Turn in ten bibliographical entries, on five leftist and five rightist sources, including at least one magazine article and one book or monograph report from the left-wing publishers and one article and book or report from the right-wing publishers in Appendices. Three and Four. Annotate the sources according to the following guidelines, and develop them into a detailed working outline keyed to citations of these entries.

1. Identify author's political position, using clues from affiliation with a particular research institute, book publisher, journal of opinion, party, or organization, and—more importantly—from arguments s/he presents that exemplify the glossary terms and the particular patterns of political rhetoric in Appendices Six and Seven; give enough quotes (or highlighted photocopies) to support your identification. In cases where the author is not arguing from an identifiable position but only reporting facts, indicate which position the reported facts support, and explain how. (Note: some newspapers, magazines, etc., have an identifiable political viewpoint in general, in their news and op-ed orientation, but also attempt to present other views at least some of the time; e.g., the *LA Times* is predominantly liberal, but often carries conservative op-ed columns, letters, etc. So you shouldn't assume that any article appearing in such a periodical will automatically have its predominant viewpoint; look for other identity clues.)
2. Apply to each source the "Semantic Calculator for Bias in Rhetoric" (Appendix Five), along with the more general principles of rhetorical analysis studied in this course.

APPENDIX TWO (A). *Political Spectrum* (Ca. 1985)*

← Left Wing   Right Wing →

Dictatorship | Political Democracy, Freedom | Dictatorship

Communism | Socialism | Capitalism | Plutocracy | Fascism

Communism: USSR, China, Cuba, North Vietnam, Cambodia, North Korea

Nicaragua (Sandinista)
Chile (Allende)

Sweden, Denmark, Norway

France, Italy, W. Germany, Spain, Canada, England

USA, Japan

Chile (Pinochet), Philippines (Marcos), South Africa, El Salvador, Nicaragua (Somoza), South Vietnam, South Korea, Taiwan
Nazi Germany, Fascist Italy, Franco Spain, Fascist Japan

American Parties
Democratic   Republican
Libertarian

---

*This version reflected the world spectrum in the mid-1980s. Regular updatings are, of course, necessary, and recent upheavals in the Communist world in particular necessitate major revision on the extreme left.

The politics of Middle Eastern countries are too complex a mix of left-wing and right-wing forces to schematize here. For example, Iran under the Shah was a plutocratic dictatorship allied with the U.S.; under the Ayatollahs it is another variety of right-wing dictatorship, a theocratic one, but is allied with some Communist and left-wing Arab forces.

APPENDIX TWO (B). *American Media and Commentators from Left to Right*

People's World	*The Nation*	*Village Voice*	*NY Times*	*Time*	*New American*
The Guardian	*In These Times*	*LA Times*	*Wash. Post*	*US News & World Report*	*Plain Truth*
	Mother Jones	*NY Review*	*Newsweek*	*Readers Digest*	*Wash. Times*
					(Insight)
	The Progressive	*Atlantic*	*New Republic*	*Wall St. Journal*	
	Z Magazine	*New Yorker*	*Harper's*	*Commentary*	
	Tikkun	PBS documentaries	*Reason*	*American Spectator*	
	Pacifica Radio	NPR	CBS news	Most newspapers, local TV & radio	
			NBC, ABC news	*National Review*	
				McLaughlin Group	
		60 Minutes			
		McNeil-Lehrer			

Alexander Cockburn				
Edward Said	Gore Vidal	Michael Kinsley	George Will	Pat Buchanan
Noam Chomsky	Barbara Ehrenreich	Anthony Lewis	Chas. Krauthammer	Phyllis Schlafly
Edward Herman	Jesse Jackson	Tom Wicker	William Safire	Pat Robertson
	Todd Gitlin	Richard Reeves	Evans & Novak	Paul Harvey
	Robert Scheer	Bill Moyers	Henry Kissinger	Jerry Falwell
	Betty Friedan	Gloria Steinem	Irving Kristol	
	Stanley Aronowitz	Seymour Hersh	Norman Podhoretz	
	Molly Ivins	David Halberstam	Midge Decter	
	Irwin Knoll	Woodward & Bernstein	Jeane Kirkpatrick	
	James Weinstein	Ted Koppel	William Buckley	
	Ralph Nader	John K. Galbraith	Michael Novak	
	Victor Navasky		Milton Friedman	
	Roger Wilkins		Thomas Sowell	
	Cornel West	Murray Rothbard (libertarian)		
		Douglas Bandow (libertarian)		

APPENDIX THREE. *Political Orientations of Publishers & Foundations*

Book Publishers

Liberal or Socialist	*Conservative or Libertarian*
Pantheon	Arlington House
Monthly Review Press	Freedom House
South End Press	Brandon Books
Praeger	Reader's Digest Books
Beacon Press	Greenhill Publishers
Seabury/Continuum Books	Laissez-Faire Books (Libertarian)
International Publishers	Paragon House
Pathfinder Press	
Routledge	
Methuen	
Schocken	
Bergin & Garvey	

Research Institutes and Foundations

Liberal or Socialist	*Conservative or Libertarian*
Institute for Policy Studies	American Enterprise Institute (Journal: *Public Opinion*—not *Public Opinion Quarterly*)
Center for Responsive Law (Journal: *Public Citizen*)	Center for Strategic and International Studies
Public Interest Research Groups	Hoover Institution (Stanford)
Common Cause (Journal: *Common Cause*)	The Media Institute
Brookings Institute	Hudson Institute
Institute for Democratic Socialism (Journals: *Democratic Left, Socialist Forum*)	Heritage Foundation (Journal: *Policy Review*)
Center for the Study of Democratic Institutions (Journal: *New Perspectives Quarterly*)	Olin Foundation
	Scaife Foundation
	Cato Foundation (Libertarian: *Cato Journal*)

APPENDIX FOUR. *Current General Periodicals*

This is a partial list intended to supplement, not replace, the more accessible, mass circulation newspapers and magazines, most of which have a center-conservative to center-liberal orientation.

American Scholar	Quarterly	Left-conservative
American Spectator	Monthly	Center-to-left conservative
Atlantic Monthly	Monthly	Center-liberal
The Black Scholar	Quarterly	Socialist
Chronicles of Culture	Monthly	Left-conservative
Commentary	Monthly	Center-conservative
Commonweal	Bi-weekly	Left-liberal Catholic

Conservative Digest	Monthly	Center-to-right conservative
Dissent	Bi-monthly	Socialist to center-liberal
Foreign Affairs	Quarterly	Center-conservative to right-liberal
The Guardian	Weekly	Socialist
Harper's	Monthly	Center-liberal to left-conservative
Human Events	Weekly	Center-to-right conservative
Insight (Washington Times)	Weekly	Center-to-right conservative
In These Times	Weekly	Socialist
Modern Age	Quarterly	Center-conservative
Mother Jones	Monthly	Socialist to left-liberal
Ms.	Monthly	Center to left-liberal
The Nation	Weekly	Socialist to left-liberal
National Review	Bi-weekly	Center-conservative
New American	Bi-weekly	Right-conservative (formerly *American Opinion*)
New Guard	Quarterly	Center-conservative
New Politics	Quarterly	Socialist
New Republic	Weekly	Right-liberal to left-conservative
New York Review of Books	Bi-weekly	Center-liberal
New York Sunday Times	Weekly	Center-liberal to left-conservative
New Yorker	Weekly	Left-to-center-liberal
People's World	Daily	Community Party USA
Progressive	Monthly	Socialist to left-liberal
Public Interest	Quarterly	Left-to-center-conservative
Public Opinion	Monthly	Center-conservative
Reason	Monthly	Conservative libertarian
Rolling Stone	Bi-weekly	Center-liberal
Social Policy	Bi-monthly	Left-liberal
Socialist Review	Quarterly	Socialist
Tikkun	Bi-monthly	Left-liberal
Utne Reader	Bi-monthly	Digest of liberal journals
Village Voice	Weekly	Left-liberal
Washington Monthly	Monthly	Center-liberal to left-conservative
World Press Review	Monthly	Digest of diverse foreign viewpoints
Z Magazine	Monthly	Socialist

APPENDIX FIVE. *A Semantic Calculator for Bias in Rhetoric**

1. What is the author's vantagepoint, in terms of social class, wealth, occupation, ethnic group, political ideology, educational level, age, gender, etc.? Is that vantagepoint apt to color her/his attitudes on the issue under discussion? Does she/he have anything personally to gain from the position she/he is arguing for, any conflicts of interest or other reasons for special pleading?

---

*This guide derives from Hugh Rank's ''Intensify-Downplay'' schema, various forms of which appear in Rank *(Persuasion, Pitch)* and in Dieterich.

2. What organized financial, political, ethnic, or other interests are backing the advocated position? Who stands to profit financially, politically, or otherwise from it?
3. Once you have determined the author's vantagepoint and/or the special interests being favored, look for signs of ethnocentrism, rationalization or wishful thinking, sentimentality, and other blocks to clear thinking, as well as the rhetorical fallacies of onesidedness, selective vision, or a double standard.
4. Look for the following semantic patterns reflecting the biases in No. 3:
   a. Playing up:
      (1) arguments favorable to his/her side,
      (2) arguments unfavorable to the other side.
   b. Playing down (or suppressing altogether):
      (1) arguments unfavorable to her/his side,
      (2) arguments favorable to the other side.
   c. Applying "clean" words (ones with positive connotations) to her/his side. Applying "dirty" words (ones with negative connotations) to the other.
   d. Assuming that the representatives of his/her side are trustworthy, truthful, and have no selfish motives, while assuming the opposite of the other side.
5. If you don't find strong signs of the above biases, that's a pretty good indication that the argument is a credible one.
6. If there *is* a large amount of one-sided rhetoric and semantic bias, that's a pretty good sign that the writer is not a very credible source. However, finding signs of the above biases does not in itself prove that the writer's arguments are fallacious. Don't fall into the *ad hominem* ("to the man") fallacy—evading the issue by attacking the character of the writer or speaker without refuting the substance of the argument itself. What the writer says may or may not be factual, regardless of the semantic biases. The point is not to let yourself be swayed by words alone, especially when you are inclined to wishful thinking on one side of the subject yourself. When you find these biases in other writers, *or in yourself,* that is a sign that you need to be extra careful to check the facts out with a variety of other sources and to find out what the arguments are on the other side of the issue.

APPENDIX SIX. *A Glossary of Political Terms and Positions*

Left wing and right wing (also see Appendix Two):

"The left wing" (adjective: "left-wing" or "leftist") is a broad term that includes a diversity of parties and ideologies (which often disagree among themselves but usually agree in their opposition to the right wing) including liberals, nearest the center of the spectrum, and—progressively toward the left—socialists and communists (the latter two are also sometimes called "radical").

"The right wing" (adjective: "right-wing" or "rightist") is a broad term that includes a diversity of parties and ideologies (which often disagree among themselves but usually agree in their opposition to the left wing) including libertarians, nearest the center of the spectrum, and—progressively toward the right—conservatives, ultraconservatives, plutocrats, and fascists.

Leftists tend to support:	*Rightists tend to support:*
The poor and working class	Middle and upper class

Labor, consumers, environmental and other controls over business	Business, management, unregulated enterprise
Equality (economic, racial, sexual)	Inequality (economic, racial, sexual)
Civil and personal liberties	Economic liberty; controls on personal liberties (e.g., sexual conduct, abortion, obscenity, drugs)
Cooperation	Competition
Internationalism	Nationalism (primary loyalty to one's own country)
Pacifism (exception: Communists)	Strong military and willingness to go to war
Questioning of authority—skepticism (exception: Communism is authoritarian)	Acceptance of authority, especially in military, police, and strong "law and order" policies
Government spending for public services like education, welfare, health care, unemployment insurance	Government spending for military, subsidies to business as incentive for profit and growth
Progressive taxes, i.e. greatest burden on wealthy individuals and corporations	Low taxes for wealthy individuals and corporations as incentive for investment ("supply-side economics" or "trickle-down theory")
Religious pluralism, skepticism, or atheism	Religious orthodoxy

Capitalism:

An economic system based on private investment for profit. Jobs and public services are provided, and public needs met, to the extent that investment in them will predictably result in a return of capital outlay. In its principles capitalism does not provide any restrictions on extremes of wealth and poverty or of social power, but its advocates (especially pure, libertarian capitalists) believe that the workings of a free market economy, unrestricted by government controls or regulation, will minimize social inequity.

Capitalism is not a political system; in principle, a capitalist economy can operate under either a democratic government or a dictatorship, as in plutocracy or fascism—see Appendix Two (A).

Socialism:

An economic system based on public investment to meet public needs, provide full employment, and reduce socioeconomic inequality. In various models of socialism, investment and industrial management are controlled either by the federal government, local governments, workers' and consumers' cooperatives, a variety of community groups, etc.

Socialism is not a political system; in principle, a socialist economy can operate under either a democratic government or a dictatorship, as in Communism—see Appendix 2 (A).

Communism:

With lower-case ''c'': Marx's ideal of the ultimate, future form of pure democratic socialism, with virtually no need for centralized government.

With upper-case ''C'' as in present-day Communist Parties: A socialist economy under undemocratic government. Historically, Communists have manipulated appeals to left-wing values like socioeconomic equality and worldwide cooperation in order to impose police-state dictatorship and military aggression.

Plutocracy:

Rule by the rich. A capitalist economy under undemocratic government.

Fascism:

A combination of capitalist and socialist economies under an undemocratic government. Historically, fascists have manipulated appeals to conservative values like patriotism, religion, competitiveness, anti-communism, respect for authority and law and order, traditional morality and the family, in order to impose police-state dictatorship.

Fascism typically is aggressively militaristic and imperialistic, and promotes racial hatred based on theories of white (or ''Pure Aryan'') supremacy and religious persecution of non-Christians. It glorifies strong authority figures with absolute power.

Conservatives, Liberals, and Socialists in America:

In the American context, conservatives are pro-capitalist. They believe the interests of business also serve the interests of labor, consumers, the environment, and the public in general—''What's good for General Motors is good for America.'' They believe that abuses by businesses can and should be best policed or regulated by business itself, and when conservatives control government, they usually appoint businesspeople to cabinet positions and regulatory agencies without perceiving any conflict of interest therein.

American liberals believe that the interests of business are frequently contrary to those of labor, consumers, the environment, and the public in general. So although they basically support capitalism, liberals think business abuses need to be policed by government regulatory agencies that are free from conflicts of interest, and that wealth should be limited.

American socialists, or radicals, believe even more strongly than liberals that the interests of business are contrary to the public interest; they believe that capitalism is basically an irrational and corrupt system where wealthy business interests inevitably gain control over government, foreign and military policy, the media, education, etc., and use the power of employment to keep the workforce and electorate under their control. They think liberal government reforms and attempts to regulate business are usually thwarted by the power of business lobbies, and that even sincere liberal reformers in government offices usually come from and represent the ethnocentric viewpoint of the upper classes. The socialist solution is to socialize at least the biggest national and international corporations, as well as the defense industry, and operate them on a nonprofit basis, and to place much higher taxes on the rich, so as to reduce the power of wealthy corporations and individuals.

APPENDIX SEVEN. *Predictable Patterns of Political Rhetoric*

Leftists will play up:	*Rightists will play up:*
Conservative ethnocentrism, wishful thinking, and sentimentality rationalizing the selfish interests of the middle and upper class and America abroad	Leftist ''negative thinking,'' ''sour grapes,'' anti-Americanism, and sentimentalizing of the lower classes and Third World rebellion
Right-wing bias in media and education	Left-wing bias in media and education
Conservative rationalization of right-wing extremism and foreign dictatorships allied with US (e.g., El Salvador, South Vietnam); rightists' use of ''Communism'' as scapegoat for rebellion against right-wing extremism	Leftist rationalization of Communist dictatorships or guerillas (e.g., Nicaragua, North Vietnam) and wishful thinking in leftists' denial of Communist influence in anti-American rebellions
US military strengths, selfish interests of the military and defense industry; right-wing scare tactics about the Russians or other adversaries being ahead	Russians' or other adversaries' military strengths; manipulation of leftist ''doves'' by Communists et al.; left-wing scare tactics about nuclear war
Rip-offs of taxpayers' money by the rich; luxury and waste in private industry and the military	Rip-offs of taxpayers' money by the poor; luxury and waste by government bureaucrats; selfish interests and inefficiency of labor, teachers, students, etc.

TEACHERS

Rosemary Deen

Notes to Stella

If my experience could help you as you begin teaching, I'd put it in the form of principles, because only principle holds in the long run. But plain principle is pretty stiff medicine, so in these notes I also suggest some practical versions of theory. Teaching needs to stay connected with generative powers that are always at work but are too simple for us to analyze.

In an attempt to head you away from problems that waste time and spirit, I want to leave you and your students free to find and engage the lively problems of how our feelings and thoughts work in words.

There's a fact I believe teachers ought to face at the start, however. Teachers hold power. Sometimes it's power over self, or skill, sometimes power over others. But power over others, as Marie said in a radio interview, is "more than anyone can bear or carry out." So all principle depends on giving up this kind of power. What kills teaching is pursuing error and weakness in others; what enables teaching is releasing students' energy by defining their success.

I. PRINCIPLES

A. Work from strength. Give writers the means to fulfill the child's motto: Self can do it!

1. Assign a concrete structure that has literary, generative power, rather than a subject, theme, or topic to inexperienced writers.
2. Use differences in the way this structure is imagined and embodied by various writers to help them all notice its underlying form.
3. Use differences in what students notice to share out the work of response.
4. Use differences, use differences, use differences. Anything you do that implies that class members are the same—for instance, reading a sample essay, or lecturing to the whole class about a common error—is a dead bore and a dead end.

B. Generate power.

1. Only literature, imaginative expression, has generative power. We are teaching composition, but we can't and would never want to separate imaginative and analytical powers. Leonard put his finger on it when he said, "Analysis is the *re*-organizer. It can't bring the body to life."
2. Develop a core of repeatable work that can't be done wrong. And repeat it. The core consists of small, discrete, variable practices.

As we all know, learning takes four steps. To begin, we have to be able to *do* the thing in some elemental form. No amount of talk and explanation will get students to do it. Most teaching founders right here in the beginning, with teachers explaining and explaining all semester what they want students to do, but never getting them to produce an elemental version of it except by chance.

The next step is to do it again. At the heart of good teaching is consistent repetition. As you well know, nothing can be mastered without patient practice, for practice is not only generative but calming. A thing your students do once, no matter how "successful" it seems, is a gimmick. Of course, you want to invent practice that's not soul-killing drill.

The next step is to do it any time we want to. At last we can do it on demand. This final stage, mastery, depends on the first three steps, on being able to begin a thing, sustain it, and practice it at will.

3. Always write with students when you ask them to write. Nothing you can say about the importance of writing will survive the evidence that you don't think it's important enough to do it yourself.

   You can help your students because you yourself are such a good writer and reader. Of course, not everyone can write in your pure, strong style. I can't. You aren't teaching students to write like you, however. What apparently separates you from inexperienced writers, your own powers, is the primer from which you learn to teach them the elements of writing to develop in their own way.

   As for teachers who can't write well, it's a commonplace that we learn by teaching. I've never had a team teacher who didn't confide to me at the end of the semester that she or he had become a better writer by writing with students.

4. Depend on sentence power. The sentence is the single, most powerful generative structure.

C. Organize and repeat incrementally.

1. Embed the order of what you know in ordered work that students can practice. You can't tell students what you know. If you talk for more than five minutes in a writing class, you've lost it.
2. The first benefit you give your students is to design a set of daily class assignments. Every time I do this elemental job, I have to remind myself of these principles as Marie articulates them. Assignments should be work that:
   - everyone does, and no one does wrong,
   - extends students' practice of what's been taught in ways they can repeat at will,
   - is useful to the group on the day it is due,
   - can be responded to quickly and with pleasure.
3. Design essays to repeat and proliferate out of a center. Anything well learned begins with the center. A student of the *Iliad* may have been struck by the image of Aphrodite threatening Helen, for example. Her study defines this as the center of her idea (not the center of the *Iliad*), and picks up the material that embodies and reflects her idea.

4. Design essays in stages so the later ones are increments of the first.

   This might mean repeating the structure of the first essay in a set of essays. Or it might mean letting students pick up an idea they have discovered in their first paper as the basis for the idea of their second paper.

   If these papers are about reading, don't worry about students' comparing very different works in different periods. Any work may be compared/contrasted with any other. We learn in an opportunistic fashion, comparing every new thing with what we already know.

D. Structure the work, not the students' response.

   Implied by the principles above, this is amplified below when we come to Practice.

II. DIALECTICS

Now comes a patch of negatives, and I don't apologize for them. We need strong negatives when we use power: "Don't drive without good brakes." "When you use electricity, don't stand in the water."

These negatives aren't guilty sins. The other side of the energy and act of teaching is, naturally, inertia and absent-mindedness, exercising power over others without intending it because we're so eager to do the right thing. I like to remember that Marie once told an anxious, brand-new mother, "Don't worry; you can't do it right." Teachers can't either, but they do get a chance to start over again every semester.

A. Recognize incorrect behavior: theirs and yours.

   Problems of "control" would appear as petty as they are if they weren't deeply ambiguous. Let me give you an image you may find useful from my first lesson in sailing. It was a small craft, a twenty-five-foot sloop, and I was grimly terrified as I tried to control wind and waves by the tiller and headsail of a boat that was only too responsive to the signals I was sending it. "If you're in trouble, let go," said my calm teacher. Let go! We'd capsize in a moment! But at his word I let go, and the boat sailed into the wind and stopped.

   Here's a minimum of negative-positive advice:

1. You'll regret it if you feature problems by going on ahead of time about what students can't do (be late, miss class or assignments, etc.).
2. Don't make many laws. Despite its beauty, law has the negative power of creating criminals. Remember Shelley's line about "Golden and sanguine laws which tempt and slay" ("England in 1819").

   Make sure the parameters are for everyone. When you keep them, their virtue will be evident to students. You won't have to tell them not to be late if you begin working promptly and keep a forward rhythm in the work. It will be obvious that work can't come in late, when work is incremental and when you return it within the week.
3. The worst breach of discipline occurs when teachers talk all the time and so imply that talk is more important than writing. I say this, though I know you would never conduct a class with the sound of your own voice. Maybe because you were the middle child or maybe because you're a good writer,

you've always practiced a beautiful tact. You'll find it a superb asset in teaching.

4. If an announcement is important, say it matter-of-factly: "I can't accept essays late because once the time I've set aside for reading them is past, I don't have that time anymore."

   You will have to repeat this to individuals (who were missing class when you told them you wouldn't accept missed classes). When transgressions appear, speak of them privately, in one sentence, and in the manner of someone mentioning an unfortunate social blunder. Treat students' errors not as terrorist explosives, but as bad form. Most people recognize that bad form is more serious than moral turpitude.

   If a childish student starts testing you, don't be drawn. Speak later and privately. Say what you have to say in one sentence: "Your conduct in class is incorrect."

5. Conferences. If conferences were videotaped and broadcast, many teachers would be taken out and hanged at once. Keep conferences productive by conferring about writing.

   In a conference you help students with their written work. But students may bring to a conference something you can't help: their absence in body or mind from the class or the assignment. If it's urgent, you could listen to about half a minute of this. You can listen courteously, or at least without getting infuriated, more easily if you know it's going to last only half a minute. Then turn to their written work, which you can help.

   When you announce your office location and hours, tell students to have their writing in hand and in mind when they come. When they come empty-handed or empty-minded, have them write for ten minutes about the most important idea of their essay. Discuss this writing. Even if a suffering student writes one sentence only, you're in business. Both of you can write observations on the sentence and read them to each other. Or you can rewrite the sentence as described below.

   There's no point in going over what you've already written on the paper, especially when authors declare themselves unable to understand your comments or know "what you want." Ask students to find the best, the most important sentence in the essay. Both of you copy it. In five minutes write three new versions of this sentence, each of which says the same thing in fresh language. Read your sentences to each other. If there's time, do it again with the second most important sentence. This gives writers eight sentences they can take home and organize into a new opening paragraph to define their idea and lay out the basis for reorganizing and rewriting the essay.

B. Avoid being the center of it all.
   1. Explaining. The more you have to explain it, the more you know you've got it wrong, and they will get it wrong. Worst are general writing assignments that take a whole period to explain.
   2. The Right Answer/The Ideal Paper. Don't ask big, Right-Answer questions, such as, "What is the main point of this article?" What follows will be a

dismal circle of Rong Answers. The Right Answer is the one you find in the Teacher's Guide to your textbook. The Ideal Paper matches the paper in your mind. You won't find it, even if you write it yourself.

3. Being the Editor and the Copy-Editor. I believe teachers edit and copy-edit students' writing because it's something they can do for students, and they don't know what else to do. In publishing, however, editors enjoy unchallenged authority. They aren't trying to teach writing, and they don't have to edit twenty-five books simultaneously. Copy-editing is a kind of refined slavery, but its skillful practitioners are respected, heeded, and often paid by the hour. When they correct "seperate," it stays corrected.

   Eliminate the attraction of your students' clichés. They stand out like beacons to you, but the students can't see them. They will pretend to see them when you point them out, if they are docile. If they are energetic, they'll fight you and demand explanations and reasons. You don't want to spend energy for that.

   When you attack writers' stereotypes, you attack sensitive blind spots. One person's news is another's cliché. And confronting students' clichés is an attack on their attachment to popular culture—probably not part of your mandate.

   Ask yourself what you expect by working on faults. Does buttressing weak points strengthen the structure? Asking writers to revise problems is an energy sink. Genuine change requires a renewal of energy by revisiting the brightness of the original idea wherever it shines through.

4. Revision to improve students' writing. The temptation to edit students' writing originates in the idea that you can teach students to write *well,* and it springs up at the prospect of revision. This is the swamp of writing and teaching. Remember Thoreau's story of the traveler who was told by a boy that the swamp ahead of him had a hard bottom. When his horse sank in up to the girths, he called back, "I thought you said this swamp has a hard bottom?"

   "So it has," said the boy, "but you haven't got half way to it yet" (*Walden,* Chapter 18).

   The swamp is the guilty muck, the airless layers of a person's failures, bad habits, and persistence in error. There's no getting to the bottom of it.

   What to do instead? In a word, turn your energy and theirs in another direction. Reward wit, humor, images, metaphor. Don't suggest, by treating literary power perfunctorily, that it is less acceptable than reasoning or that reason enters by another gate. Humans think in images. Wit is energizing. Teaching students to begin writing with topic generalizations is the shortcut to the swamp.

5. Avoid the town of Morality. Remember how Pilgrim's burden got heavier as he got into the town of Morality? We can help students past most problems without even naming the problem and certainly without wasting time analyzing it. Prolific writing brings about fluency. Rewriting from strength takes care of most errors, except copy-editing errors, the kind you can't

hear when the essay is read aloud, like spelling. But spellers are born, not made—or they were made way back, when they learned handwriting.

Chronic misspellers may be dyslexic. They need another person to correct their work and write out for them the way the word looks. I don't believe you can teach spelling and writing in the same class. But at any rate, have nothing to do with the Morality of Spelling. It aims to make writers conscious of errors so they will look up the fatal words next time. They won't. All you will have given them is the Consciousness of Sin.

6. Shock tactics. A sure sign of the desperation of an enterprise is the way teachers apply shock to induce it. The best example is revision. Teachers want writers to revise, but they don't know how to make them do it, so they apply aggressive pressure. One notion is to make students begin with "chaos," "perplexity," "anxiety," "uncertainty." (I'm quoting from an essay on the subject.) These folks want to make students conscious of their "process" or their "protocols." They confront writers with their faults by "peer critiquing." They keep talking about the "agony" of writing. Students who enjoy or might enjoy writing can see that their sort of writing isn't serious.

   Behind these practices is the realization that even incomplete ideas jell when they are written down. Teachers think it will help students revise later to make them begin orally ("brainstorming") or with some kind of writing that isn't really writing, that denies form and seems fluid enough to go into "revision." Students get stuck with one or two essays kept perpetually in flux all semester with much supervision and managing by the teacher-editor.

   These practices induce self-consciousness and bring on the "writer's block" the teacher had predicted. "Writer's block" is a function of the idea that writing is heroic. It occurs when writers are made to focus on themselves instead of on what they have in mind.

7. Later I suggest some practices that teach rewriting (as distinguished from revision). Now let me say some truths contrary to the errors embedded in shock tactics.

   Writing is not heroic. It's an ordinary act, like learning to ride a bicycle when you know how to walk, turning one form of bi-pedal motion into another for the pleasure of it.

   The whole of an act is in its beginning. Writing begins with writing; what begins orally is going to go on orally.

   The first literary pleasure is the sense of closure. To complete something small and shapely is the elemental or source experience of writing.

   The author is the authority. You can teach students to write, but not to write well, though they can learn to write well. You can't teach students to revise ("see it again"), though you can teach them to rewrite. When you teach any skill, the best you can do is let writers practice it often and in ways that manifest its power.

III. PRAXIS

A. Master elemental skills.

I spoke earlier of having a repeatable core of things writers can't do wrong.

However you devise such a useful repertory for any of your classes, I suggest that there are five elemental writing skills and that they are the nucleus of all practice in writing:

- Writing prolifically (timed, nonstop writing),
- Writing from generative structures,
- Writing both concretely and abstractly,
- Writing observations,
- Rewriting.

Prolific writing and structured writing are polar and necessary to each other. The first promotes fluency, which is primary. Without fluency there is no writing. By itself, however, prolific writing tends to be amorphous. Structure enables writers to sustain and think about writing.

B. Repeat, "You Can't Do It Wrong."

Say this every time you ask students to write in class. Make sure it's true by expecting a version of something elemental. If you ask them to write a sentence, ask for it in at least three versions; that way the writer gets at least one good one. And one good one is all you need for starters.

After one of my team teachers had gone off to the university, he sent me this note: "Please find enclosed, as authors' royalties, $1.36, for the phrase 'You can't do it wrong,' used by me 272 times during the first three weeks of teaching composition. On a TA's salary, I estimate each repetition to be worth about 1/2¢. Please divide accordingly with Marie Ponsot."

The dollar is still in the envelope, and the quarter, dime, and penny still taped to the letter. Wit, after all, is priceless.

C. Assign structures.

1. Structure leads a student to discovery.

If your assignment is a one-time topic or subject, express its structure concretely. Instead of a general topic, like "Family Customs," let students write a set of concrete sentences after you give them a sample like this:

- In my family we all eat dinner together.
- In my family we hate exercise and love junk food.
- In my family, the family car belongs to Father.

Let them read these to each other, write some more, and take them home to choose the best one to begin the essay.

If your assignment is analytical and must be expressed abstractly, then repeat the abstract structure in a series of essays.

The simplest structure I can describe larger than a sentence turns on a center. It's the structure of plot: an action begins, turns, and ends. A repeatable analytic assignment for an essay about any kind of reading is to answer the question, "What is the turning point and why do you think so?"

This essay teaches the elemental move in analysis: to begin with the center—that is, to re-order the order of what you have read. There's no one Right Answer to this question, of course, because it's not where the student locates the turning point that counts, but how she or he defines it, then finds and orders evidence.

2. Structure pulls material out of the mind.

Shortly after I first met Marie, she happened to tell me the story about sonnets. As a young writer, she noted, you look for good advice. She had written verse since she was a child, but at sixteen or so she was ready for something serious. She read somewhere, "If you write a sonnet every day for a year, you'll be a poet." So she did.

I was writing my first sonnet at the time, and I did not find it easy. So my reaction to her story was to say, "But how did you have time to write anything else?"

She laughed and said that in fact after the daily sonnet, other poems seemed to spring out ready formed. She was right. After I'd struggled dutifully with my third terrible sonnet and finished it, I got a good one "free." The reason is simple. Structure pulls material out of the mind. You pluck something out of the mind's plenty that you would not otherwise pay attention to because you need it to complete your structure. That's how language enables us to express our minds in speech, after all.

Students naïvely think structure is restricting instead of enabling, although not a one of them would tolerate a baseball game with any number of outs. Luckily you know better.

D. Encourage observations, not critiques.

1. Teach students to write observations about their colleagues' essays and about their reading. Observations are descriptions of how something is written or designed.

   Observations are useful because they are concrete, simple, discrete, and public. They are the most direct way you can record real study material. As an example, here's an observation of Leonard's about Wordsworth's "Three Years She Grew." He noticed that the poem begins in the past tense, moves to the future, and ends in the past, without having ever gone through a present. As you can see, this is concrete, simple, and obvious—once someone has noticed it.

2. Observations teach students the language of description, so much more useful in all their classes than the language of evaluation. They are evidence from which many different inferences may be drawn. They are written efficiently as a list, nonstop in two or three minutes if necessary, and may be read aloud by each student, without comment, for the edification of all writers present.

   Observations are preferable to opinions because they direct students' attention to the writing as literature and give students practice in paying attention. Most people don't need practice in having opinions.

3. Give students the benefit of your literary training. Find students' literary power with your own observations. Describe and define the brightest aspect of the writing read in class especially. Your aim is not praise, but definition. Praise is cheap, but the definition of a power is a gift. It gives writers something to think about and language in which to think about it.

E. Practice rewriting.

1. Begin with the best.

   Rewriting is not remedial. No real writer ever went to the weakest part of a piece of writing and started fooling around with it. Don't send your students down this dead end out of a mistaken sense of pedantic duty to make them aware of their faults. When a part is bad, as you know, a real writer tosses it out and looks for the strongest and brightest part to expand and develop by writing more about it.

Rewriting unpacks an idea impacted in language. That is its sovereign, its blessed power. Practice it in class.

Choose the best sentence in an essay read in class—perhaps the best first or last sentence. In five minutes let everyone write three versions of it. Read these around to hear how some writers open up surprises in the idea, transform it concretely or abstractly, or simply reproduce it in elegant, witty sentences that are the rhetoric of the idea.

2. Begin by rereading.

   For years it never occurred to me to tell students that rewriting begins with rereading. I hadn't noticed that I reread in order to rewrite, but took it for granted, like breathing. Finally a team-teacher, who had been my student and whom I knew for a crackerjack rewriter, showed me my unconsciousness when I asked her to give the class some advice about rewriting. "Well," she said, "you begin by rereading." The thunderbolt of the obvious struck me. In fact, I've since realized, the power to rewrite is proportional to the amount of rereading.
3. Structure is self-correcting. If the assignment has a structure or clear shape, it will be easier for writers to rewrite. Students in the three-genre writing course tell me they never understood rewriting till they wrote their sonnets. Then they discovered they had to rewrite *as* they were writing to engage the play between idea and form.
4. When students have written many essays in a course, they enjoy the chance to look back, choose two or three of their best, and rewrite them to make them elegant, original expressions of an idea.

F. Recognize a student's work.

Students like to tell you how long it took them to do an assignment and how hard they worked. They've been trained to think of marks simply as the reward for work. Teachers who often can't see how much work a student actually does and who do see that some students are more gifted than others, tend, I think, to underrate the claim of work.

The basis of this claim, however, is certainly reasonable. In fairness, our job is to design a course in which all the work pays off, and no student who completes it (who hands in all the work assigned) can fail.

In Marie's elemental writing course, students write six fables, four parables, and fourteen essays. In an introduction to poetry course, my students write fourteen essays, each accompanied by two sets of sentences and a set of observations. I read all this with pleasure and I read it quickly. The essays vary in length and excellence, of course. Some students write five-page essays; some can barely keep up. But barely keeping up counts; and in the long run, the whole body of work has taught them method: learning by writing to put what they've studied into their heads.

In courses like these the difference between working faithfully and being a gifted writer isn't crucial. It may make the difference between an "A" and a "B," but not between an "A" and a "C." Good writers get to exercise the habit of writing well. Faithful workers get to extend their imaginative and literary powers. And teachers have earned the right to give good marks to workers who deserve them.

Theoria, the Greeks said, is contemplation. *Praxis* is doing. Teaching is both, although like most worldly work, it spends most of its time transmuting principle into act. It ought to be honored more than it is. I remember that my greatest teacher had no office—the college was very poor. But when I began to teach, I found her office in me. You'll find your best self in you when you begin to teach, faithful worker and good writer that you are.

Peter Elbow

Embracing Contraries in the Teaching Process

My argument is that good teaching seems a struggle because it calls on skills or mentalities that are actually contrary to each other and thus tend to interfere with each other. It was my exploration of writing that led me to look for contraries in difficult or complex processes. I concluded that good writing requires on the one hand the ability to conceive copiously of many possibilities, an ability which is enhanced by a spirit of open, accepting generativity; but on the other hand good writing also requires an ability to criticize and reject everything but the best, a very different ability which is enhanced by a tough-minded critical spirit. I end up seeing in good writers the ability somehow to be extremely creative and extremely critical, without letting one mentality prosper at the expense of the other or being halfhearted in both. (For more about this idea see my *Writing With Power* [Oxford University Press, 1981], especially Chapter 1.)

In this frame of mind I began to see a paradoxical coherence in teaching where formerly I was perplexed. I think the two conflicting mentalities needed for good teaching stem from the two conflicting obligations inherent in the job: we have an obligation to students but we also have an obligation to knowledge and society. Surely we are incomplete as teachers if we are committed only to what we are teaching but not to our students, or only to our students but not to what we are teaching, or halfhearted in our commitment to both.

We like to think that these two commitments coincide, and often they do. It happens often enough, for example, that our commitment to standards leads us to give a low grade or tough comment, and it is just what the student needs to hear. But just as often we see that a student needs praise and support rather than a tough grade, even for her weak performance, if she is really to prosper as a student and a person—if we are really to nurture her fragile investment in her studies. Perhaps we can finesse this conflict between a "hard" and "soft" stance if it is early in the semester or we are only dealing with a rough draft; for the time being we can give the praise and support we sense is humanly appropriate and hold off strict judgment and standards till later. But what about when it is the end of the course or a final draft needs a grade? It is comforting to take as our paradigm that first situation where the tough grade was just right, and to consider the trickier situation as somehow anomalous, and thus to assume that we always serve students best by serving knowledge, and vice versa. But I now think I can throw more light on the nature of teaching by taking our conflicting loyalties as paradigmatic.

From *College English* 45 (April 1983): 327-39.

Our loyalty to students asks us to be their allies and hosts as we instruct and share: to invite all students to enter in and join us as members of a learning community—even if they have difficulty. Our commitment to students asks us to assume they are all capable of learning, to see things through their eyes, to help bring out their best rather than their worst when it comes to tests and grades. By taking this inviting stance we will help more of them learn.

But our commitment to knowledge and society asks us to be guardians or bouncers: we must discriminate, evaluate, test, grade, certify. We are invited to stay true to the inherent standards of what we teach, whether or not that stance fits the particular students before us. We have a responsibility to society—that is, to our discipline, our college or university, and to other learning communities of which we are members—to see that the students we certify really understand or can do what we teach, to see that the grades and credits and degrees we give really have the meaning or currency they are supposed to have.[1]

A pause for scruples. Can we give up so easily the paradigm of teaching as harmonious? Isn't there something misguided in the very idea that these loyalties are conflicting? After all, if we think we are being loyal to students by being extreme in our solicitude for them, won't we undermine the integrity of the subject matter or the currency of the credit and thereby drain value from the very thing we are supposedly giving them? And if we think we are being loyal to society by being extreme in our ferocity—keeping out *any* student with substantial misunderstanding—won't we deprive subject matter and society of the vitality and reconceptualizations they need to survive and grow? Knowledge and society only exist embodied—that is, flawed.

This sounds plausible. But even if we choose a middle course and go only so far as fairness toward subject matter and society, the very fact that we grade and certify at all—the very fact that we must sometimes flunk students—tempts many of them to behave defensively with us. Our mere fairness to subject matter and society tempts students to try to hide weaknesses from us, "psych us out," or "con us." It is as though we are doctors trying to treat patients who hide their symptoms from us for fear we will put them in the hospital.

Student defensiveness makes our teaching harder. We say, "Don't be afraid to ask questions," or even, "It's a sign of intelligence to be willing to ask naive questions." But when we are testers and graders, students too often fear to ask. Towards examiners they must play it safe, drive defensively, not risk themselves. This stunts learning. When they trust the teacher to be wholly an ally, students are more willing to take risks, connect the self to the material, and experiment. Here is the source not just of learning but also of genuine development or growth.

Let me bring this conflict closer to home. A department chair or dean who talks with us about our teaching and who sits in on our classes is our ally insofar as she is trying

---

1. I lump "knowledge and society" together in one phrase but I acknowledge the importance of the potential conflict. For example, we may feel *society* asking us to adapt our students to it, while we feel *knowledge*—our vision of the truth—asking us to unfit our students for that society. Socrates was convicted of corrupting the youth. To take a more homely example. I may feel institutions asking me to teach students one kind of writing and yet feel impelled by my understanding of writing to teach them another kind. Thus where this paper paints a picture of teachers pulled in two directions, sometimes we may indeed be pulled in three.

to help us teach better; and we can get more help from her to the degree that we openly share with her our fears, difficulties, and failures. Yet insofar as she makes promotion or tenure decisions about us or even participates in those decisions, we will be tempted not to reveal our weaknesses and failures. If we want the best help for our shortcomings, someone who is merely fair is not enough. We need an ally, not a judge.

Thus we can take a merely judicious, compromise position toward our students only if we are willing to settle for being *sort of* committed to students and *sort of* committed to subject matter and society. This middling or fair stance, in fact, is characteristic of many teachers who lack investment in teaching or who have lost it. Most invested teachers, on the other hand, tend to be a bit passionate about supporting students or else passionate about serving and protecting the subject matter they love—and thus they tend to live more on one side or the other of some allegedly golden mean.

But supposing you reply, "Yes, I agree that a compromise is not right. Just middling. Muddling. Not excellence or passion in either direction. But that's not what I'm after. My scruple had to do with your very notion of *two directions*. There is only one direction. Excellence. Quality. The very conception of conflict between loyalties is wrong. An inch of progress in one direction, whether toward knowledge or toward students, is always an inch in the direction of the other. The needs of students and of knowledge or society are in essential harmony."

To assert this harmony is, in a sense, to agree with what I am getting at in this paper. But it is no good just asserting it. It is like asserting, "Someday you'll thank me for this," or, "This is going to hurt me worse than it hurts you." I may say to students, "My fierce grading and extreme loyalty to subject matter and society are really in your interests," but students will still tend to experience me as adversary and undermine much of my teaching. I may say to knowledge and society, "My extreme support and loyalty to all students is really in your interests," but society will tend to view me as a soft teacher who lets standards down.

It is the burden of this paper to say that a contradictory stance is possible—not just in theory but in practice—but not by pretending there is no tension or conflict. And certainly not by affirming only one version of the paradox, the "paternal" version, which is to stick up for standards and firmness by insisting that to do so is good for students in the long run, forgetting the "maternal" version which is to stick up for students by insisting that to do so is good for knowledge and society in the long run. There is a genuine paradox here. The positions are conflicting and they are true.

Let me turn this structural analysis into a narrative about the two basic urges at the root of teaching. We often think best by telling stories. I am reading a novel and I interrupt my wife to say, "Listen to this, isn't this wonderful!" and I read a passage out loud. Or we are walking in the woods and I say to her, "Look at that tree!" I am enacting the pervasive human itch to share. It feels lonely, painful, or incomplete to appreciate something and not share it with others.[2]

But this urge can lead to its contrary. Suppose I say, "Listen to this passage," and

2. Late in life, I realize I must apologize and pay my respects to that form of literary criticism that I learned in college to scorn in callow fashion as the "Ah lovely!" school: criticism which tries frankly to share a perception and appreciation of the work rather than insist that there is some problem to solve or some complexity to analyze.

my wife yawns or says, "Don't interrupt me." Suppose I say, "Look at that beautiful sunset on the lake," and she laughs at me for being so sentimental and reminds me that Detroit is right there just below the horizon—creating half the beauty with its pollution. Suppose I say, "Listen to this delicate irony," and she can't see it and thinks I am neurotic to enjoy such bloodless stuff. What happens then? I end up *not* wanting to share it with her. I hug it to myself. I become a lone connoisseur. Here is the equally deep human urge to protect what I appreciate from harm. Perhaps I share what I love with a few select others—but only after I find a way somehow to extract from them beforehand assurance that they will understand and appreciate what I appreciate. And with them I can even sneer at worldly ones who lack our taste or intelligence or sensibility.

Many of us went into teaching out of just such an urge to share things with others, but we find students turn us down or ignore us in our efforts to give gifts. Sometimes they even laugh at us for our very enthusiasm in sharing. We try to show them what we understand and love, but they yawn and turn away. They put their feet up on our delicate structures; they chew bubble gum during the slow movement; they listen to hard rock while reading *Lear* and say, "What's so great about Shakespeare?"

Sometimes even success in sharing can be a problem. We manage to share with students what we know and appreciate, and they love it and eagerly grasp it. But their hands are dirty or their fingers are rough. We overhear them saying, "Listen to this neat thing I learned," yet we cringe because they got it all wrong. Best not to share.

I think of the medieval doctrine of poetry that likens it to a nut with a tough husk protecting a sweet kernel. The function of the poem is not to disclose but rather to conceal the kernel from the many, the unworthy, and to disclose it only to the few worthy (D. W. Robertson, *A Preface to Chaucer* [Princeton, N.J.: Princeton University Press, 1963], pp. 61 ff.). I have caught myself more than a few times explaining something I know or love in this tricky double-edged way: encoding my meaning with a kind of complexity or irony such that only those who have the right sensibility will hear what I have to say—others will not understand at all. Surely this is the source of much obscurity in learned discourse. We would rather have readers miss entirely what we say or turn away in boredom or frustration than reply,"Oh, I see what you mean. How ridiculous!" or, "How naive!" It is marvelous, actually, that we can make one utterance do so many things: communicate with the right people, stymie the wrong people, and thereby help us decide who *are* the right and the wrong people.

I have drifted into an unflattering portrait of the urge to protect one's subject, a defensive urge that stems from hurt. Surely much bad teaching and academic foolishness derive from this immature reaction to students or colleagues who will not accept a gift we tried generously to give (generously, but sometimes ineffectually or condescendingly or autocratically). Surely I must learn not to pout just because I can't get a bunch of adolescents as excited as I am about late Henry James. Late Henry James may be pearls, but when students yawn, that doesn't make them swine.

But it is not immature to protect the integrity of my subject in a positive way, to uphold standards, to insist that students stretch themselves till they can do justice to the material. Surely these impulses are at the root of much good teaching. And there is nothing wrong with these impulses in themselves—only *by themselves*. That is, there is nothing wrong with the impulse to guard or protect the purity of what we cherish so long as that act is redeemed by the presence of the opposite impulse also to give it away.

In Piaget's terms learning involves both assimilation and accommodation. Part of the job is to get the subject matter to bend and deform so that it fits inside the learner (that is, so it can fit or relate to the learner's experiences). But that's only half the job. Just as important is the necessity for the learner to bend and deform himself so that he can fit himself around the subject without doing violence to it. Good learning is not a matter of finding a happy medium where both parties are transformed as little as possible. Rather both parties must be maximally transformed—in a sense deformed. There is violence in learning. We can not learn something without eating it, yet we can not really learn it either without letting it eat us.

Look at Socrates and Christ as archetypal good teachers—archetypal in being so paradoxical. They are extreme on the one hand in their impulse to share with everyone and to support all learners, in their sense that everyone can take and get what they are offering; but they are extreme on the other hand in their fierce high standards for what will pass muster. They did not teach gut courses, they flunked "gentleman C" performances, they insisted that only "too much" was sufficient in their protectiveness toward their "subject matter." I am struck also with how much they both relied on irony, parable, myth, and other forms of subtle utterance that hide while they communicate. These two teachers were willing in some respects to bend and disfigure and in the eyes of many to profane what they taught, yet on the other hand they were equally extreme in their insistence that learners bend or transform themselves in order to become fit receptacles.

It is as though Christ, by stressing the extreme of sharing and being an ally—saying "suffer the little children to come unto me" and praising the widow with her mite—could be more extreme in his sternness: "unless you sell all you have," and, "I speak to them in parables, because seeing they do not see and hearing they do not hear, nor do they understand" (saying in effect, "I am making this a tough course *because* so many of you are poor students"). Christ embeds the two themes of giving away and guarding—commitment to "students" and to "subject matter"—in the one wedding feast story: the host invites in guests from the highways and byways, anybody, but then angrily ejects one into outer darkness because he lacks the proper garment.

Let me sum up the conflict in two lists of teaching skills. If on the one hand we want to help more students learn more, I submit we should behave in the following four ways:

1. We should see our students as smart and capable. We should assume that they *can* learn what we teach—all of them. We should look through their mistakes or ignorance to the intelligence that lies behind. There is ample documentation that this "teacher expectation" increases student learning (Robert Rosenthal, "Teacher Expectation and Pupil Learning," in R. D. Strom, ed., *Teachers and the Learning Process* [Englewood Cliffs, N.J.: Prentice-Hall, 1971], pp. 33–60).

2. We should show students that we are on their side. This means, for example, showing them that the perplexity or ignorance they reveal to us will not be used against them in tests, grading, or certifying. If they hide their questions or guard against us they undermine our efforts to teach them.

3. Indeed, so far from letting their revelations hurt them in grading, we should be as it were lawyers for the defense, explicitly trying to help students do better against the judge and prosecuting attorney when it comes to the "trial" of testing and

grading. (''I may be able to get you off this charge but only if you tell me what you really were doing that night.'') If we take this advocate stance students can learn more from us, even if they are guilty of the worst crimes in the book: not having done the homework, not having learned last semester, not *wanting* to learn. And by learning more—even if not learning perfectly—they will perform better, which in turn will usually lead to even better learning in the future.

4. Rather than try to be perfectly fair and perfectly in command of what we teach—as good examiners ought to be—we should reveal our own position, particularly our doubts, ambivalences, and biases. We should show we are still learning, still willing to look at things in new ways, still sometimes uncertain or even stuck, still willing to ask naive questions, still engaged in the interminable process of working out the relationship between what we teach and the rest of our lives. Even though we are not wholly peer with our students, we can still be peer in this crucial sense of also being engaged in learning, seeking, and being incomplete. Significant learning requires change, inner readjustments, willingness to let go. We can increase the chances of our students being willing to undergo the necessary anxiety involved in change if they see we are also willing to undergo it.

Yet if, on the other hand, we want to increase our chances of success in serving knowledge, culture, and institutions I submit that we need skill at behaving in four very different ways:

1. We should insist on standards that are high—in the sense of standards that are absolute. That is, we should take what is almost a kind of Platonic position that there exists a ''real world'' of truth, of good reasoning, of good writing, of knowledge of biology, whatever—and insist that anything less than the real thing is not good enough.
2. We should be critical-minded and look at students and student performances with a skeptical eye. We should assume that some students cannot learn and others will not, even if they can. This attitude will increase our chances of detecting baloney and surface skill masquerading as competence or understanding.
3. We should not get attached to students or take their part or share their view of things; otherwise we will find it hard to exercise the critical spirit needed to say, ''No, you do not pass,'' ''No, you cannot enter in with the rest of us,'' ''Out you go into the weeping and gnashing of teeth.''
4. Thus we should identify ourselves primarily with knowledge or subject matter and care more about the survival of culture and institutions than about individual students—even when that means students are rejected who are basically smart or who tried as hard as they could. We should keep our minds on the harm that can come to knowledge and society if standards break down or if someone is certified who is not competent, rather than on the harm that comes to individual students by hard treatment.

Because of this need for conflicting mentalities I think I see a distinctive distribution of success in teaching. At one extreme we see a few master or genius teachers, but they are striking for how differently they go about it and how variously and sometimes surprisingly they explain what they do. At the other extreme are people who teach very badly, or who have given up trying, or who quit teaching altogether: they are debilitated by the conflict between trying to be an ally as they teach and an adversary as they grade. Between these two extremes teachers find the three natural ways of making peace between contraries: there are ''hard'' teachers in whom loyalty to knowledge or society has won out; ''soft'' teachers in whom loyalty to students has won out; and middling, mostly dispirited teachers who are sort of loyal to students and sort of loyal to knowledge or society. (A few of this last group are not dispirited at all but live on a kind of knife edge of almost palpable tension as they insist on trying to be scrupulously fair both to students and to what they teach.)

This need for conflicting mentalities is also reflected in what is actually the most traditional and venerable structure in education: a complete separation between teaching and official assessment. We see it in the Oxford and Cambridge structure that makes the tutor wholly an ally to help the student prepare for exams set and graded by independent examiners. We see something of the same arrangement in many European university lecture-and-exam systems which are sometimes mimicked by American PhD examinations. The separation of teaching and examining is found in many licensing systems and also in some new competence-based programs.

Even in conventional university curricula we see various attempts to strengthen assessment and improve the relationship between teacher and student by making the teacher more of an ally and coach. In large courses with many sections, teachers often give a common exam and grade each others' students. Occasionally, when two teachers teach different courses within each other's field of competence, they divide their roles and act as ''outside examiner'' for the other's students. (This approach, by the way, tends to help teachers clarify what they are trying to accomplish in a course since they must communicate their goals clearly to the examiner if there is to be any decent fit between the teaching and examining.) In writing centers, tutors commonly help students improve a piece of writing which another teacher will assess. We even see a hint of this separation of roles when teachers stress collaborative learning: they emphasize the students' role as mutual teachers and thereby emphasize their own pedagogic role as examiner and standard setter.

But though the complete separation of teacher and evaluator is hallowed and useful I am interested here in ways for teachers to take on both roles better. It is not just that most teachers are stuck with both; in addition I believe that opposite mentalities or processes can enhance each other rather than interfere with each other if we engage in them in the right spirit.

How can we manage to do contrary things? Christ said, ''Be ye perfect,'' but I don't think it is good advice to try being immensely supportive and fierce in the same instant, as he and Socrates somehow managed to be. In writing, too, it doesn't usually help to try being immensely generative and critical-minded in the same instant as some great writers are—and as the rest of us sometimes are at moments of blessed inspiration. This is the way of transcendence and genius, but for most of us most of the time there is too much interference or paralysis when we try to do opposites at once.

But it is possible to make peace between opposites by alternating between them so that you are never trying to do contrary things at any one moment. One opposite leads naturally to the other; indeed, extremity in one enhances extremity in the other in a positive, reinforcing fashion. In the case of my own writing I find I can generate more and better when I consciously hold off critical-minded revising till later. Not only does it help to go whole hog with one mentality, but I am not afraid to make a fool of myself since I know I will soon be just as wholeheartedly critical. Similarly, I can be more fierce and discriminating in my critical revising because I have more and better material to work with through my earlier surrender to uncensored generating.

What would such an alternating approach look like in teaching? I will give a rough picture, but I do so hesitantly because if I am right about my theory of paradox, there will be widely different ways of putting it into practice.

In teaching we traditionally end with the critical or gatekeeper function: papers, exams, grades, or less institutionalized forms of looking back, taking stock, and evaluating. It is also traditional to start with the gatekeeper role: to begin a course by spelling out all the requirements and criteria as clearly as possible. We often begin a course by carefully explaining exactly what it will take to get an A, B, C, etc.

I used to be reluctant to start off on this foot. It felt so vulgar to start by emphasizing grades, and thus seemingly to reinforce a pragmatic preoccupation I want to squelch. But I have gradually changed my mind, and my present oppositional theory tells me I should exaggerate, or at least take more seriously than I often do, my gatekeeper functions rather than run away from them. The more I try to soft-pedal assessment, the more mysterious it will seem to students and the more likely they will be preoccupied and superstitious about it. The more I can make it clear to myself and to my students that I do have a commitment to knowledge and institutions, and the more I can make it specifically clear how I am going to fulfill that commitment, the easier it is for me to turn around and make a dialectical change of role into being an extreme ally to students.

Thus I start by trying to spell out requirements and criteria as clearly and concretely as possible. If I am going to use a midterm and final exam, it would help to pass out samples of these at the beginning of the course. Perhaps not a copy of precisely the test I will use but something close. And why not the real thing? If it feels as though I will ruin the effectiveness of my exam to "give it away" at the start, that means I must have a pretty poor exam—a simple-minded task that can be crammed for and that does not really test what is important. If the exam gets at the central substance of the course then surely it will help me if students see it right at the start. They will be more likely to learn what I want them to learn. It might be a matter of content: "Summarize the three main theories in this course and discuss their strengths and weaknesses by applying them to material we did not discuss." Or perhaps I am more interested in a process or skill: "Write an argumentative essay on this (new) topic." Or, "Show how the formal characteristics of this (new) poem do and do not reinforce the theme." I might want to give room for lots of choice and initiative: "Write a dialogue between the three main people we have studied that illustrates what you think are the most important things about their work." Passing out the exam at the start—and perhaps even samples of strong and weak answers—is an invitation to make a tougher exam that goes more to the heart of what the course is trying to teach. If I don't use an exam, then it is even more crucial that I say how I will determine the grade—even if I base it heavily on slippery factors: e.g., "I will count half your grade on my impression of how well you motivate and invest yourself," or "how well you work collaboratively with your peers.'' Of course this kind of announcement makes for a tricky situation, but if these are my goals, surely I want my students to wrestle with them all term—in all their slipperiness and even if it means arguments about how unfair it is to grade on such matters—rather than just think about them at the end.

When I assign papers I should similarly start by advertising my gatekeeper role, by clearly communicating standards and criteria. That means not just talking theoretically about what I am looking for in an A paper and what drags a paper down to B or C or F, but rather passing out a couple of samples of each grade and talking concretely about what makes me give each one the grade I give it. Examples help because our actual grading sometimes reflects criteria we do not talk about, perhaps even that we are not aware of. (For example, I have finally come to admit that neatness counts.) Even if our practice fits our preaching, sometimes students do not really understand preaching without examples. Terms like ''coherent'' and even ''specific'' are notoriously hard for students to grasp because they do not read stacks of student writing. Students often learn more about well-connected and poorly-connected paragraphs or specificity or the lack of it in examples from the writing of each other than they learn from instruction alone, or from examples of published writing.

I suspect there is something particularly valuable here about embodying our commitment to knowledge and society in the form of documents or handouts: words on palpable sheets of paper rather than just spoken words-in-the-air. Documents heighten the sense that I do indeed take responsibility for these standards: writing them forces me to try to make them as concrete, explicit, and objective as possible (if not necessarily fair). But most of all, having put all this on paper I can more easily go on to separate myself from them in some way—leave them standing—and turn around and schizophrenically start being a complete ally of students. I have been wholehearted and enthusiastic in making tough standards, but now I can say, ''Those are the specific criteria I will use in grading; that's what you are up against, that's really me. But now we have most of the semester for me to help you attain those standards, do well on those tests and papers. They are high standards but I suspect all of you can attain them if you work hard. I will function as your ally. I'll be a kind of lawyer for the defense, helping you bring out your best in your battles with the other me, the prosecuting-attorney me when he emerges at the end. And if you really think you are too poorly prepared to do well in one semester, I can help you decide whether to trust that negative judgment and decide now whether to drop the course or stay and learn what you can.''

What is pleasing about this alternating approach is the way it naturally leads a teacher to higher standards yet greater supportiveness. That is, I feel better about being really tough if I know I am going to turn around and be more on the student's side than usual. And contrarily I do not have to hold back from being an ally of students when I know I have set really high standards. Having done so, there is now no such thing as being ''too soft,'' supportive, helpful, or sympathetic—no reason to hold back from seeing things entirely from their side, worrying about their problems. I can't be ''cheated'' or taken advantage of.

In addition, the more clearly I can say what I want them to know or be able to do, the better I can figure out what I must provide to help them attain those goals. As I make progress in this cycle, it means I can set my goals even higher—ask for the deep knowledge and skills that are really at the center of the enterprise.

But how, concretely, can we best function as allies? One of the best ways is to be a kind of coach. One has set up the hurdle for practice jumping, one has described the strengths and tactics of the enemy, one has warned them about what the prosecuting attorney will probably do: now the coach can prepare them for these rigors. Being an ally is probably more a matter of stance and relationship than of specific behaviors. Where a professor of jumping might say, in effect, "I will explain the principles of jumping," a jumping coach might say, in effect, "Let's work on learning to jump over those hurdles; in doing so I'll explain the principles of jumping." If we try to make these changes in stance, I sense we will discover some of the resistances, annoyances, and angers that make us indeed reluctant genuinely to be on the student's side. How can we be teachers for long without piling up resentment at having been misunderstood and taken advantage of? But the dialectical need to be in addition an extreme adversary of students will give us a legitimate medium for this hunger to dig in one's heels even in a kind of anger.

This stance provides a refreshingly blunt but supportive way to talk to students about weaknesses. "You're strong here, you're weak there, and over here you are really out of it. We've got to find ways to work on these things so you can succeed on these essays or exams." And this stance helps reward students for volunteering weaknesses. The teacher can ask, "What don't you understand? What skills are hard for you? I need to decide how to spend our time here and I want it to be the most useful for your learning."

One of the best ways to function as ally or coach is to role-play the enemy in a supportive setting. For example, one can give practice tests where the grade doesn't count, or give feedback on papers which the student can revise before they count for credit. This gets us out of the typically counterproductive situation where much of our commentary on papers and exams is really justification for the grade—or is seen that way. Our attempt to help is experienced by students as a slap on the wrist by an adversary for what they have done wrong. No wonder students so often fail to heed or learn from our commentary. But when we comment on practice tests or revisable papers we are not saying, "Here's why you got this grade." We are saying, "Here's how you can get a better grade." When later we read final versions as evaluator we can read faster and not bother with much commentary.[3]

It is the spirit or principle of serving contraries that I want to emphasize here, not any particular fleshing out in practice such as above. For one of the main attractions of this theory is that it helps explain why people are able to be terrific teachers in such diverse ways. If someone is managing to do two things that conflict with each other, he

---

3. Since it takes more time for us to read drafts and final versions too, no matter how quickly we read final versions, it is reasonable to conserve time in other ways—indeed I see independent merits. Don't require students to revise every draft. This permits you to grade students on their best work and thus again to have higher standards, and it is easier for students to invest themselves in revising if it is on a piece they care more about. And in giving feedback on drafts, wait till you have two drafts in hand and thus give feedback only half as often. When I have only one paper in hand I often feel, "Oh dear, everything is weak here; nothing works right; where can I start?" When I have two drafts in hand I can easily say, "This one is better for the following reasons; it's the one I'd choose to revise: see if you can fix the following problems." With two drafts it is easier to find genuine strengths and point to them and help students consolidate or gain control over them. Yet I can make a positive utterance out of talking about what *didn't* work in the better draft and how to improve it.

is probably doing something mysterious: it's altogether natural if his success involves slipperiness, irony, or paradox. For example, some good teachers look like they are nothing but fierce gatekeepers, cultural bouncers, and yet in some mysterious way—perhaps ironically or subliminally—they are supportive. I think of the ferocious Marine sergeant who is always cussing out the troops but who somehow shows them he is on their side and believes in their ability. Other good teachers look like creampuffs and yet in some equally subtle way they embody the highest standards of excellence and manage to make students exert and stretch themselves as never before.

For it is one's spirit or stance that is at issue here, not the mechanics of how to organize a course in semester units or how to deal in tests, grading, or credits. I do not mean to suggest that the best way to serve knowledge and society is by having tough exams or hard grading—or even by having exams or grades at all. Some teachers do it just by talking, whether in lectures or discussions or conversation. Even though there is no evaluation or grading, the teacher can still demonstrate her ability to be wholehearted in her commitment to what she teaches and wholehearted also in her commitment to her students. Thus her talk itself might in fact alternate between attention to the needs of students and flights where she forgets entirely about students and talks over their head, to truth, to her wisest colleagues, to herself.[4]

The teacher who is really in love with Yeats or with poetry will push harder, and yet be more tolerant of students' difficulties because his love provides the serenity he needs in teaching: he knows that students cannot hurt Yeats or his relationship with Yeats. It is a different story when we are ambivalent about Yeats or poetry. The piano teacher who mean-spiritedly raps the fingers of pupils who play wrong notes usually harbors some inner ambivalence in his love of music or some disappointment about his own talent.

In short, there is obviously no one right way to teach, yet I argue that in order to teach well we must find *some* way to be loyal both to students and to knowledge or society. Any way we can pull it off is fine. But if we are teaching less well than we should, we might be suffering from the natural tendency for these two loyalties to conflict with each other. In such a case we can usually improve matters by making what might seem an artificial separation of focus so as to give each loyalty and its attendant skills and mentality more room in which to flourish. That is, we can spend part of our teaching time saying in some fashion or other, "Now I'm being a tough-minded gatekeeper, standing up for high critical standards in my loyalty to what I teach"; and part of our time giving a contrary message: "Now my attention is wholeheartedly on trying to be your ally and to help you learn, and I am not worrying about the purity of standards or grades or the need of society or institutions."

It is not that this approach makes things simple. It confuses students at first because

---

4. Though my argument does not imply that we need to use grades at all, surely it implies that if we do use them we should learn to improve the way we do so. I used to think that conventional grading reflected too much concern with standards for knowledge and society, but now I think it reflects too little. Conventional grading reflects such a single-minded hunger to *rank* people along a single scale or dimension that it is willing to forego any communication of what the student really knows or can do. The competence-based movement, whatever its problems, represents a genuine attempt to make grades and credits do justice to knowledge and society. (See Gerald Grant, et al., *On Competence. A Critical Analysis of Competence-Based Reform in Higher Education* [San Francisco: Jossey-Bass, 1979]. See also my "More Accurate Evaluation of Student Performance," *Journal of Higher Education,* 40 [1969], 219-230.)

they are accustomed to teachers being either ''hard'' or ''soft'' or in the middle—not both. The approach does not take away any of the conflict between trying to fulfill two conflicting functions. It merely gives a context and suggests a structure for doing so. Most of all it helps me understand better the demands on me and helps me stop feeling as though there is something wrong with me for feeling pulled in two directions at once.

I have more confidence that this conscious alternation or separation of mentalities makes sense because I think I see the same strategy to be effective with writing. Here too there is obviously no one right way to write, but it seems as though any good writer must find some way to be both abundantly inventive yet tough-mindedly critical. Again, any way we can pull it off is fine, but if we are not writing as well as we should—if our writing is weak in generativity or weak in tough-minded scrutiny (not to mention downright dismal or blocked)—it may well be that we are hampered by a conflict between the accepting mentality needed for abundant invention and the rejecting mentality needed for tough-minded criticism. In such a case too, it helps to move back and forth between sustained stretches of wholehearted, uncensored generating and wholehearted critical revising to allow each mentality and set of skills to flourish unimpeded.

Even though this theory encourages a separation that could be called artificial, it also points to models of the teaching and writing process that are traditional and reinforced by common sense: teaching that begins and ends with attention to standards and assessment and puts lots of student-directed supportive instruction in the middle; writing that begins with exploratory invention and ends with critical revising. But I hope that my train of thought rejuvenates these traditional models by emphasizing the underlying structure of contrasting mentalities which is central rather than merely a mechanical sequence of external stages which is not necessary at all.

In the end, I do not think I am just talking about how to serve students and serve knowledge or society. I am also talking about developing opposite and complementary sides of our character or personality: the supportive and nurturant side and the tough, demanding side. I submit that we all have instincts and needs of both sorts. The gentlest, softest, and most flexible among us really need a chance to stick up for our latent high standards, and the most hawk-eyed, critical-minded bouncers at the bar of civilization among us really need a chance to use our nurturant and supportive muscles instead of always being adversary.

Lad Tobin

Reading Students, Reading Ourselves: Revising the Teacher's Role in the Writing Class

At the end of each semester I ask my students to write an essay on writing, to identify and comment on some significant feature of their own writing process. The idea is to help them better understand how they have written in the past so that they will have more control over how they write in the future. Most of my students find this assignment tedious and end up writing a fairly perfunctory self-study, but I keep giving this assignment for two reasons: first, I am really curious about how students view the writing process and, second, when these "process papers" are good, they are remarkably good.

Recently I was telling two of my colleagues about a particularly insightful essay one of my students wrote about the relationship between thought and language. In her essay, Nicki argues that a writer can only think clearly when she is allowed to use a voice and a style that she has mastered. In my course, she felt that she had been able to think through important issues in original ways; however, in her humanities class, she had trouble developing and organizing her ideas about Homer, Cicero, and the Hebrew prophets. She accounted for the difference not by the difficulty of the material—she took on complicated problems in my course—but rather by the encouragement I gave her to explore ideas that mattered to her in personal and informal language. Her humanities professor, she complained, had denied her this access by insisting on numerous references to the text and "impeccable English prose."

Her point was not simply that her expression became more awkward in her humanities papers; instead, she was arguing that in the translation from her own form of expression to the academic language required in that course, her actual ideas were lost or distorted. The irony, she concludes, is that although her humanities teacher claims to value creativity and logic, he insists that students write in a form which virtually guarantees detachment and confusion. "But what is best about her essay," I told my colleagues, "is that it is so well written. At the end she writes something like, 'The essay I am writing right now proves my point. I am comfortable and I am able to use "I" and "you" which allows me to tell you clearly and directly what I think. But when I try to write "impeccable English prose," I lose sight of my audience and I disappear as a writer.' "

They seemed impressed, maybe even won over by the idea of this assignment. But

From *College English* 53 (March 1991): 333-48.

as I walked back to my office, I started worrying that I had overstated the value of the assignment and the quality of Nicki's essay. When I re-read it, I was embarrassed to discover how much I had organized and focused her argument in my re-telling. It is not as if her essay was without thought or skill. In fact, the section that I singled out for its rhetorical sophistication was actually much better in Nicki's paper than in my memory and re-telling:

> In Humanities, I have to remember a certain format and I have to back up every general statement with specific examples. Oh, and that word "I," I just used. You would never see that word in one of my Humanities papers. Neither would you see "you." It would be marked with red ink and a comment, "You who???" or "To Whom do you think you are referring?"

But in general the writing seemed much flatter and more prosaic that I had remembered it:

> Though it is good to be able to write for different audiences, I do not want to have to change my preference in writing because of some particular "format" I am supposed to follow. There is no law that states that I must write in a certain way. When I write I like to feel as if I have gotten across what I want to say.

But it wasn't just the writing. My discomfort grew as I began to see how much her whole argument echoed my own ideas—I, also, believe that a student should be allowed to write in her own voice, that she should be able to choose topics, that writing is a mode of thinking, and so on—all ideas to which I have a strong ideological and personal commitment. For years I have argued with colleagues who believe that students should not be allowed to write in first person or from personal experience, who insist on impeccable prose, correctness, and perfect one-inch margins. So it only makes sense that I would be pleased and excited to see that my student's writing supported and even validated my own positions and, therefore, that I would make her argument more eloquent and sophisticated than it actually was.

But there were other reasons for my misreading. This was not the first essay of Nicki's I had read. All semester I had seen her work: I read this final essay in terms of all of our interactions. From our conferences, I knew that her parents were first-generation Greek-Americans and that she was a first-generation college student from a small, working class town in Massachusetts. From her previous essays, I knew that during her last two years of high school she had been involved with a man in his twenties who was addicted to cocaine, who cheated on her with other women, and who once beat her up at a party. I also knew that throughout that whole relationship her worst fear was that he would break up with her. I knew that she considered herself a "good Catholic" but was shocked and angry at the Church for "never telling her the truth about God."

So when I read Nicki's essay on writing and personal voice, I was also reading Nicki herself and imagining—rightly or wrongly—that this first term of college was a crucial time in her development. I was thinking about how she ended her essay on that self-destructive high-school relationship:

> To this day, I am not sure why I loved someone like that. Why was I drawn to a person who treated me so badly? I guess you could say he was my drug. He was my

> high and my addiction. It was hard to "just say no," but I finally did. I've been clean for almost six months now and I plan to stay that way.

And I was thinking about how upset she was when her humanities teacher dismissed her argument as superficial and that the God in Exodus and the Book of Job who was sometimes "vengeful, jealous, and merciless," was "more realistic" than the all-loving, perfect God that the nuns had described.

But to make matters still more complicated, I was also reading myself. I had a vested interest in thinking that my teaching and my course had provided Nicki something she did not get in her humanities class. I had an interest in thinking that my teaching helped her feel confident about her abilities and her potential. By reading Nicki's text in such a way that it reached a self-confident and successful resolution, by making *her* into a text with a happy ending, I could congratulate myself not only for helping another writer succeed but also for helping another student establish her identity. And, perhaps most complicated of all, by reading her in a particularly imaginative and integrated way, I could use her (as I am trying to do right now) for my own benefit in my writing and research.

Obviously the specific circumstances of my reading or, more accurately, misreading are unique—and that is part of my point. But I am also suggesting that, in many ways, my misreading illustrates common issues and problems. As teachers, we play a crucial—but generally misunderstood—role in our students' writing processes. While we have begun to understand how students compose and to develop a more comprehensive and flexible view of the unconscious forces which shape their composing, we continue to oversimplify the teacher's reading or interpretative processes. Or to put it another way, while we have come to see writing as socially constructed, we have failed to understand the teacher's role in the construction of that meaning. We need to develop a theory of reading student texts which takes into account our reading of the students themselves, of our own unconscious motivations and associations, and, finally, of the interactive and dialectical nature of the teacher-student relationship.

Reading and Misreading Student Essays

The most significant relationship in any writing course is the one between the writer and her text. But if reading and writing are reciprocal or transactional processes, as Louise Rosenblatt believes, we also need to develop the teacher's relation to a text. That I misread Nicki's essay in certain ways is not significant in itself. After all, most of us in English studies have grown relatively comfortable with the notion that our readings are not simple or literal decodings of texts, that when we read we create and recreate, deconstruct and reconstruct. While this fact seems to cause shock and anguish in old-fashioned New Critics and neo-Aristotelians, most writing teachers are comfortable with the idea that meaning is found not solely in the text nor solely in the reader but rather in the interaction between the two. In fact, that process is at the very center of our work as writing teachers: we *must* misread every student text in order to help students say what we think they really mean. It is this sort of generous and deliberate misreading—readings in which we go beyond the words' literal meanings to try to draw out possibilities in a text, to imagine what the text might be trying to become—that is at the basis of Shaughnessy's analysis of error, Elbow's believing game, and Bartholomae and Petrosky's plan to integrate reading and writing.

So far, so good. But the next step causes resistance: few writing teachers want to go so far as to admit that we actually *create* the meaning of our students' texts, particularly if this creative act is largely the result of our unconscious biases and associations. The problem with admitting our role as co-author is that it violates most of our fundamental beliefs about the objectivity of the teacher, the integrity of the text, the rights of the individual author. And yet that next step seems unavoidable, a fact not lost on those interested in the application of critical theory to the composition classroom: if great literary works are unstable and subject to multiple readings and interpretations, then how unstable is the evolving draft of an inexperienced composition student (Harris 158)? If every reading of Chaucer and Shakespeare is a re-writing, then how can teachers avoid becoming authors of our students' drafts (Eagleton 12)? Or, to put it another way, if a teacher is reading a text that was written specifically for her, with revisions that are a direct result of her suggestions, how can she possibly have any clear sense of where the text stops and her reading begins?

But in spite of these nagging realities, I sense that in practice most of us cling to the notion that our readings of student essays are somehow "objective"; in spite of our knowledge of reader response theory and deconstruction, we believe that when we read student essays we are responding to some objective reality in—or noticeably missing from—the text itself rather than to a text we have unconsciously revised or even created. We are not unaware that we bring to our teaching of writing and our reading of student essays strong beliefs and biases. We know, for example, how we feel about abortion and gun control, how we respond more favorably to some rhetorical strategies than to others, even how we like some students much better than others. But we conveniently forget those issues and pretend that we can willingly suspend those beliefs and disbeliefs. We see ourselves as neutral, objective, open-minded. We give each student an equal chance. We are ready to like essays on any topic in any mode. We just want students to find their own voices, to find themselves.

This paradigm of the teacher-as-objective-reader fails to do justice to the complexity of the reading and writing processes and to our relationship to our students. When we read an essay on abortion or a presidential election, most of us go out of our way to be fair, to try to evaluate the writing for its own sake, if such a thing is even possible. But what happens when we read an essay on a seemingly "unpolitical" issue or topic about which we have powerful (and often unconscious) associations? Consider, for example, this exchange during a discussion I had a few weeks ago with two other writing teachers. First teacher: "If I get one more essay on 'how I won the big high school football game,' I'll scream. I mean these guys describe each play in great detail and then show how they saved the day at the end. Yuck. They are so self-serving and so trivial." Other teacher: "You're missing the point. Those aren't trivial at all. For an adolescent male, those games can be his most significant experiences." In part this is a gender issue: the writers of most of these sports essays and the second teacher are male, while the first teacher is female. In part it is personal: the male teacher went on to explain that he remembers high school sports as perhaps the one "pure thing" in his life, while I went on to admit that because my memories of high school sports include failed expectations—mine and my father's—it is for me one of the most *impure* things in my life.

Of course, it's not true that every reading is equally idiosyncratic and personal or that student texts do not exist until we de- and then reconstruct them. I am not suggesting that all student papers are Rorschach tests or random ink blots on the page. Clearly there is a text in the class, and it is even a text for which we can—and have—developed shared criteria for evaluation. Sometimes this "interpretative community" is consciously and deliberately created, such as the training of teachers participating in holistic scoring sessions; more often, though, it is the result of shared unconscious preferences or, as Lester Faigley's study of teacher preferences demonstrates, shared "unstated cultural definitions" (410). There is even a certain type of essay (I will call it the autobiographical narrative of a self-actualizing event) that most of us in this interpretative community prefer. But the fact that we agree a text exists and that we agree about some of the criteria for evaluation should not make us underestimate our own creative and often idiosyncratic role in the process.

My point (similar to one that Louise Phelps makes on the various stages of teacher response) is that we need to develop a theory of reading student drafts that reflects these issues, that allows us to acknowledge—to our students and to ourselves—that we play a central role in the composing process, not only when we give our students guidelines and heuristics, not only when we suggest changes in conferences, but also when we read the essays themselves. We need a theory of reading that takes into account the "intertextual" nature of our work; that is, a theory that takes into account the fact that we cannot read any student essay without unconsciously and simultaneously reading a number of other texts as well. And, finally, we need a theory that allows us to recognize our limitations, to say first to ourselves, and then directly to a student, "I am not going to be a good reader of an essay on this topic. You should know that going in."

In part, then, this is a process for which we need to use and extend what we have learned about reading and analysis from critical theory. But it is more—and less—than that. The evolving student draft is not identical to the published literary work and thus requires, as Phelps and others have argued, new theories of reading and response. Our readings of student essays are contextualized in ways that readings of literary texts are not. We know the authors of these texts, we work with them, we suggest changes to them, we have something to gain if they succeed, or—if we dislike the students involved—something to gain if they fail. None of this is static or linear or unilateral but changes with each teacher and each student. Therefore, in order to develop a more dialectical theory of reading and interpretation we need to consider how actual readers and writers—teachers and students—interact. We will not come to understand this interaction by decontextualizing context (as I believe Linda Flower and other experimentalists often do in their research on this subject) but rather by examining our readers within the student-teacher relationship.

Towards Redefining the Teacher's Role

If we are to respond differently in the teacher-student relationship, we need to re-define our role and self-image. Ironically, as much as the teacher's role seems to have changed in the great paradigm shift from product to process, one thing remains the same: we still have written ourselves relatively minor and unfulfilling parts to play in the writing process. In the traditional class the writing teacher played several roles—provider of information, lecturer, upholder of standards, corrector—but each was relatively static, unilateral: the teacher provided the students with rules and models of good writing and then graded them according to how closely the results approximated those rules and models. Not only did this role fail to reflect the complexity and pleasure of the writer, it failed to acknowledge the intelligence, creativity, and interests of the teacher. In fact our role in the traditional classroom seems to me a little like the tyrant's rule in Orwell's "Shooting an Elephant." By denying our students power, we actually limited our own freedom. Although we did most of the talking, although we told the students the rules and gave them the models, although we believed that we were in control, there was actually very little room for the sort of originality, risk-taking, and inquiry that Cynthia Onore and others have argued is essential if a writing relationship is to be successful (240).

When I say that our role is still dull and one-dimensional, I am not suggesting that there has been no significant change over the last two decades. Nor am I ignoring some current examples of more innovative and interactive teacher-student relationships. It's just that the new role most process teachers have adopted is in many respects as narrow and rigid as the old one. I'm referring to teachers who describe themselves as "facilitators" (as if they have no agenda of their own, or rather, as if their agenda is not important) or as "just another member of the writing workshop." The concept of the de-centralized writing classroom is based on the following logic (or illogic): "all we really have to do is get out of our students' way and let them write." I realize that I am creating something of a caricature here of the process teacher and classroom, but I think that there has been an element of naiveté in this approach.

Many writing teachers deny their role as co-authors and their tremendous authority in the classroom because it does not fit with the image they would like to project. Most of us are uncomfortable admitting that we are the center of a "de-centered" classroom, that we hold so much power, that we are largely responsible for success and, even worse, for failure. But while there are good reasons for our discomfort—many of us would like for political reasons to think of our classroom as democratic, supportive, and non-hierarchical—there are even better reasons to face the truth: from a student's perspective a writing teacher *is* an authority figure, even—or especially—in process classrooms. (In fact, as Tom Newkirk has argued, the teacher in composition classes in which students are asked to write about their personal feelings and to meet in one-to-one conferences actually holds more authority, because the stakes are higher.)

I suspect that the notion of teacher-as-non-authority developed as a necessary stage or antithesis to the thesis offered by traditional classroom teachers. The synthesis is to move beyond either/or thinking—either we have authority or they do, either we own the text or they do, either the meaning is in the writer or in the reader—towards a more dialectical definition. Rather than dichotomizing the teacher's and the student's roles, we need to see how they are inseparably related. Just as Janet Emig argued that traditional models of the composing process failed because they ignored the role and uses of the writer's unconscious, most of our current views fail because they ignore the role and uses of the teacher's unconscious. Until we have a clearer and more realistic notion of how we shape and influence student writing and how, in return, that writing shapes and influences us, we will continue to limit our student's potential development.

And to limit our own. One reason many English teachers dislike teaching composition is that they feel they are supposed to dislike it and then set out to prove it. The teaching of writing should not be fun, they feel, and a writing course certainly should not be tailored to a teacher's individual taste or preference. This sense of composition as a teacher's unpleasant duty or burden runs deep in our profession and is one of the reasons so many people distrust, resent, and envy those writing teachers who talk about their work in intensely personal and positive terms. I know for a fact that my colleagues are more than a little skeptical when Toby Fulwiler gloats that Freshman Writing is the "Best Course in the University to Teach" or when Don Murray muses, "There must be something wrong with a fifty-four-year-old man who is looking forward to his thirty-fifth conference of the day" ("Listening" 232). This kind of enthusiasm for composition does not seem possible to teachers who have scrupulously sought to remove themselves and their own interests from the course. By remaining detached in this way, by refusing to misread essays in personal and playful ways, we make composition an unpleasant duty—for our students *and* for ourselves.

do!

don't!

do!

The Lure, Lore, and Leery(ness) of Therapeutic Models

So how do we write more interesting and satisfying roles for ourselves to play in the writing class? And how do we develop a clearer and more realistic notion of the way that our responses and non-responses shape student writing? My own suggestion—and it is one that may not be particularly popular or politically correct—is that we pay more careful attention to the research and experience of psychotherapists. I am not equating composition and therapy nor am I suggesting that psychotherapeutic relationships are free from the power politics and self-deceptions that I am criticizing in the writing class. I am simply saying that it makes no sense to ignore lessons from the field in which the workings of the unconscious and the subtle dynamics of dyad relationships have been carefully and systematically analyzed. I think that most writing teachers know that therapeutic models can help us explain and explore the teacher-student relationship, but because they find this comparison threatening they publicly deny it. That may also explain why so many composition theorists offer instructive models from and comparisons to psychotherapy which they then immediately disown. Take, for example, this paragraph by James Moffett:

> The processes of psychotherapy and writing both require maximum synthesizing of firsthand and secondhand knowledge into a full, harmonious expression of individual experience. This calls for the removal of spells to which the person has not agreed and of which he is unconscious. Freud asked the patient to start talking about anything that came into his head—in other words, to attempt to verbalize his stream of consciousness or externalize his inner speech. This technique presupposes that from the apparent chaos of all this disjointed rambling will emerge for analyst and patient an order, eventually "betrayed" by motifs, by sequencing, by gradual filling in of personal cosmology. Thus, if successful, the subject's cosmologizing processes, the idiosyncratic ways of structuring and symbolizing experience, stand more clearly revealed and presumably more amenable to deliberate change, if desired. The most important thing a writer needs to know is how she does think and verbalize and how

> he or she might. . . . Not for a moment do I suggest that the teacher play psychiatrist. The therapeutic benefits from writing are natural fallout and nothing for a school to strive for. (100-101)

I think Moffett is saying that, "Writing and psychotherapy are similar processes, but composition teachers and therapists have nothing in common." In other words, although he is unquestionably drawn to—and willing to draw from—the experience of psychotherapists, he is determined to distance himself from this model as quickly as possible. In fact, Moffett's statement is only the clearest example of the schizophrenic response that most writing teachers have to the composition-as-therapy metaphor. For example, Thomas Carnicelli, concerned about the kinds of questions and clues that promote self-discovery, suggests first that Rogerian questioning might help, but then quickly offers an artificial distinction: "The teacher's function is to lead students to adopt the teacher's values, the common criteria of good writing shared by the teacher, and the English profession, and, with certain wide variations, educated people in general. The therapist's function is to lead clients to clarify or develop their own individual values" (116). Similarly, Stephen Zelnick, in writing about conferences, admits, "I am afraid that whether we wish it or not, we become role models for our students" and "there is the romantic/sexual vibration. If it is in any way possible, conferences set going a buzz and flutter of fantasies" (49), but then he dismisses the therapeutic model altogether: "Translating student conferences into other, simpler paradigms of efficient, smooth client relations, or psychotherapeutic self-exploration impoverishes education. We can do better than that" (58).

Oddly enough, Don Murray, the writing teacher whose work seems most heavily influenced by psychotherapeutic goals and methods, is perhaps the most outspoken critic of this analogy. While Murray talks again and again about reading "my other self," about "writing to learn," about writing conferences in which the teacher listens and the student speaks, about a process which, in fact, sounds suspiciously similar to making the unconscious conscious, he finds the comparison ludicrous:

> Responsive teaching is often confused with a stereotypical therapeutic role in which the teacher always nods, always encourages, always supports, and never intervenes. That is ridiculous. . . . The conference isn't a psychiatric session. Think of the writer as an apprentice at the workbench with the master workman. (*Writer* 154)

I can't help but wonder why these writing teachers are going so far to deny a connection that they actually brought up themselves. No one claims that conference teaching equals therapy; but the fact that there are significant differences between teaching writing and doing psychotherapy is hardly the point. Carnicelli, Zelnick, Murray, and others seem to admit that there is role-modeling, sexual tension, even transference, in the teaching of writing and the teacher-student relationship, but because these things make them uncomfortable (which they should) they deny their significance and suggest that we focus on the writing process and product as if it existed in a decontextualized situation and relationship.

Still, these early attempts to link composition and therapy were valuable because they called attention to important aspects of the teacher-student relationship and paved the way for more recent essays which unapologetically take advantage of therapeutic models. I want to mention two of these that focus on the unconscious drives and associations that shape the way our students respond to us as teachers. Robert Brooke, relying heavily on Lacan, suggests that students in "response" classrooms of the type that Murray and Elbow describe improve their writing because they identify with—and want desperately to please—the teacher, the "Subject Who is Supposed to Know" (Brooke 680). The student then projects or transfers emotions and associations from his own early-life relationships, particularly with his parents, onto the teacher. Ann Murphy, relying more heavily on Freud, extends Brooke's argument by demonstrating how transference can also account for our students' occasional resistance to us, to writing, to self-knowledge, to education. Murphy argues:

> Despite their many obvious and important differences, both psychoanalysis and teaching writing involve an intensely personal relationship in which two people painstakingly establish trust beyond the apparent limitations of their institutional roles, in order that both might learn and one might achieve a less marginal, more fully articulated life. (181)

While I think these essays go a long way in explaining classroom dynamics, I want to go still further and suggest that counter-transference—our unconscious responses to our students or, more significantly, our unconscious responses to their unconscious responses to us—also shapes the reading and writing processes. Freud's explanation of counter-transference has important implications for writing teachers:

> We have become aware of the 'counter-transference,' which arises in [the analyst] as a result of a patient's influence on his unconscious feelings, and we are almost inclined to insist that he shall recognize this counter-transference in himself and overcome it. Now that a considerable number of people are practising psychoanalysis and exchanging their observations with one another, we have noticed that no psychoanalyst goes further than his own complexes and internal resistances permit; and we consequently require that he shall begin his activity with self-analysis and continually carry it deeper while he is making observations on his patients. Anyone who fails to produce results in a self-analysis of this kind may at once give up any idea of being able to treat patients by analysis. (145)

As teachers, we also can go no further than our own complexes and internal resistances permit, and thus we, too, need to begin with self-analysis. We, too, need to identify the extent to which our responses to our students and their writing are not neutral or objective, the extent to which counter-transference responses interfere with our ability to help students improve their writing.

If writing teachers react negatively to the suggestion that they play therapist, I assume that my recommendations—that we analyze ourselves, that we consider our own neuroses in the reading and teaching processes, that we also play patient—seem even more irrelevant and threatening. Again it's not that writing teachers are unaware that our own unconscious issues often obscure and shape our actions; it's just that we hope if we don't talk about this, it will go away. For instance, Rosenblatt acknowledges that when students read and write personally, they often reveal some of their "conflicts and obsessions," thereby tempting teachers to deal directly with these psychological issues (207). Although she points out some instances in which students have benefitted from this sort of interaction, she ends up warning teachers against "officious meddling with the emotional life of their students" (207) because teachers cannot be trusted in this sort of relationship:

> Unfortunately, like members of any other group, many teachers are themselves laboring under emotional tensions and frustrations. Given the right to meddle in this way, they would be tempted to find solutions for their own problems by vicariously sharing the student's life. They might also project upon the student their own particular preoccupations and lead him to think that he was actually suffering difficulties and frustrations that were the teacher's. Assuredly even worse than the old indifference to what is happening psychologically to the student is the tampering with personality carried on by well-intentioned but ill-informed adults. The wise teacher does not attempt to be a psychiatrist. (208)

Rosenblatt is right to point out that teachers have the power to impose themselves on their students in dangerous ways, but it is not always so easy to distinguish between a teacher who is guilty of projecting his "own particular preoccupations" onto his students and "tampering with personality" from one who is emotionally engaged in his teaching and honestly interested in influencing his students' values and ideas. By attempting to edit feelings, unconscious associations, and personal problems out of a writing course, we are fooling ourselves and shortchanging our students. The teaching of writing is about solving problems, personal and public, and I don't think we can have it both ways: we cannot create intensity and deny tension, celebrate the personal and deny the significance of the personalities involved. In my writing courses, I *want* to meddle with my students' emotional lives, and I want their writing to meddle with *mine*. Transference and counter-transference emotions are threatening because they are so powerful, but they are most destructive and inhibiting in the writing class when we fail to acknowledge and deal with them.

Reading Myself Reading My Students: A Classroom Example

Let me try to illustrate this process of identifying and using counter-transference emotions with an example from my own teaching. Last fall I taught two sections of freshman composition; from the very first week, one section went extremely well while the other was a nightmare. I had trouble getting the students involved in the discussions or in their own writing, and I grew increasingly irritated during class. I was especially bothered by the four 18-year-old male students who sat next to each other, leaning back in their desks against the wall. They usually wore sunglasses; they always wore sneakers with untied laces. Whenever I tried to create drama or intensity, they joked or smirked. Whenever I tried to joke, they acted aggressively bored, rolling their eyes or talking to each other. At first, I tried to ignore them, trying not to let them get to me. But I found that it was a little like trying not to think about an elephant. I was always aware of them, even when they were not acting out.

After two weeks, I decided that everyone was being distracted by these students, that they were responsible for the unproductive mood of the classroom. But for some reason, I was not able to confront them directly about their aggressive behaviors in class or their passive efforts outside of class. It was as if in confronting them I would be acknowledging that they were bothering me, and I refused to do that, partly because I always prided myself on my relationships with students and the comfortable, relaxed atmosphere in my classrooms. So instead of confronting them directly, I stewed inside and—I am embarrassed to admit—fantasized about revenge: "Be patient," I told myself. "Grading time will come along eventually and then you can get even. You can fail them all."

I suppose the other reason I did not confront them was that when they came for their first individual conferences, they were polite, even a bit deferential. They were emotionally detached, but they answered my questions, accepted most of my suggestions, and, except for one, even seemed somewhat grateful. Still their writing was relatively weak, and I made little effort to help them improve. I read their texts looking more for problems than for possibilities. I had essentially written them off: I had decided that these four were just insecure, adolescent boys trying to act tough in class in front of the other students; that they were not secure enough with their roles, with their masculinity, to be independent, serious, or mature; that if they wanted to get nothing out of this class, then that was fine by me, and, finally, that I would just concentrate on the other students in the class and ignore them.

But that noble plan failed miserably. It seemed that every time I would accommodate their acting out, they would raise the stakes. For example, during small-group peer response times, they would choose to work together and then spend the time talking about football or dorm parties. Even worse, if I assigned groups, they would talk about writing for a few minutes and then call to each other across groups. I retaliated (note the aggressive language) by indirectly threatening them. I interrupted the class one day to give an angry and sarcastic speech on how anyone who was not taking the class seriously would fail and end up taking it again. I told them how sorry I would feel if that happened, but that I had no choice. Although I knew that these four students would not fail—their essays were not that bad—I looked at them when I made the threat.

Finally, one day, I snapped. I walked into class, saw them together, laughing and leaning against the wall, and in a voice that conveyed much too much anger and disgust I said, "I have never had to do this in ten years of college teaching; in fact, I left high school specifically so I wouldn't have to deal with shit like this, but you guys are completely out of control. I don't want you to sit together any more." There was an awkward silence and then one of the boys said in a mocking voice, "Completely out of control? Fine, I'll move." Another asked, "That's why you left high school?" It was an embarrassing moment because it was clear—to them and to me—that I was the one who felt out of control.

What was going on? I was usually relaxed and comfortable with students. I was reasonable. I was well liked. So the problem had to be with them. They were threatened by me, I told myself, so insecure that they had to stick together and act tough. They saw me as an authority figure and were rebelling, not only against me but against authority figures from their pasts. And those explanations were partially true. But that still did not explain why my response was so angry. I had allowed myself to get caught up in a macho competition with these students, and I was losing. Clearly this had as much or more to do with my insecurities and unconscious responses as it did with theirs.

That's when I realized the significance of my slip about high school. I had meant to say, "That's why I left high school *teaching*," but I had referred accidentally to my own experience as a high school student. I remembered periods when I acted like these students and later periods when they were the type I felt I was competing against. And I realized how much, for whatever reasons, I was still bothered by the group behavior of adolescent males. The realization helped: by recognizing and somehow naming the source of my anger, it dissipated and became more manageable. I'm not saying I suddenly felt comfortable with these students or with their texts, but the situation now seemed within my own realm, somehow within my control.

Although this example may have more to do with my neuroses than with composition theory, the point is that this knowledge changed the way that I read these students and their texts; it helped me in my teaching and, indirectly, helped these students in their writing. I began to confront them directly, asking them if they agreed with certain points, inviting them to criticize my readings, giving them room and invitation in conference and class to challenge me in (what I took to be) constructively indirect ways: I encouraged them to freewrite about the course and me. I asked them to write metaphorically about writing. I told them to push back when they felt I was pushing them too hard in a conference.

Although one of the students continued to write essays that showed little effort or commitment, the other three made significant progress. One wrote an essay in which he used the metaphor of writing as playing the drums to argue against my emphasis on revision: a writer has to revise just as a drummer has to tune his kit, but "sometimes you just have to let me play." Another wrote a satiric essay on "productive procrastination," suggesting not only that I took writing too seriously but also that my view of the process was limited and limiting. He ended his essay by saying,

> If you begin writing too early, the pressure may not be great enough. If you begin too late, your ideas will not have time to take shape. Procrastination is the key because it triggers your unconscious ideas. Oh, by the way, it is now 3:27 a.m. And you probably thought I wouldn't have time to write a good essay.

The fact that these challenges to my authority came in conventional forms that supported my authority neutralized my anger or defensiveness; the fact that I allowed and encouraged these challenges neutralized theirs.

But the third student, Jack, provided the best example of this sort of interaction. From the beginning of the year, he had seemed the angriest and the least cooperative. I was irritated that the first essay he brought into conference, "The Advantages and Disadvantages of Biotechnology," was clearly written as a report for high-school class. When I asked him to write something new, he brought in "How to Make a Peanut Butter and Jelly Sandwich." There were attempts at humor ("A true P, B, and J expert takes this science a step further by experimenting with exotic varieties of peanut butters and jellies."), but for the most part it was a flat description of the process.

As I was reading it, he spoke up, "Remember in class what you said? You said that there are no good or bad topics, that someone could write a trivial essay on something profound, like nuclear war, or a profound one on something trivial, like making a peanut butter and jelly sandwich. So I tried it." Again I felt irritated, and couldn't quite figure out how to respond, so I asked him the purpose of the essay. "To tell the reader how to make peanut butter and jelly sandwich. Why? Isn't that OK?"

"But doesn't a reader already know that?"

"Yeah. So are you saying that something is missing . . . but what else can you say about this topic?"

When I asked him if he meant the essay to be funny, he said, "Sort of," so I suggested he try to locate and develop the humor in a revision.

After he left, I knew I had been too aggressive in my responses to him and too passive in my readings of his texts. I was not making any effort to read or rather misread meaning or possibility or potential into his writing because I felt convinced that he was trying to get away with something; he was provoking and mocking me. Still I was frustrated with myself: rather than calling him on anything directly, saying "I don't want dredged-up high school essays" or "Why waste time making fun of the assignment?" I was still operating at a stage in which I did not want Jack or any of the others to know they were getting to me.

It was during the next week that I began to realize why I was so upset by these four students. It was also the time that I realized I had to confront their resistance more directly while at the same time giving them more room to channel it. So when Jack came back with a revision of the peanut butter and jelly sandwich essay, I responded differently. He had made a few minor changes, but nothing striking. When we discussed it, he said he tried to make it funnier by making the instructions "more ridiculous." When I asked him why he was writing a comic essay on making a peanut butter and jelly sandwich, he had no idea. I suggested that if the essay were meant to be satiric, he ought to think about who or what was being satirized. He seemed totally confused and asked for an example. I said that the essay could, for example, be making fun of technical writers who complicate simple processes. He looked irritated. "Or, maybe you are making fun of teachers who give foolish assignments." He looked surprised for a second, then laughed. I had not planned to confront him in that way, but as soon as I did I was convinced it was the right move.

"I decided to drop the peanut butter and jelly essay," Jack told me in his next conference. "You kept asking me what I learned from writing it and what I wanted the reader to learn and my answer was always 'I don't know, probably nothing.' So I decided that if I couldn't learn anything from it, the reader can't be expected to either. So I wrote an essay about why this wasn't a good topic." Now it could certainly be argued that Jack had simply quit resisting or that he was now putting me on in a new way, but at the time I only focused on how this new essay was an interesting discussion on the role and difficulty of topic selection in the writing process. His main point was that a "simpler topic is actually harder to work with than a more complicated and in-depth one." He tried to prove that point by comparing his peanut butter and jelly essay to a classmate's essay on the death of his father. He argued that he had struggled to generate ideas because his topic was so simple, while his classmate "had many avenues and moral implications to explore." I encouraged him, pointing out that I thought this essay had more potential than his earlier ones. I raised questions about certain nuances of his argument. And I talked a little about what kind of topics *I* found easier and harder to write about. In short, I finally tried to misread one of his essays in ways that would open up the topic for him and for me. After Jack revised his essay, we both agreed that it was by far the strongest piece he had written all semester; not coincidentally, it was also the first one in which we both felt an investment.

Until I recognized the fact that my unconscious responses were creating much of the resistance, Jack stalled as a writer. After that recognition, we both were more productive in our respective roles. The essay on the relative difficulty of certain topics may have begun as the same kind of dare as the first peanut butter and jelly paper, but it is clear that in writing that essay, he and I both became interested in the topic, more connected to the text and to each other. In fact, until I could recognize how much my anger and defensiveness were shaping my responses to all four of these student writers, I was not an effective writing teacher for them or the other students in the class.

The Personal Is Pedagogical

Of course, these students may have had difficulty as writers in my class for all sorts of reasons that have nothing to do with my personal hangups or limitations. In fact, I'm certain that there were a combination of explanations for their problems early in the year. But the fact remains that I may have contributed to their problems by responding to them and to their writing in ways that limited our relationship. The same is true in Nicki's case. It's possible that she was able to write effectively in my course partially because of her transference emotions and identification with me. But it is also true that I may have failed to push her as hard as I might have if I were not caught up in feeling proud of myself. Nicki's writing directly and indirectly validated my teaching and, as a result, I was flattered; I read the early drafts and behavior of these four males as threatening and critical and I, in return, was defensive and punishing.

Of course, a sexual component is in this: we cannot ignore gender as a factor in the way students respond to their teachers and the way teachers respond to their students. But beyond the sexual tension—most of which is unconscious—there is simply the problem that I respond more favorably to students—male or female—who make me feel secure than to those who threaten me. And that is what I need to monitor: as soon as I find myself giving up on a student or, on the other hand, feeling tremendous personal pride in a student's work, I need to question my own motives. I need to discover in what ways my biases and assumptions—both conscious and unconscious—are shaping my teaching.

Now I suspect that this concentration on my own feelings and associations seems self-indulgent and misguided to composition specialists who believe in more "scholarly" research. I further suspect that they would advise me to quit thinking so much about myself and to focus instead on the tropes and conventions of academic discourse, or on the problems of task representation, or on new ways to empower student writers. But this is not an either/or choice, not a decision to study myself rather than my students or their texts; my point is that we can never fully separate one from the other. If we want to find less constrained and constraining ways of responding as writing teachers, we have to examine our responses within the contexts of the relationships in which they occur. By engaging in ongoing self-analysis, by becoming more self-conscious about the source of our misreadings, by recognizing that our unconscious associations are a significant part of a writing course, we can become more creative readers and more effective teachers. By avoiding this process, we will never know in what ways we are limiting our students, their writing, and ourselves.

Works Cited

Bartholomae, David, and Anthony Petrosky. *Facts, Artifacts, and Counterfacts: Theory and Method for a Reading and Writing Course*. Portsmouth, NH: Heinemann, 1986.

Brooke, Robert. "Lacan, Transference, and Writing Instruction." *College English* 49 (1987): 679-691.

Carnicelli, Thomas. "The Writing Conference: A One-to-One Conversation." *Eight Approaches to Composition*. Ed. Timothy R. Donovan and Ben W. McClelland. Urbana, IL: NCTE, 1980. 101-132.

Eagleton, Terry. *Literary Theory: An Introduction*. Minneapolis: U of Minnesota P, 1983.

Elbow, Peter. *Writing Without Teachers*. New York: Oxford UP, 1973.

Emig, Janet. "The Uses of the Unconscious in Composing." *The Web of Meaning: Essays on Writing, Teaching, Learning, and Thinking*. Ed. Dixie Goswami and Maureen Butler. Upper Montclair, NJ: Boynton/Cook, 1983. 44-53.

Faigley, Lester. "Judging Writing, Judging Selves." *College Composition and Communication* 40 (1989): 395-413.

Flower, Linda. "Cognition, Context, and Theory Building." *College Composition and Communication* 40 (1989): 282-311.

Freud, Sigmund. "The Future Prospects of Psycho-Analytic Therapy." *Standard Edition of the Complete Works of Sigmund Freud*. Ed. James Strachey. Vol. 11. London: Hogarth, 1957. 144-145. 24 vols.

Fulwiler, Toby. "Freshman Writing: It's the Best Course in the University to Teach." *Composition and Literature: Exploring the Human Experience*. Ed. Jesse Jones, Veva Vonler, and Janet Harris. San Diego: Harcourt, 1987. 17-20.

Harris, Joseph. "The Plural Text/The Plural Self: Roland Barthes and William Coles." *College English* 49 (1987): 158-170.

Moffett, James. "Writing, Inner Speech, and Mediation." *Coming on Center: Essays in English Education*. 2nd ed. Portsmouth, NH: Heinemann, 1988. 133-181.

Murphy, Ann. "Transference and Resistance in the Basic Writing Classroom: Problematics and Praxis." *College Composition and Communication* 40 (1989): 175-187.

Murray, Donald. "The Listening Eye: Reflections on the Writing Conference." *The Writing Teacher's Sourcebook*. Ed. Gary Tate and Edward Corbett. New York: Oxford UP, 1968. 232-237.

———. *A Writer Teaches Writing*. 2nd ed. Boston: Houghton, 1985.

Onore, Cynthia. "The Student, The Teacher, and the Text: Negotiating Meanings Through Response and Revision." *Writing and Response: Theory, Practice, and Research*. Ed. Chris M. Anson. Urbana, IL: NCTE, 1989. 231-260.

Phelps, Louise Wetherbee. "Images of Student Writing: The Deep Structure of Teacher Response." *Writing and Response: Theory, Practice, and Research*. Ed. Chris M. Anson. Urbana, IL: NCTE, 1989. 37-67.

Rosenblatt, Louise. *Literature as Exploration*. New York: MLA, 1938.

Shaughnessy, Mina. *Errors and Expectations: A Guide for the Teacher of Basic Writing*. New York: Oxford UP, 1977.

Zelnick, Stephen. "Student Worlds in Student Conferences." *Writing Talks: Views on Teaching Writing From Across the Professions*. Ed. Muffy E. A. Siegal and Toby Olson. Upper Montclair, NJ: Boynton/Cook, 1983. 47-58.

Susan C. Jarratt

Teaching Across and Within Differences

After trying for several years to finesse my feminist position in the writing class, I finally decided last fall to conduct an "overtly" feminist composition class. By that I meant most of all that I would use the word "feminism" from the very beginning to describe my pedagogy—in my syllabus, in daily discussion of pedagogical decisions with students, and in a final letter to the students describing and evaluating our work together. Both national and local factors worked against a discourse of feminism at Miami University. The popular media have been gloating for some time over the decline of feminism, along with other progressive social movements, during the conservative eighties. Susan Faludi's *Backlash* documents this phenomenon in painstaking detail. In tune with the national chilling of feminist fervor, some students at our Oxford campus reserve a special distaste for the "feminism" that disrupted the normal social life of their university within recent memory. Five years ago an undergraduate women's group staged a courageous sit-in at the administration building to protest the forced hiring of a man in the all-woman escort service. That year, protests of a sorority date auction and of a beauty contest held by the men's glee club further alienated, even disgusted, many students who have no experience of social protest as a legitimate form of participation in public affairs.

For newer students several years away from those local events, "feminism" still comes from an alien language. Like "marxism" and "socialism," it names an alternative worldview often demonized by schools, churches, and the media in the attenuated public discourse in our society. Even after I've convinced my students to complicate their assumption that anyone calling herself (or even himself) a feminist seeks the end of love, intimate heterosexual relations, marriage, family, and the male of the species, there remains the genuinely threatening critique feminism offers of those gendered social conventions they've worked so hard to master and that serve as their very survival skills in the frightening new world of the university outside the classroom.

Despite these impediments, my 17 first-year students, my undergraduate teaching fellow, and I had a fine semester that fall. Class discussions were lively but respectful, the writing was engaging, student evaluations were high. It seems that we experienced this success through a paradoxical opening and closing of multiple differences. Though the gesture of naming myself and my class "feminist" opened a chasm between me and my students, various kinds of solidarity were forged behind and within our work together.

From *College Composition and Communication* 43 (October 1992): 297-322.

I think my own self-assurance was an important personal difference from past teaching experiences. Though I engaged in activism in the sixties, it was more counterculturally than politically motivated. Without a history in the feminist movement, I've often felt like a late-comer without the necessary credentials. In the past, I most often introduced social issues within a rhetorically based pedagogy and sometimes expressed my own views on issues students chose to take up, but I did not name myself with a particular ideological position. They labeled me with their own terms—former hippie, liberal—but I kept an ironic distance from those labels. To announce myself as something was a completely different strategy, a different rhetoric of the classroom, one I now felt comfortable with. In closing with the name "feminist," I modeled for my students a committed but provisional social identity, good only until the withering away of sexism (not to be confused with the withering away of men)! The problem for both conservative students and also, paradoxically, for some radical teachers is that "feminism" names only one site of social difference: gender. But defining feminism for myself and my students as a theoretical vehicle for naming multiple social identities and for analyzing the play of language and power at those various sites moves it beyond a narrow identity politics. More simply put, I convinced them that there was more to "feminism" for us than my own gender.

Having just completed the Feminist Sophistics seminar and conference made a more communal difference. It marked a moment of connection for the participants between composition/rhetoric as a field and feminism. Certainly feminism is far from the first political intervention in the writing classroom. From long before the beginning of the recent creation of composition as an academic field, teachers and scholars have engaged politics in and of the classroom. Greg Myers, James Berlin, Richard Ohmann and many others have chronicled the junctures of radical politics and the teaching of writing during this century. The connection with feminism has been later and perhaps more tentative (see Flynn, "Composing"), but seeing the growth of a group of feminist fellow travelers specifically in the teaching of writing was significant for me as a teacher. As I stepped into the classroom, I felt supported by a group. The inevitably agonistic experience of counterhegemonic cultural work was ameliorated.

In another more material way I wasn't alone in the classroom last fall. I was assisted throughout the term by an undergraduate Teaching Fellow. Miami offers to English majors interested in education the opportunity to assist a faculty member in teaching a lower-division class. Theresa Squires had been in a class I taught the previous spring on women and writing. As a sophomore, Theresa was beginning to discover a feminist consciousness and was passionately interested in social issues in the classroom. Her presence as an older student, a self-described "brown girl" (with parents of different racial backgrounds), and a social being in process had a powerful influence on the class. She attended all the classes, read all the students' papers and wrote comments on them, contributed to decisions about writing and reading, and joined in class discussion. Though the two of us shared some general sympathies, we sometimes had different opinions on issues discussed in class. So Theresa never functioned simply as a "yes" person, just reinforcing a party line. Rather, she served a significant role for the students in mediating my authority as an older, tenured faculty member offering a counterhegemonic pedagogy. For me, she was a source of support, a sounding board, and a reality check on student concerns and lives. Her presence bridged the student/teacher difference.

"How many feminists does it take to change a light bulb?"
"One—and it's *not* funny!"

Another element of our experience last fall was my serious decision to use humor in my feminist classroom. I may be more sensitive than others to the bad rap feminists get about being humorless because I am generally a pretty serious person. Because students both male and female feel that their most deeply valued personal interests are threatened by feminism, I made an effort to lighten the burden of relentless critique (see Shor). By using humor, I communicated that I could laugh at myself and other feminists, and that I cared enough about my students to share the pleasure of laughter with them. But I didn't suggest that feminism wasn't something I took seriously. Laughing together made us co-conspiring cultural critics, bound us together as a group, cutting across the social boundaries we were simultaneously working to demarcate through our writing and reading about difference. After I started the semester with the joke above, we fell into a pattern of sharing jokes at the beginning of class each day. We noticed that lots of jokes were funny at someone else's expense, so we tried to find ones that weren't or figure out why people's laughter hurts others. Too often, though, I found myself analyzing for my students why the jokes others brought in were sexist, racist, classist, etc. That role gave me too much power and, of course, killed the humor of the joke. Power shifts when the students themselves have control of language tools—a principle at the heart of the composition revolution.

I think the most powerful fusion across difference came from sharing with my students tools of sociolinguistic analysis—tools the students both applied and critiqued. By naming my pedagogy and myself feminist, I wasn't communicating that the writing class should be a platform for my views about specific issues. It meant that becoming a responsible language user demands an understanding of the ways language inscribes difference. My students had only to agree to use the analytic tools I provided (see Bizzell). In this way, my pedagogy may resemble a more instrumentalist composition theory. For example, instead of taking my word that adding concrete details would improve a personal narrative, I asked my students to take my word that naming their own social locations would ground their stories in socially specific, and thus more socially responsible, accounts of personal history. Instead of persuading my students that "information" is socially constructed, I made them graph the data from a national news source along two axes of social difference—race and gender. When we looked at tv and film, I set them the task of focusing not only on the medium as text but on its production and reception as well, questions generated by work in cultural studies. I didn't tell them what the outcome of these processes would be because I didn't know myself, nor did I tell them what to make of the outcomes. But it was clear to all of us that the politics are in the techniques of analysis. Requiring students to perform the practices above is different both from demanding that they share my views and from teaching them to write unified, coherent five-paragraph themes—all political acts.

Again, the difference for me in this moment of feminist teaching was a reduction of difference between me and my students. We were collectively engaged in a process of cultural critique and production, an engagement reflected in their evaluations: "I'm addicted to this analysis stuff!" Very few, if any, of my 17 students were ready to sign up as feminists at the end of the semester, but that was not my goal. Most of them were ready to describe themselves as language users engaged in a socially located and responsible practice of analysis and production. The point of my account is not that differences should disappear; it is that socially oriented and personally satisfying composition teaching can happen across and within different kinds of difference.

Works Cited

Berlin, James A. *Rhetoric and Reality: Writing Instruction in American Colleges, 1900-1985*. Carbondale: Southern Illinois UP, 1987.

Bizzell, Patricia. "Power, Authority, and Critical Pedagogy." *Journal of Basic Writing* 10 (1991): 54-70.

Faludi, Susan. *Backlash: The Undeclared War Against American Women*. New York: Crown, 1991.

Flynn, Elizabeth A. "Composing as a Woman." *College Composition and Communication* 39 (1988): 423-435.

Myers, Greg. "Reality, Consensus, and Reform in the Rhetoric of Composition Teaching." *College English* 48 (1986): 154-174.

Ohmann, Richard. *Politics of Letters*. Middletown: Wesleyan UP, 1987.

Shor, Ira. *Critical Teaching and Everyday Life*. Boston: South End, 1980.

Donald M. Murray

The Listening Eye: Reflections on the Writing Conference

It was dark when I arrived at my office this winter morning, and it is dark again as I wait for my last writing student to step out of the shadows in the corridor for my last conference. I am tired, but it is a good tired, for my students have generated energy as well as absorbed it. I've learned something of what it is to be a childhood diabetic, to raise oxen, to work across from your father at 115 degrees in a steel-drum factory, to be a welfare mother with three children, to build a bluebird trail, to cruise the disco scene, to be a teen-age alcoholic, to salvage World War II wreckage under the Atlantic, to teach invented spelling to first graders, to bring your father home to die of cancer. I have been instructed in other lives, heard the voices of my students they had not heard before, shared their satisfaction in solving the problems of writing with clarity and grace. I sit quietly in the late afternoon waiting to hear what Andrea, my next student, will say about what she accomplished on her last draft and what she intends on her next draft.

It is nine weeks into the course and I know Andrea well. She will arrive in a confusion of scarves, sweaters, and canvas bags, and then produce a clipboard from which she will precisely read exactly what she has done and exactly what she will do. I am an observer of her own learning, and I am eager to hear what she will tell me.

I am surprised at this eagerness. I am embedded in tenure, undeniably middle-aged, one of the gray, fading professors I feared I would become, but still have not felt the bitterness I saw in many of my own professors and see in some of my colleagues. I wonder if I've missed something important, if I'm becoming one of those aging juveniles who bound across the campus from concert to lecture, pleasantly silly.

There must be something wrong with a fifty-four-year-old man who is looking forward to his thirty-fifth conference of the day. It is twelve years since I really started teaching by conference. I average seventy-five conferences a week, thirty weeks a year, then there's summer teaching and workshop teaching of teachers. I've probably held far more than 30,000 writing conferences, and I am still fascinated by this strange, exposed kind of teaching, one on one.

It doesn't seem possible to be an English teacher without the anxiety that I will be exposed by my colleagues. They will find out how little I do; my students will expose me to them; the English Department will line up in military formation in front of Hamil-

From *College English* 41 (September 1979): 13-18.

ton Smith Hall and, after the buttons are cut off my Pendleton shirt, my university library card will be torn once across each way and let flutter to the ground.

The other day I found myself confessing to a friend, "Each year I teach less and less, and my students seem to learn more. I guess what I've learned to do is to stay out of their way and not to interfere with their learning."

I can still remember my shock years ago when I was summoned by a secretary from my classroom during a writing workshop. I had labored hard but provoked little discussion. I was angry at the lack of student involvement and I was angry at the summons to the department office. I stomped back to the classroom and was almost in my chair before I realized the classroom was full of talk about the student papers. My students were not even aware I had returned. I moved back out to the corridor, feeling rejected, and let the class teach itself.

Of course, that doesn't always happen, and you have to establish the climate, the structure, the attitude. I know all that, and yet . . .

I used to mark up every student paper diligently. How much I hoped my colleagues would see how carefully I marked my student papers. I alone held the bridge against the pagan hordes. No one escaped the blow of my "awk." And then one Sunday afternoon a devil bounded to the arm of my chair. I started giving purposefully bad counsel on my students' papers to see what would happen. "Do this backward," "add adjectives and adverbs," "be general and abstract," "edit with a purple pencil," "you don't mean black you mean white." Not one student questioned my comments.

I was frightened my students would pay so much attention to me. They took me far more seriously that I took myself. I remembered a friend in advertising told me about a head copywriter who accepted a piece of work from his staff and held it overnight without reading it. The next day he called in the staff and growled, "Is this the best you can do?"

They hurried to explain that if they had more time they could have done better. He gave them more time. And when they met the new deadline, he held their copy again without reading it, and called them together again and said, "Is *this* the best you can do?"

Again they said if only they had more time, they could . . . He gave them a new deadline. Again he held their draft without reading it. Again he gave it back to them. Now they were angry. They said, yes, it was the best they could do and he answered, "I'll read it."

I gave my students back their papers, unmarked, and said, make them better. And they did. That isn't exactly the way I teach now, not quite, but I did learn something about teaching writing.

In another two-semester writing course I gave 220 hours of lecture during the year. My teaching evaluations were good; students signed up to take this course in advance. Apparently I was well-prepared, organized, entertaining. No one slept in my class, at least with their eyes shut, and they did well on the final exam. But that devil found me in late August working over my lecture notes and so, on the first day of class, I gave the same final exam I had given at the end of the year. My students did better before the 220 hours of lectures than my students had done afterwards. I began to learn something about teaching a non-content writing course, about under-teaching, about not teaching what my students already know.

The other day a graduate student who wanted to teach writing in a course I supervise indicated, "I have no time for non-directive teaching. I know what my students need to know. I know the problems they will have—and I teach them."

I was startled, for I do not know what my students will be able to do until they write without any instruction from me. But he had a good reputation, and I read his teaching evaluations. The students liked him, but there was a minor note of discomfort. "He does a good job of teaching, but I wish he would not just teach me what I already know" and "I wish he would listen better to what we need to know." But they liked him. They could understand what he wanted, and they could give it to him. I'm uncomfortable when my students are uncomfortable, but more uncomfortable when they are comfortable.

I teach the student not the paper but this doesn't mean I'm a "like wow" teacher. I am critical and I certainly can be directive but I listen before I speak. Most times my students make tough—sometimes too tough evaluations—of their work. I have to curb their too critical eye and help them see what works and what might work so they know how to read evolving writing so it will evolve into writing worth reading.

I think I've begun to learn the right questions to ask at the beginning of a writing conference.

"What did you learn from this piece of writing?"

"What do you intend to do in the next draft?"

"What surprised you in the draft?"

"Where is the piece of writing taking you?"

"What do you like best in the piece of writing?"

"What questions do you have of me?"

I feel as if I have been searching for years for the right questions, questions which would establish a tone of master and apprentice, no, the voice of a fellow craftsman having a conversation about a piece of work, writer to writer, neither praise nor criticism but questions which imply further drafts, questions which draw helpful comments out of the student writer.

And now that I have my questions, they quickly become unnecessary. My students ask these questions of themselves before they come to me. They have taken my conferences away from me. They come in and tell me what has gone well, what has gone wrong, and what they intend to do about it.

Some of them drive an hour or more for a conference that is over in fifteen minutes. It is pleasant and interesting to me, but don't they feel cheated? I'm embarrassed that they tell me what I would hope I would tell them, but probably not as well. My students assure me it is important for them to prepare themselves for the conference and to hear what I have to say.

"But I don't say anything," I confess. "You say it all."

They smile and nod as if I know better than that, but I don't.

What am I teaching? At first I answered in terms of form: argument, narrative, description. I never said comparison and contrast, but I was almost as bad as that. And then I grew to answering, "the process." "I teach the writing process." "I hope my students have the experience of the writing process." I hear my voice coming back from the empty rooms which have held teacher workshops.

That's true, but there's been a change recently. I'm really teaching my students to react to their own work in such a way that they write increasingly effective drafts. They write; they read what they've written; they talk to me about what they've read and what the reading has told them they should do. I nod and smile and put my feet up on the desk, or down on the floor, and listen and stand up when the conference runs too long. And I get paid for this?

Of course, what my students are doing, if they've learned how to ask the right questions, is write oral rehearsal drafts in conference. They tell me what they are going to write in the next draft, and they hear their own voices telling me. I listen and they learn.

But I thought a teacher had to talk. I feel guilty when I do nothing but listen. I confess my fear that I'm too easy, that I have too low standards, to a colleague, Don Graves. He assures me I am a demanding teacher, for I see more in my students than they see in themselves. I certainly do. I expect them to write writing worth reading, and they do—to their surprise, not mine.

I hear voices from my students they have never heard from themselves. I find they are authorities on subjects they think ordinary. I find that even my remedial students write like writers, putting down writing that doesn't quite make sense, reading it to see what sense there might be in it, trying to make sense of it, and—draft after draft—making sense of it. They follow language to see where it will lead them, and I follow them following language.

It is a matter of faith, faith that my students have something to say and a language in which to say it. Sometimes I lose that faith but if I regain it and do not interfere, my students do write and I begin to hear things that need saying said well.

This year, more than ever before, I realize I'm teaching my students what they've just learned.

They experiment, and when the experiment works I say, "See, look what happened." I put the experiment in the context of the writing process. They brainstorm, and I tell them that they've brainstormed. They write a discovery draft, and I point out that many writers have to do that. They revise, and then I teach them revision.

When I boxed I was a counterpuncher. And I guess that's what I'm doing now, circling my students, waiting, trying to shut up—it isn't easy—trying not to interfere with their learning, waiting until they've learned something so I can show them what they've learned. There is no text in my course until my students write. I have to study the new text they write each semester.

It isn't always an easy text to read. The student has to decode the writing teacher's text; the writing teacher has to decode the student's writing. The writing teacher has to read what hasn't been written yet. The writing teacher has the excitement of reading unfinished writing.

Those papers without my teacherly comments written on them haunt me. I can't escape the paranoia of my profession. Perhaps I should mark up their pages. There are misspellings, comma splices, sentence fragments (even if they are now sanctified as "English minor sentences.") Worse still, I get papers that have no subject, no focus, no structure, papers that are underdeveloped and papers that are voiceless.

I am a professional writer—a hired pen who ghostwrites and edits—yet I do not know how to correct most student papers. How do I change the language when the student writer doesn't yet know what to say? How do I punctuate when it is not clear what the student must emphasize? How do I question the diction when the writer doesn't know the paper's audience?

The greatest compliment I can give a student is to mark up a paper. But I can only mark up the best drafts. You can't go to work on a piece of writing until it is near the end of the process, until the author has found something important to say and a way to say it. Then it may be clarified through a demonstration of professional editing.

The student sits at my right hand and I work over a few paragraphs, suggesting this change, that possibility, always trying to show two, or three, or four alternatives so that the student makes the final choice. It is such satisfying play to mess around with someone else's prose that it is hard for me to stop. My best students snatch their papers away from my too eager pen but too many allow me to mess with their work as if I knew their world, their language, and what they had to say about their world in their language. I stop editing when I see they really appreciate it. It is not my piece of writing; it is not my mind's eye that is looking at the subject; not my language which is telling what the eye has seen. I must be responsible and not do work which belongs to my students, no matter how much fun it is. When I write it must be my own writing, not my students'.

I realize I not only teach the writing process, I follow it in my conferences. In the early conferences, the prewriting conferences, I go to my students; I ask questions about their subject, or if they don't have a subject, about their lives. What do they know that I don't know? What are they authorities on? What would they like to know? What would they like to explore? I probably lean forward in these conferences; I'm friendly, interested in them as individuals, as people who may have something to say.

Then, as their drafts begin to develop and as they find the need for focus, for shape, for form, I'm a bit removed, a fellow writer who shares his own writing problems, his own search for meaning and form.

Finally, as the meaning begins to be found, I lean back, I'm more the reader, more interested in the language, in clarity. I have begun to detach myself from the writer and from the piece of writing which is telling the student how to write it. We become fascinated by this detachment which is forced on student and teacher as a piece of writing discovers its own purpose.

After the paper is finished and the student starts on another, we go back through the process again and I'm amused to feel myself leaning forward, looking for a subject with my student. I'm not coy. If I know something I think will help the student, I share it. But I listen first—and listen hard (appearing casual)—to hear what my student needs to know.

Now that I've been a teacher this long I'm beginning to learn how to be a student. My students are teaching me their subjects. Sometimes I feel as if they are paying for an education and I'm the one getting the education. I learn so many things. What it feels like to have a baby, how to ski across a frozen lake, what rights I have to private shoreline, how complex it is to find the right nursery school when you're a single parent with three children under six years old.

I expected to learn of other worlds from my students but I didn't expect—an experienced (old) professional writer—to learn about the writing process from my students.

But I do. The content is theirs but so is the experience of writing—the process through which they discover their meaning. My students are writers and they teach me writing most of the time.

I notice my writing bag and a twenty-page paper I have tossed towards it. Jim has no idea what is right or wrong with the paper—and neither do I. I've listened to him in conference and I'm as confused as he is. Tomorrow morning I will do my writing, putting down my own manuscript pages, then, when I'm fresh from my own language, I will look at Jim's paper. And when he comes back I will have at least some new questions for him. I might even have an answer, but if I do I'll be suspicious. I am too fond of answers, of lists, of neatness, of precision; I have to fight the tendency to think I know the subject I teach. I have to wait for each student draft with a learning, listening eye. Jim will have re-read the paper and thought about it too and I will have to be sure I listen to him first for it is his paper, not mine.

Andrea bustles in, late, confused, appearing disorganized. Her hair is totally undecided; she wears a dress skirt, lumberjack boots, a fur coat, a military cap. She carries no handbag, but a canvas bag bulging with paper as well as a lawyer's briefcase which probably holds cheese and bread.

Out comes the clipboard when I pass her paper back to her. She tells me exactly what she attempted to do, precisely where she succeeded and how, then informs me what she intends to do next. She will not work on this draft; she is bored with it. She will go back to an earlier piece, the one I liked and she didn't like. Now she knows what to do with it. She starts to pack up and leave.

I smile and feel silly; I ought to do something. She's paying her own way through school. I have to say something.

"I'm sorry you had to come all the way over here this late."

Andrea looks up surprised. "Why?"

"I haven't taught you anything."

"The hell you haven't. I'm learning in this course, really learning."

I start to ask Andrea what she's learning but she's out the door and gone. I laugh, pack up my papers, and walk home.

CLASSROOMS

Terry Dean

Multicultural Classrooms, Monocultural Teachers

Remember gentlemen, John Chrysostom's exquisite story about the day he entered the rhetorician Libanius' school in Antioch. Whenever a new pupil arrived at his school, Libanius would question him about his past, his parents, and his country.

Renan, *La Réforme intellectuelle et morale*

Sometimes more than others, I sense the cultural thin ice I walk on in my classrooms, and I reach out for more knowledge than I could ever hope to acquire, just to hang on. With increasing cultural diversity in classrooms, teachers need to structure learning experiences that both help students write their way into the university and help teachers learn their way into student cultures. Now this is admittedly a large task, especially if your students (like mine) are Thai, Cambodian, Vietnamese, Korean, Chinese, Hmong, Laotian (midland Lao, lowland Lao), Salvadoran, Afro-American, Mexican, French, Chicano, Nicaraguan, Guatemalan, Native American (Patwin, Yurok, Hoopa, Wintu), Indian (Gujarati, Bengali, Punjabi), Mexican-American, Jamaican, Filipino (Tagalog, Visayan, Ilocano), Guamanian, Samoan, and so on. It may take a while for the underpaid, overworked freshman composition instructor to acquire dense cultural knowledge of these groups. But I have a hunch that how students handle the cultural transitions that occur in the acquisition of academic discourse affects how successfully they acquire that discourse. The very least we can do, it seems to me, is to educate ourselves so that when dealing with our students, in the words of Michael Holzman, "We should stop doing harm if we can help it" (31).

Some would question how much harm is being done. If enough students pass exit exams and the class evaluations are good, then everything is OK. Since we want our students to enter the mainstream, all we need worry about is providing them the tools. Like opponents of bilingual education, some would argue that we need to concern ourselves more with providing student access to academic culture, not spending time on student culture. But retention rates indicate that not all students are making the transition into academic culture equally well. While the causes of dropout are admittedly complex, cultural dissonance seems at the very least to play an important role. If indeed we are going to encounter "loss, violence, and compromise" (142) as David Bartholomae describes the experience of Richard Rodriguez, should we not be directing students to the counseling center? And if the attainment of biculturalism in many cases is painful and

From *College Composition and Communication* 40 (February 1989): 23-37.

difficult, can we be assured, as Patricia Bizzell suggests, that those who do achieve power in the world of academic discourse will use it to argue persuasively for preservation of the language and the culture of the home world view? (299). This was not exactly Richard Rodriguez's response to academic success, but what if, after acquiring the power, our students feel more has been lost than gained? I think as teachers we have an obligation to raise these issues. Entering freshmen are often unaware of the erosion of their culture until they become seniors or even later. Like Richard Rodriguez, many students do not fully realize what they have lost until it is too late to regain it. Let me briefly outline the problem as I see it and offer some possible solutions.

The Problem

A lot is being asked of students. David Bartholomae describes the process: "What our beginning students need to learn is to extend themselves, by successive approximations, into the commonplaces, set phrases, rituals and gestures, habits of mind, tricks of persuasion, obligatory conclusions, and necessary connections that determine 'what might be said' and constitute knowledge within the various branches of our academic community" (146). "Rituals and gestures, tricks of persuasion" mean taking on much more than the surface features of a culture. Carried to an extreme, students would have to learn when it is appropriate to laugh at someone slipping on a banana peel. When we teach composition, we are teaching culture. Depending on students' backgrounds, we are teaching at least academic culture, what is acceptable evidence, what persuasive strategies work best, what is taken to be a demonstration of "truth" in different disciplines. For students whose home culture is distant from mainstream culture, we are also teaching how, as a people, "mainstream" Americans view the world. Consciously or unconsciously, we do this, and the responsibility is frightening.

In many situations, the transitions are not effective. Several anthropological and social science studies show how cultural dissonance can affect learning. Shirley Brice Heath examines the ways in which the natural language environments of working-class Black and white children can interfere with their success in schools designed primarily for children from middle-class mainstream culture. The further a child's culture is from the culture of the school, the less chance for success. Classroom environments that do not value the home culture of the students lead to decreased motivation and poor academic performance (207-272). In a study of Chicano and Black children in Stockton, California, schools, John Ogbu arrives at a similar conclusion. Susan Urmston Philips analyzes the experiences of Warm Springs Native American children in a school system in Oregon where the administrators, teachers, and even some parents thought that little was left of traditional culture. But Philips shows that "children who speak English and who live in a material environment that is overwhelmingly Western in form can still grow up in a world where by far the majority of their enculturation experience comes from their interaction with other Indians. Thus school is still the main source of their contact with mainstream Anglo culture" (11). Philips describes the shock that Warm Springs Native American children experience upon entering a school system designed for the Anglo middle-class child. Because of differences in the early socialization process of Native American children (especially in face-to-face interaction), they feel alienated in the classroom and withdraw from class activities (128).

Pierre Bourdieu and Jean-Claude Passeron examine how the cultural differences of social origin relate directly to school performance (8-21). Educational rewards are given to those who feel most at home in the system, who, assured of their vocations and abilities, can pursue fashionable and exotic themes that pique the interest of their teachers, with little concern for the vocational imperatives of working-class and farm children. Working-class and farm children must struggle to acquire the academic culture that has been passed on by osmosis to the middle and upper classes. The very fact that working-class and farm children must laboriously acquire what others come by naturally is taken as another sign of inferiority. They work hard because they have no talent. They are remedial. The further the distance from the mainstream culture, the more the antipathy of mainstream culture, the more difficulty students from outside that culture will have in acquiring it through the educational system (which for many is the only way):

> Those who believe that everyone would be given equal access to the highest level of education and the highest culture, once the same economic means were provided for all those who have the requisite "gifts," have stopped halfway in their analysis of the obstacles; they ignore the fact that the abilities measured by the scholastic criteria stem not so much from natural "gifts" (which must remain hypothetical so long as the educational inequalities can be traced to other causes), but from the greater or lesser affinity between class cultural habits and the demands of the educational system or the criteria which define success within it. [Working-class children] must assimilate a whole set of knowledge and techniques which are never completely separable from social values often contrary to those of their class origin. For the children of peasants, manual workers, clerks, or small shopkeepers, the acquisition of culture is acculturation. (Bourdieu and Passeron 22)

Social and cultural conditions in the United States are not the same as in France, but the analyses of Bourdieu, Passeron, Heath, Ogbu, and others suggest interesting lines of inquiry when we look at the performance of students from different cultures and classes in U.S. schools. Performance seems not so much determined by cultural values (proudly cited by successful groups), but by class origins, socio-economic mobility, age at time of immigration, the degree of trauma experienced by immigrants or refugees, and the acceptance of student culture by the mainstream schools. Stephen Steinberg argues in *The Ethnic Myth* that class mobility precedes educational achievement in almost all immigrant groups (131-132). I really do not believe that Black, Native American, and Chicano cultures place less emphasis on the importance of education than Chinese, Jewish, Vietnamese, or Greek cultures do. We do not have over one hundred Black colleges in the United States because Blacks don't care about education. I have never been to a Native American Studies Conference or visited a rancheria or reservation that did not have newsletters, workshops, and fund-raisers in support of education. Bourdieu and Passeron's analysis suggests that educational success depends to a large extent on cultural match, and if an exact match is not possible, there must at least be respect and value of the culture children bring with them. Acculturation (assimilation) is possible for some, but it is not viable for all.

Acculturation itself poses problems. Jacquelyn Mitchell shows how cultural conflict affects the preschool child, the university undergraduate, the graduate student, and the faculty member as well. Success brings with it, for some people, alienation from the values and relationships of the home culture. ''In fulfilling our academic roles, we interact increasingly more with the white power structure and significantly less with members of our ethnic community. This is not without risk or consequence; some minority scholars feel in jeopardy of losing their distinctive qualities'' (38). The question Mitchell poses is, ''How can blacks prepare themselves to move efficiently in mainstream society and still maintain their own culture?'' (33). Jacqueline Fleming, in a cross-sectional study of Black students in Black colleges and predominantly white colleges, found Black colleges more effective despite the lack of funding because Black colleges are more ''supportive'' of students (194). Long before the recent media coverage of racial incidents on college campuses, Fleming noted that ''all is not well with Black students in predominantly white colleges'' (162). And in California, the dropout rate of ''Hispanics'' (a term that obscures cultural diversity much as the term ''Asian'' does) is greater than that of any other group except possibly Native Americans. But despite gloomy statistics, there is hope.

Theoretical Models for Multicultural Classrooms

Several theoretical models exist to help students mediate between cultures. In ''Empowering Minority Students: A Framework for Intervention,'' James Cummins provides one:

> The central tenet of the framework is that students from ''dominated'' societal groups are ''empowered'' or ''disabled'' as a direct result of their interactions with educators in the schools. These interactions are mediated by the implicit or explicit role definitions that educators assume in relation to four institutional characteristics of schools. These characteristics reflect the extent to which (1) minority students' language and culture are incorporated into the school program; (2) minority community participation is encouraged as an integral component of children's education; (3) the pedagogy promotes intrinsic motivation on the part of students to use language actively in order to generate their own knowledge; and (4) professionals involved in assessment become advocates for minority students rather than legitimizing the location of the ''problem'' in the students. For each of these dimensions of school organization the role definitions of educators can be described in terms of a continuum, with one end promoting the empowerment of students and the other contributing to the disabling of students. (21)

Like Bourdieu, Cummins sees bicultural ambivalence as a negative factor in student performance. Students who have ambivalence about their cultural identity tend to do poorly whereas ''widespread school failure does not occur in minority groups that are positively oriented towards both their own and the dominant culture, [that] do not perceive themselves as inferior to the dominant group, and [that] are not alienated from their own cultural values'' (22). Cummins argues that vehement resistance to bilingual education comes in part because ''the incorporation of minority languages and cultures into the school program confers status and power (jobs for example) on the minority group'' (25). But for Cummins, it is precisely this valuing of culture within the school that leads to academic success because it reverses the role of domination of students by the school.

Shirley Brice Heath's model is similar to Cummins'. The main difference is that she focuses on ethnography as a way for both teachers and students to mediate between home and school cultures. A consideration of home culture is the only way students can succeed in mainstream schools, increase scores on standardized tests, and be motivated to continue school: "Unless the boundaries between classrooms and communities can be broken, and the flow of cultural patterns between them encouraged, the schools will continue to legitimate and reproduce communities of townspeople who control and limit the potential progress of other communities who themselves remain untouched by other values and ways of life" (369). Like Cummins, Heath aims for cultural mediation. As one student stated: "Why should my 'at home' way of talking be 'wrong' and your standard version be 'right'? . . . Show me that by adding a fluency in standard dialect, you are adding something to my language and not taking something away from me. Help retain my identity and self-respect while learning to talk 'your' way" (271). Paulo Freire, on the other hand, wants those from the outside to totally transform mainstream culture, not become part of it: "This, then, is the great humanistic and historical task of the oppressed: to liberate themselves and their oppressors as well" (*Pedagogy* 28). Yet in almost all other respects, Freire's model, like Cummins' and Heath's, is grounded in a thorough knowledge of the home culture by teachers, and actively learned, genuine knowledge by the student. Each of these models has different agendas. Teachers should take from them whatever suits their teaching style, values, and classroom situations. I would simply encourage an inclusion of the study of the wide diversity within student cultures.

Teaching Strategies for Multicultural Classes

Cultural Topics. Culturally oriented topics are particularly useful in raising issues of cultural diversity, of different value systems, different ways of problem solving. Several successful bridge programs have used comparison of different cultural rituals (weddings, funerals, New Year's) as a basis for introducing students to analytic academic discourse. Loretta Petrie from Chaminade University in Hawaii has a six-week summer-school curriculum based on this. Students can use their own experience, interview relatives, and read scholarly articles. Reading these papers to peer response groups gives students additional insights into rituals in their own culture as well as making them aware of similarities and differences with other groups. I have used variations of Ken Macrorie's I-Search paper (Olson 111-122) to allow students to explore part of their cultural heritage that they are not fully aware of. One Vietnamese student, who was three years old when she came to America, did a paper on Vietnam in which she not only interviewed relatives to find out about life there but sought out books on geography and politics; she literally did not know where Vietnam was on the map and was embarrassed when other students would ask her about life there. Several students whose parents were from the Philippines did research that was stimulated by the desire to further understand family customs and to explain to themselves how the way they thought of themselves as "American" had a unique quality to it. One student wrote:

> For over eighteen years I have been living in the United States. Since birth I have been and still am a citizen of this country. I consider myself a somewhat typical

> American who grew up with just about every American thing you can think of; yet, at home I am constantly reminded of my Filipino background. Even at school I was reminded of my Filipino culture. At my previous school, the two other Filipinos in my class and I tried to get our friends to learn a little about our culture.

But classmates were not always open to cultural diversity, and their rejection raised the central question of just how much you have to give up of your culture to succeed in the mainstream society:

> During the Philippine presidential election, there were comments at my school that we Filipinos were against fair democracy and were as corrupt as our ex-President Marcos. Also it was said that Filipinos are excessively violent barbaric savages. This is partly due to our history of fighting among ourselves, mostly one group that speaks one dialect against another group of a different language within the islands. Also maybe we are thought savages because of the food we eat such as ''chocolate meat'' and ''balut,'' which is sort of a salted egg, some of which may contain a partially developed chicken. So, to be accepted into society you must give up your old culture.

''You must give up your old culture'' is misleading. The student had to be careful about sharing home culture with peers, but he isn't giving up his culture; he is gaining a greater understanding of it. His essay ended with:

> I now have a better understanding of why I was doing a lot of those things I didn't understand. For example, whenever we visited some family friends, I had to bow and touch the older person's hand to my forehead. My mother didn't really explain why I had to do this, except it was a sign of respect. Also my mother says my brother should, as a sign of respect, call me ''manong'' even though he is only two years younger than I. At first I thought all this was strange, but after doing research I found out that this practice goes back a long way, and it is a very important part of my Filipino culture.

Richard Rodriguez's widely anthologized ''Aria'' (from *Hunger of Memory*) allows students to analyze his assertion that loss of language and culture is essential to attain a ''public voice.'' Although the student above seems to agree in part with Rodriguez, most students find Rodriguez's assertions to be a betrayal of family and culture:

> I understand Rodriguez's assertion that if he learned English, he would lose his family closeness, but I think that he let paranoia overcome his senses. I feel that the lack of conversation could have been avoided if Rodriguez had attempted to speak to his parents instead of not saying anything just because they didn't. I am sure that one has to practice something in order to be good at it and it was helpful when his parents spoke in English to them while carrying on small conversation. To me, this would have brought the family closer because they would be helping each other trying to learn and grow to function in society. Instead of feeling left out at home and in his society, Rodriguez could've been included in both.

I have used this topic or variations on it for a half dozen years or so, and most of my freshmen (roughly 85% of them) believe they do not have to give up cultural and home values to succeed in the university. I find quite different attitudes among these students when they become juniors and seniors. More and more students graduate who feel that they have lost more than they gained. Raising this issue early provides students more choices. In some cases it may mean deciding to play down what seem to be unacceptable parts of one's culture (no balut at the potluck); in other cases it may lead to the assertion of positive values of the home culture such as family cohesiveness and respect for parents and older siblings. Courageous students will bring the balut to the potluck anyway and let mainstream students figure it out. I often suspect that some of the students who drop out of the university do so because they feel too much is being given up and not enough is being received. Dropping out may be a form of protecting cultural identity.

Cultural topics are equally important, if not more so, for students from the mainstream culture. Many mainstream students on predominantly white campuses feel inundated by Third World students. Their sense of cultural shock can be as profound as that of the ESL basic writer. One student began a quarter-long comparison/contrast essay on the immigration experiences of his Italian grandparents with the experiences of Mexicans and Vietnamese in California. As the quarter went on, the paper shifted focus as the student became aware that California was quickly becoming non-white. It scared him. The essay was eventually titled "Shutting the Doors?" and ended with:

> I have had some bad experiences with foreigners. On a lonely night in Davis, three other friends and I decided to go to a Vietnamese dance. When we got there, I couldn't believe how we were treated. Their snobbishness and arrogance filled the air. I was upset. But that was only one incident and possibly I am over reacting. I often reflect on my high school teacher's farewell address. He called for our acceptance of the cultural and religious background of each other. But after long days thinking, I, like many others, am unable to answer. My only hope is that someone has a solution.

The student had grown up in Richmond, California, a culturally diverse East Bay community, and was friendly with students from many cultures (it was his idea to go to the dance). The very recognition on the part of the student of what it feels like to be surrounded by difference at the dance is a beginning step for him to understand what it means to try to be who he is in the midst of another culture. I see this student quite often. He is not racist. Or if he is, he at least has the courage to begin asking questions. The solution for which the student yearned was not immediately forthcoming: his yearning for one is worth writing about.

Language Topics. It is not unusual for ESL errors to persist in the writing or the pronunciation of highly educated people (doctors, lawyers, engineers, professors) because, consciously or unconsciously, those speech patterns are part of the person's identity and culture. The same can be true of basic writers. Language-oriented topics are one way to allow students to explore this kind of writing block. Assignments that require students to analyze their attitudes toward writing, their writing processes, and the role that writing plays in their lives can make these conflicts explicit. For example, one student wrote:

> Moreover, being a Chinese, I find myself in a cultural conflict. I don't want to be a cultural betrayer. In fact, I want to conserve my culture and tradition. It would be

> enjoyable if I wrote my mother language. For example I like to write letter in Chinese to my friends because I can find warmth in the letter. The lack of interest in writing and the cultural conflict has somehow blocked my road of learning English.

Overcoming this block may, however, cause problems in the home community. In response to Rodriguez, one student wrote:

> I can relate when Rodriguez say that his family closeness was broken. Even though I speak the language that is understood and is comfortable at home, when I speak proper English around my friends at home, they accuse me of trying to be something I'm not. But what they don't realize is that I have to talk proper in order to make it in the real world.

Jacquelyn Mitchell writes of returning to her community after college:

> My professional attire identified me in this community as a "middle-class," "siddity," "uppity," "insensitive" school teacher who had made her way out of the ghetto, who had returned to "help and save," and who would leave before dark to return to the suburbs. My speech set me up as a prime candidate for suspicion and distrust. Speaking standard English added to the badges that my role had already pinned on me. (31)

In some way problems like this affect us all. In the small town that I grew up in, simply going to college was enough to alienate you from your peers. Although my parents encouraged me to get an education, they and their peers saw college as producing primarily big egos—people who thought they knew everything. No easy choices here. It took my uncle, who made the mistake of becoming a Franciscan priest, forty years to be accepted back into the family by my mother.

Another way to help students with cultural transitions is to make the home language the subject of study along with the different kinds of academic discourse they will be required to learn. Suzy Groden, Eleanor Kutz, and Vivian Zamel from the University of Massachusetts have developed an extensive curriculum in which students become ethnographers and analyze their language patterns at home, at school, and in different social situations, using techniques developed by Shirley Brice Heath. This approach takes time (several quarters of intensive reading and writing), but such a curriculum has great potential for helping students acquire academic discourse while retaining pride and a sense of power in the discourse they bring with them.

Peer Response Groups. Peer response groups encourage active learning and help students link home and university cultures. The Puente Project, in affiliation with the Bay Area Writing Project, combines aggressive counseling, community mentors, and English courses that emphasize active peer response groups. The Puente Project has turned what used to be a 50-60% first-year dropout rate into a 70-80% retention rate in fifteen California community colleges. All of the students are academically high risk (meaning they graduated from high school with a D average), they are Chicanos or Latinos, and all have a past history of avoidance of English classes and very low self-confidence when it comes to writing. Writing response groups give the students a sense of belonging on campus. As students make the transition from home to school, the groups become, in the students' own words, "una familia":

> Now after two quarters of Puente, it's totally different. My writing ability has changed to about 110%. I might not be the best speller in the world, but I can think of different subjects faster and crank out papers like never before . . . having Latinos in a class by themselves is like a sun to a rose. This is the only class where I know the names of every student and with their help I decide what to write. (*Puente* 20)

Joan Wauters illustrates how structured non-confrontational editing can make peer response more than a support group. Students work in pairs on student essays with specific training and instructions on what to look for, but the author of the essay is not present. The author can later clarify any point she wishes with the response group, but Wauters finds that the non-confrontational approach allows students to be more frank about a paper's strengths and weaknesses and that it is "especially valuable for instructors who work with students from cultures where direct verbal criticism implies 'loss of face' " (159). Wauters developed these techniques with Native American students, but they apply equally well to other cultures.

Response groups do not have to be homogeneous. Any small group encourages participation by students who may not feel comfortable speaking up in class for whatever reason. They provide a supportive environment for exploring culturally sensitive issues that students might hesitate to bring up in class or discuss with the teacher. The following paragraph was read by a Black student to a group consisting of a Filipino, a Chinese, and two Chicano students:

> I am black, tall, big, yet shy and handsome. "I won't hurt you!" Get to know who I am first before you judge me. Don't be scared to speak; I won't bite you. My size intimidates most people I meet. I walk down my dormitory hallway and I can feel the tension between me and the person who's headed in my direction. A quick "Hi!" and my response is "Hello, how are you doing?" in a nice friendly way. It seems that most of the guys and girls are unsure if they should speak to me. I walk through the campus and eyes are fixed on me like an eagle watching its prey. A quick nod sometimes or a half grin. Do I look like the devil? No, I don't. Maybe if I shrink in size and lightened in color they wouldn't be intimidated. Hey, I'm a Wild and Crazy Guy too!

This small group discussion of what it felt like to be an outsider spilled over into the class as a whole, and students that normally would not have participated in class discussion found themselves involved in a debate about dorm life at Davis. I know that peer response groups have limitations, need structure, and can be abused by students and teachers alike. But I have never heard complaints from teachers using peer groups about how difficult it is to get ESL students to participate. In some cases the problem is to shut them up.

Class Newsletters. Class newsletters encourage students to write for an audience different from the teacher, and they generate knowledge about multicultural experiences. I use brief 20-minute in-class writing assignments on differences between the university and home, or how high school is different from the university, or ways in which the university is or is not sensitive to cultural differences on campus. Sometimes I simply have students finish the statement, "The university is like. . . ." These short paragraphs serve as introductions to issues of cultural transition, and when published, generate class discussions and give ideas for students who are ready to pursue the topic in more detail. Newsletters can be done in a variety of formats from ditto masters to desktop publishing. Students who feel comfortable discussing ethnic or cultural tensions establish a forum for those students whose initial response would be one of denial. For example, the story of a Guamanian and a Black student who thought they were not invited to a white fraternity party led to an extended class discussion of whether this kind of experience was typical and whether they had not been invited for ethnic reasons. The next time I had students write, a Chicana student articulated her awakening sense of cultural conflict between the university and her family:

> I was so upset about leaving home and coming to Davis. I was leaving all my friends, my boyfriend, and my family just to come to this dumb school. I was angry because I wanted to be like all my other friends and just have small goals. I was resentful that I had to go away just to accomplish something good for me. I felt left out and angry because it seemed that my family really wanted me to go away and I thought it was because they did not love me. I was not studying like I should because I wanted to punish them. My anger grew when I realized I was a minority at Davis. My whole town is Mexican and I never thought of prejudice until I came to Davis.

Family is central to Chicano/Latino/Mexican-American students. The pull toward home can create ambivalence for students about their school commitments. In this case, the family was aware of this pull, and encouraged the student to give college the priority. The daughter interpreted this as a loss of love. The student's ambivalence about home and school put her on probation for the first two quarters at Davis. She is now a junior, doing well as a premed student, and her chances for a career in medicine look good. I don't think just writing about these issues made the difference. The class discussion generated by her article helped her realize that her situation was shared by others who were experiencing the same thing but had not quite articulated it. She was not alone.

Bringing Campus Events into the Classroom. I recently assigned a paper topic for a quarter-long essay that made reading of the campus newspapers mandatory. I was surprised to find that many students did not read the campus newspapers on a daily basis and in many cases were quite unaware of campus issues that directly concerned them, for example, the withdrawal of funding from the Third World Forum (a campus newspaper that deals with Third World issues), compulsory English examinations for international graduate teaching assistants in science and math classes, and increasing incidents of hostility toward Asians (as reflected in bathroom graffiti, "Lower the curve: kill a chink"). Admission policies at UC Berkeley, particularly as reported in the press, have pitted Blacks and Chicanos against Asians, and quite often students find themselves in dorm discussions without having enough specific knowledge to respond. The more articulate students can be about these issues, the greater the chance the students will feel integrated in the university.

Assignments can make mandatory or strongly encourage students to attend campus events where cultural issues will be discussed within the context of campus life—issues such as the self-images of women of color or how Vietnamese students feel they are perceived on campus. A panel entitled "Model Minority Tells the Truth" called for Vietnamese students to become more involved in campus life partly in order to overcome misperceptions of some students:

> Vietnamese students feel inferior to Americans because Americans do not understand why we are here. We are refugees, not illegal immigrants. There are a lot of unspoken differences between Vietnamese and Americans because the memory of the Vietnam war is so fresh, and it is difficult for Americans to be comfortable with us because we are the conflict. Vietnamese students also suffer an identity crisis because the Vietnamese community has not established itself yet in America as other Asian groups have.

Anecdotes. Anecdotes about oneself and former class experiences are another way to generate discussion and raise issues of cultural transition and identity. The teacher's own curiosity and experience of cultural diversity will often give students ideas for other topics. Cultural identity does not depend on a Spanish surname alone nor does it reside in skin color. Richard Rodriguez, for example, does not consider himself a Chicano and was insulted when he was so identified at Berkeley during the sixties. I mentioned this in class and described the experiences of several former students whose parents were from different cultures (so-called "rainbow children"). Several weeks later, a student whose mother was Mexican and whose father was Anglo wrote:

> Someone once told me that I'd have to fight everyday to prove to everyone that I wasn't "another stupid Mexican." He was convinced that the whole American population was watching his every move, just waiting for him to slip and make a mistake. Having it emphasized that he is a minority certainly won't help his attitude any. It will just remind him that he is different. All of my life I never considered myself a minority. I didn't speak Spanish, I didn't follow Mexican customs, and I hung around with "American" kids. It is real hard for me to understand why minorities get so much special treatment.

This student found the existence of different student cultural groups on campus to be disturbing:

> When I came to Davis, instead of seeing a melting pot like the one I expected, I saw distinct cultural groups. When I first heard about CHE (Chicanos in Health Education) I was furious. I could not see any need for a special club just for Hispanic students interested in health careers. It seemed that the students in CHE were segregating themselves from the real world. They should actually be interacting with everyone else proving that they weren't different.

After interviewing members of CHE and VSA (Vietnamese Student Association), this student was able to see how some people benefited from these clubs:

> I didn't think I could find some positive aspects about these clubs, but I found some. Some clubs help immigrants assimilate into the Western culture. They provide the member with a sense of pride about who they are and they strengthen cultural bonds. If students attend classes and become discouraged by lower grades than they ex-

> pected, they can go to a CHE meeting and see "one of their own" explain how they made it through the bad times and how they came back to beat the odds. I asked Trinh why she joined the VSA. She said she joined to learn more about her culture and to improve her language. But others join to help themselves assimilate into Western culture. I was wondering what was so important about her culture that Trinh couldn't retain unless she went to these meetings. She said, "I can't explain it." But there is an atmosphere there that she can't get anywhere else. And if this gives her a good feeling, then more power to her.

This student still has reservations about cultural differences on campus (primarily because she does not want to see herself as "different"), but the movement from "I was furious about CHE" to "more power to her" is a step toward recognizing her own cultural diversity. Cultural identity is not always simple. I have seen second-generation Vietnamese, Indian, Korean, and Chinese students who saw themselves primarily as "American" (no hyphens), and in some cases as white. One Chinese student, from a Black neighborhood in Oakland, grew up wanting to be Black. It was the cool thing to be.

For some students, examining home culture and the culture of the university can cause anxiety. Teenagers often do not relish the idea of being "different." They have enough difficulty keeping their grades up, forming peer relationships, adjusting to being in a new environment. My sixteen-year-old stepson argues constantly that he is just like his peers. He certainly tries to be. But all you have to do is walk on the school grounds, look around a few minutes, and it is impossible to miss the six-foot-tall, dark-skinned, handsome boy bobbing up and down amongst a sea of white faces: definitely Indian (other Indians can identify him on sight as Telegu). It is becoming increasingly difficult on predominantly white campuses for students to deny differences in culture. What is important to learn is that while differences between home and school can lead to conflict, differences in themselves do not inherently cause conflict. The home culture can be a source of strength which can enable the student to negotiate with the mainstream culture. One of the major factors of success of students coming from cultures least valued by society is the ability of the family to help the student maintain a positive self-image that allows her to withstand rejection and insensitivity of mainstream peers. Occasionally, I have discovered some parents who did not want their children in school at all and did everything in their power to deter the education of their children. But most often, it is not the home culture that causes problems, but a fear on the part of students that elements of that culture will not be accepted in the university environment.

Implications for Teacher Training and Classroom Research

These topics and assignments not only help students mediate between school and home cultures, they provide windows for the teacher into the diversity within each of the cultures that students bring with them. They can serve as a base for ongoing teacher research into the ways in which home and university cultures interact. There simply is no training program for teachers, and can be no definitive research study that will ever account for the realities our students bring with them. Change is constant. Each generation is different. Given the lack of homogeneity in our classes, given the incredible diversity of cultures we are being exposed to, who better to learn from than our students? The culture and language topics I have described here comprise roughly 30-50% of the assignments I give in a ten-week course that meets for two hours twice a week. Some quarters I find myself using more cultural topics than others; it depends on the students. The course is English A, a four-unit course (two units counting toward graduation) with English Department administered, holistically-graded diagnostic, midterm, and final exams. The point is that if one can begin to integrate cultures under these constraints, one should be able to do it anywhere.

The cultural transitions we ask of our students are by no means easy. Cultural transition is ultimately defined by the student, whether she decides to assimilate and leave her culture behind, or attempts to integrate her world view with the academic world view. As composition teachers we are offered a unique opportunity to make these transitions easier for students, and at the same time increase our skills in moving between cultures. Clifford Geertz puts it this way: "The primary question, for any cultural institution anywhere, now that nobody is leaving anybody else alone and isn't ever again going to, is not whether everything is going to come seamlessly together or whether, contrariwise, we are all going to exist sequestered in our separate prejudices. It is whether human beings are going to be able, in Java or Connecticut, through law, anthropology, or anything else, to imagine principled lives they can practicably lead" (234). But we may find, and this has been my experience, that in helping students make cultural transitions, we learn from them how to make transitions ourselves.

Works Cited

Bartholomae, David. "Inventing the University." *When a Writer Can't Write*. Ed. Mike Rose. New York: Guilford, 1985. 134-165.

Bizzell, Patricia. "What Happens When Basic Writers Come to College?" *College Composition and Communication* 37 (1986): 294-301.

Bourdieu, Pierre, and Jean-Claude Passeron. *The Inheritors: French Students and Their Relation to Culture*. Chicago: U of Chicago P, 1979.

Cummins, James. "Empowering Minority Students: A Framework for Intervention." *Harvard Educational Review* 56 (1986): 18-36.

Fleming, Jacqueline. *Blacks in College: A Comparative Study of Students' Success in Black and in White Institutions*. San Francisco: Jossey-Bass, 1985.

Freire, Paulo. *Pedagogy of the Oppressed*. New York: Continuum, 1982.

———. *The Politics of Education: Culture, Power, and Liberation*. South Hadley, MA: Bergin & Garvey, 1985.

Geertz, Clifford. *Local Knowledge*. New York: Basic Books, 1983.

Groden, Suzy, Eleanor Kutz, and Vivian Zamel. "Students as Ethnographers: Investigating Language Use as a Way to Learn Language." *The Writing Instructor* 6 (Spring-Summer 1987): 132-140.

Heath, Shirley Brice. *Ways with Words: Language, Life, and Work in Communities and Classrooms*. New York: Cambridge UP, 1983.

Holzman, Michael. "The Social Context of Literacy Education." *College English* 48 (1986): 27-33.

Mitchell, Jacquelyn. "Reflections of a Black Social Scientist: Some Struggles, Some Doubts, Some Hopes." *Harvard Educational Review* 52 (1982): 27-44.

Ogbu, John. *The Next Generation: An Ethnography of Education in an Urban Neighborhood.* New York: Academic Press, 1974.

Olson, Carol Booth, ed. *Practical Ideas for Teaching Writing as a Process.* Sacramento: California State Department of Education, 1986.

Petrie, Loretta. "Pulling Together the Multicultural Composition Class." CCCC Convention. New Orleans, March 1986.

Philips, Susan Urmston. *The Invisible Culture: Communication in the Classroom and Community on the Warm Springs Indian Reservation.* New York: Longman, 1983.

The Puente Project: Building Bridges. Berkeley: Bay Area Writing Project, 1985.

Rodriguez, Richard. *Hunger of Memory*. Boston: Bantam Books, 1982.

Steinberg, Stephen. *The Ethnic Myth: Race, Ethnicity, and Class in America.* New York: Atheneum, 1981.

Wauters, Joan. "Non-Confrontational Critiquing Pairs: An Alternative to Verbal Peer Response Groups." *The Writing Instructor* 7 (Spring/Summer 1988): 156-66.

Rae Rosenthal

Male and Female Discourse: A Bilingual Approach to English 101

With each passing year that I have taught freshman composition, I have become increasingly uncomfortable with the course. Until recently, I had identified the source of that discontent as boredom, boredom with similar students being taught similar tasks, semester after semester. Lately, however, I have come to the notion that perhaps I am bored not with the teaching of writing, but rather that I am bored with the way in which I have been teaching writing. I'm not here describing the type of malaise which prompts me every summer to revamp my syllabus and discard old assignments in favor of new, fresher ones. This is a larger malaise which has led me to question certain fundamentals of composition which I have accepted blindly for so long.

Gradually, I have become suspicious of traditional rhetoric, with its emphasis on objectivity, logic, facts, conciseness, and clarity. More and more, I have come to see composition, as we now teach it, as a masculine form of communication. And once having contemplated that possibility, I have begun to fear that as composition specialists, even those of us with strong interests in feminist theory, gender, and women's literature, we have been advocating the adoption of a discourse which has created the very marginalization against which we struggle. I now wonder whether I have not, unknowingly, been perpetuating a male-dominated discourse at the expense of female modes of writing.

Yet, I find myself pausing over the notion of female writing. What is female writing? Are there indeed distinctly male and female modes of writing? Do men and women actually write differently? I began first to reconsider the findings of Robin Lakoff, Carol Gilligan, and Dale Spender, three women who have demonstrated to us through extensive research that male and female speech differs and that these differences may be attributed to the clear and demonstrable effect which gender has on language. Lakoff, Gilligan, and Spender all point out, in addition, that not only do male and female conversations differ in terms of content, but that the ways in which men and women express themselves differ as well, even when the subject matter remains constant.

I then began to consider similar studies in the area of literary research. More and more literary scholars have begun to examine the relationship between gender and literary production. For example, Robyn R. Warhol notes in a 1986 *PMLA* article that nineteenth-century women novelists (she cites Gaskell, Stowe, and Eliot as examples)

From *Focuses* 3 (Fall 1990): 99-113. Reprinted by permission of the publisher.

tended to choose what she terms "engaging" narrators who endeavor "to close the gaps between the narratee, the addressee, and the receiver" (811), while nineteenth-century male novelists (here she cites Fielding and Thackeray) tended rather to invent "distancing" narrators who create a "distance between the actual reader and the inscribed 'you' in the text" (811). Warhol's argument then goes on to suggest that these choices were culturally determined according to gender. Other literary scholars have also established connections between gender and genre, as in the work of Shari Benstock and Sidonie Smith, both of whom have uncovered a distinctly female tradition of autobiography. What these studies have in common is their suggestion that literary choices, like oral communication, are strongly influenced by gender.

Of even larger significance, though, is the value system which has been imposed on these differences. Men and women have not simply chosen to speak and to write differently; this would imply an unrealistic neutrality. Instead, in both instances, the male version has been deemed standard, normal, that against which all others will be judged. Or, to put it more simply, what is male has been valued and what is valued has been male. This may sound perhaps hyperbolic, but consider, for example, our literary canon. Certainly literary studies have historically valued writing which is male and that which is male is what has been valued. Canon revisionists have documented the many ways in which women writers and female forms of writing have been either denied, ignored, or trivialized. Consider, as Nina Baym notes in her article, "Melodramas of Beset Manhood: How Theories of American Fiction Exclude Women Authors," that "as late as 1977, [the American] canon did not include any women novelists" (123). Similarly, the Harcourt and Brace two-volume anthology *Major British Writers,* a text which was commonly used in introductory literature courses when I was an undergraduate in the late seventies and which began with Chaucer and ended with T. S. Eliot, includes not one woman writer. They simply do not exist. So after two semesters of English literature, I had not read a single word written by a woman (nor had I had a female professor, but that of course is another issue). And today, despite protestations to the contrary, particularly on the part of publishers and the innumerable "book reps" we all see in our offices every semester, the situation has improved only marginally; if you examine, for example, the 1985 edition of the Macmillan anthology of American literature which begins with "The Literature of Colonial America" and ends with "20th-Century Literature," you will find that out of 103 authors, only 18 are women, approximately 17%.

And what is true for the canon has been true for female speech. There is a taboo against the voice of women in patriarchal society because, according to Spender, "the yardstick against which women's talk is, in fact, measured is that of SILENCE" (148). And as Spender goes on to point out, "Women are socially devalued so it is not inconsistent to devalue them linguistically" (169). Gilligan came to similar conclusions when she examined various linguistic and psychological studies; what she discovered was that inevitably "the male model is [presumed] the better one" (10). The result of such thinking is that whatever differs is necessarily deemed erroneous, and of course to be female is, by definition, to be different. Consequently, as Gilligan notes, "when women do not conform to the standards . . . , the conclusion has generally been that something is wrong with the women" (14). Faced with the pervasiveness of such fallacious thinking, it is not surprising that the women who do strive to express themselves, who struggle to escape their bond of inarticulateness, regardless of the medium which they choose, encounter almost universal discouragement.

But none of this is news. What is news here, at least for me, are the questions which these issues raise, by implication, for the field of composition, whether it be the study of writing, the tutoring of writing, or the teaching of writing. Having recognized the presence of gender-determined speech and gender-determined literary choices, what does this mean for composition studies? More and more it is seeming to me unlikely that the field of composition could be exempt, aloof from the gender influences which appear to be ever more pervasive. And not surprisingly, once having decided to apply theories of gender determination to the study of writing and to consider what kinds of writing are labeled male and which female and which are rewarded and which devalued, a pattern emerges.

In Thomas J. Farrell's 1979 *College English* essay, "The Female and Male Modes of Rhetoric," he details thoroughly the differences which constitute this gender-differentiated pattern. According to Farrell, the female writing style tends to be "eidetic, methectic, open-ended, and generative," whereas the male style often "appears framed, contained, more preselected, and packaged" (910). "The male mode of rhetoric," Farrell continues, "seems to assume that antagonism is all right because intellectual life presumably proceeds agonistically. . . . The female mode, on the other hand, seems to avoid unnecessary antagonism or differentiation" (916).[1] Although historically there has been a hierarchical element to this pattern which has elevated the male over the female, Farrell strives, throughout his discussion, to avoid statements of evaluation, preferring to be, as he terms it, "descriptive and exploratory" rather than "definitive" (920). Yet in his closing remarks, Farrell draws several disturbing conclusions which show that he too has been influenced by the hierarchical thinking which he had hoped to avoid; Farrell states that "one must first master the male mode of rhetoric before attempting the female mode, because the female mode requires an even greater degree of control" and that, therefore, "students in college composition courses [should] be required to master the male mode of rhetoric" (920); later, he also states that "the male mode of rhetoric is probably better suited than the female mode for written discourse" (920). Such comments are troublesome in and of themselves, but when made by a composition specialist who appears to have as one of his goals the promotion of female writing, the dangers multiply.

Margaret Pigott's essay, "Sexist Roadblocks in Inventing, Focusing, and Writing," which appears in the same volume of *College English,* continues Farrell's discussion of gender and writing differences, and like Farrell, Pigott sees a pattern of differentiation in the writing styles of her male and female students. However, Pigott begins from the premise that the female mode, by virtue of its emphasis on the personal and its reliance on intuition, is inherently less valuable. According to Pigott, her female students "rely *solely* on personal experience in their thinking, in their focusing on topics, and, most notably, in

---

1. Farrell ascribes these contrasting styles to biological differences: "[Walter] Ong maintains that differences in male and female patterns of contesting are rooted in biological differences—a position that has been accepted in the formulation of the present paper" (917). Farrell later states that "the female and male modes of rhetoric are related to larger psychological developments in the evolution of human consciousness and to the individual person's recapitulation of those processes" (918). This, of course, is a position which most feminists reject.

their writing, scribbling a world of particulars within which the incident is isolated, individual, non-generalizable, and stunted because it does not relate to universal concepts which could give it meaning'' (922). The result of such thinking, Pigott contends, is that her female students ''have unique problems in thinking, focusing, and writing'' (927) and that they will ''limit, even thwart, their intellectual development'' (922).

For Pigott, the crucial question which these issues raise is ''what can the composition instructor, especially the woman composition instructor, do to *aid those females* who cannot generalize?'' (927; emphasis added). In response to her own query, Pigott suggests various classroom exercises which are designed to stretch or eliminate what she terms the ''limitations of the woman student'' (927) and to encourage female students to write in a more masculine mode. Seeking to strengthen her female students (i.e. make them more masculine), she rejects the female in favor of the male and mistakes what are female strengths for weaknesses. This distrust of the ''intuitive analysis'' (922) which her female students employ and this privileging of the public over the private is equivalent to a distrust of the female mode and, ultimately, of all that is feminine. Pigott's conclusion—that female students need to learn to reject female thinking in favor of more masculine modes of thought and discourse—is based on that very fallacy which Gilligan finds so pervasive in our culture: the tendency to deem whatever is female and/or feminine deficient because it differs from the male ''norm.'' Gilligan actually found that most psychological theorists began their studies by ''adopting the male life as the norm,'' and from there, ''they have tried to fashion women out of a masculine cloth'' (6). And even though Pigott seems to view her discussion as being feminist in its approach, she has not been able to avoid this male-norm trap. Rather, she has unknowingly adopted the patriarchal tenet which states that what is male is ''normal'' and that what is different should try not to be. In this way, despite intentions to the contrary, Pigott's essay suffers from the same hierarchical, anti-female thinking which plagues Farrell.

Still, both Farrell and Pigott appear to regard their work as being supportive of women's issues and as making a positive contribution to the murky waters of gender studies and composition. And to a certain extent they fulfill their intentions; they raise significant, yet largely untouched issues, and they help to demonstrate the strong connection between gender and composition. My point is not that Farrell and Pigott were kindly but erroneous; rather, I see them as examples of the ease with which feminist-minded scholars can fall victim to the patriarchal thinking which surrounds us. Indeed, it is painfully easy to do so. And as I have begun to examine more carefully the type of composition class which I have been teaching and to hold that class up against the feminist theories which I have simultaneously been studying and teaching, I have recognized some of the same disturbing inconsistencies which plague Farrell and Pigott. I too have been somewhat lacking in self-awareness, and furthermore, I suspect that my failing is not uncommon. I now suspect that as composition specialists, many of us, including those who are critical of the literary canon and its restrictions, those who resist male dominated discourse, we have been creating a similarly oppressive hierarchy in our own writing, in our composition classes, and in our writing centers; we tend to teach what is male, and we tend to value what we teach.

In doing so, we have provided additional evidence of the powerful connections between composition studies, the field of rhetoric, and patriarchy. And further examination of the three only reveals greater interconnectedness. For instance, both composition specialists and rhetoricians have traditionally valued the same traits which characterize male speech: strong assertion, clear logic, and irrefutable facts. Also, as Elisabeth Daumer and Sandra

Runzo point out in their essay "Transforming the Composition Classroom," what has been termed "good" writing corresponds close to what are generally classified as positive male characteristics. In fact, Daumer and Runzo note that the "description of an accomplished writing style, found in virtually any rhetoric textbook, coincides, interestingly, with Carol Gilligan's description of what she calls the masculine mode of thought and morality" (52). As an example, consider the traditional preference for objectivity; students of composition and rhetoric have long been taught that well written prose should not be clouded by bias or preference. This maxim is firmly stated in *The Elements of Style* by Strunk and White, that "little book" which has had such a vast influence on composition instruction. This text advises us, in its markedly decisive and confident voice, "do not inject opinion into a piece of writing" (80).[2] Not coincidentally, this is an injunction which Gilligan and Daumer and Runzo see as being particularly well suited to a male orientation. In Gilligan's study of male and female development, she notes that "masculinity is defined through separation" (8). Daumer and Runzo then argue that as a result of this separation, "males learn to resolve ethical dilemmas by depersonalizing situations—by detaching themselves from others" (52). Consequently, males are predisposed by early training towards detachment and objectivity.[3] The male developmental pattern thus neatly coincides with the development of good writing skills.

This complicated interplay between composition, rhetoric, and patriarchy has led many of us, whether composition teachers, writing program administrators, or writing center tutors, to be the unwitting endorsers of patriarchal values. In the name of composition instruction, we have cooperated with the dominant culture by accepting its values and by instructing our students to do the same. And even while keeping up with contemporary composition theory—even while advocating process over product, while encouraging revision, while requiring journals—we still have often held, albeit unconsciously, traditional aims for our students. We have urged them to write prose which is assertive, logical, factual, objective. We have advised them to exhaust their subject matter. We have urged them to write male prose, and we have rewarded them for doing so.[4]

What we now must ask ourselves is to what extent is this really objectionable and what are the desirable alternatives? Until recently, I had never thought to question the definition of good writing, but now, I find myself wondering—what exactly is good writing? In the past, when asked to define good writing, I had always answered with confidence, "good writing is clear and concise." Now though, I have to wonder, what

2. Gilligan notes, interestingly, that the examples used in *The Elements of Style* tend to be highly reflective of traditional sex roles: "A recent discovery of this sort pertains to the apparently innocent classic *The Elements of Style* by William Strunk and E. B. White. A Supreme Court ruling on the subject of sex discrimination led one teacher of English to notice that the elementary rules of English usage were being taught through examples which counterposed the birth of Napoleon, the writings of Coleridge, and statements such as 'He was an interesting talker. A man who had traveled all over the world and lived in half a dozen countries,' with 'Well, Susan, this is a fine mess you are in' or, less drastically, 'He saw a woman, accompanied by two children, walking slowly down the road' " (6).

3. Feminist theory has, for the most part, rejected the concept of objectivity as an impossibility and has endorsed subjectivity as a more honest, more legitimate approach; however, the feminist position does not alter the high value which composition studies have traditionally granted objectivity and the continued widespread belief that one can and should be objective in one's thoughts and writings.

4. Wendy Goulston and Olivia Frey both note the distinctly feminine nature of the recent pedagogical innovations which are currently in use in so many composition classes; however, a feminine teaching style does not guarantee that one is teaching feminine writing. Indeed, I suspect that the one often blinds instructors to their failure to accomplish the other.

makes clear and concise so desirable? Are ambiguity and abundance somehow less? They seem to me to be more. Yet, for years, I have blindly accepted what were presented to me as the universal standards of good writing. But this argument of universality, of quality as a definable, discoverable entity, beyond the prejudices of gender, beyond the reaches of politics, is of course the same argument used by those who support the traditional literary canon: we should simply read what is good—we should adhere to universal standards of quality regardless of gender, even though the reality has been that "universal" rarely includes women. Indeed, what is "quality" all too often translates to what is male. For this reason, quality itself, as many of us have had to recognize, is highly debatable matter.

The problem which we are thus left with is being able to discover and define a writing mode which is feminine and positively so, a task of significant magnitude and one which is compounded by the prevalence of what Joan Bolker terms the "Griselda syndrome." In her discussion of the writing deficiencies of two bright, female students, Bolker suggests that her "Griseldas" have a writing problem unique to women: they are so concerned with being "good girls," so devoted to "blatant goodness" (907) that they cannot write well. They are actually paralyzed in terms of voice and style by their desire to "please all and offend none" (907). Because being " 'a good girl' means learning what pleases those around you, and acting that way" (907), they constantly, in their writing, obliterate themselves in service to their reader. Ironically, this effort to be "good" leads to the exact opposite traits from those which describe "good" writing. By writing as they have been taught to behave, the Griseldas succeed only in developing a style which " 'smiles' all the time, shows very little of a thought process, but strives instead to produce a neat package tied with a ribbon" (907). They avoid risk and strong assertion, and consequently, they avoid censure, but also success. Thinking to succeed as they always have, by behaving well, they doom themselves to failure. In fact, these students are caught in a classic Catch-22 situation: if they write essays which accord with the standard definition of good writing—forceful, reasoned, factual, argumentative essays—they will be guilty of unfeminine behavior, but if they adopt a feminine mode, they will have written "poorly."[5]

Daumer and Runzo discuss this dilemma at some length, noting the impossibility of choosing to one's advantage:

> The qualities of 'good' writing as they are advocated in textbooks and rhetoric books—directness, assertiveness and persuasiveness, precision and vigor—collide with what social conventions dictate proper femininity to be. Even should a woman succeed at becoming a 'good' writer she will still have to contend with either being considered too masculine because she does not speak 'like a Lady,' or, paradoxically, too feminine and hysterical because she is, after all, a woman. The belief that the qualities that make good writing are somehow 'neutral' conceals the fact that their meaning and evaluation changes depending on whether the writer is a man or woman. (52)

---

5. In her discussion of women and oral language, Robin Lakoff notes a similar bind: "So a girl is damned if she does, damned if she doesn't. If she refuses to talk like a lady, she is ridiculed and subjected to criticism as unfeminine; if she does learn, she is ridiculed as unable to think clearly, unable to take part in a serious discussion; in some sense, as less than fully human. These two choices which a woman has—to be less than a woman or less than a person—are highly painful" (6).

In an attempt to escape their bind, Bolker's students—the Griseldas—seek a compromise position: they try to imitate the masculine writing style which they have been taught to admire, while simultaneously striving to maintain a feminine decorum. However, this admirable effort, like so many compromises, pleases neither camp and results only in essays which are technically correct but "thoroughly dull" (Bolker 906), void of vigor, assertion, and original thought.

In an effort to free her Griseldas from this bind, Bolker encourages a more female style of writing. She suggests to her students that "most readers are more pleased by the sloppy sound of the human voice in a piece of writing than they are by neatness and goodness" (908) and that they develop a personal voice. In order to assist in the development of these skills, Bolker assigns journal writing and free writing, and when selecting readings, she seeks works with a "strong and even outrageous voice" (908). The results, according to Bolker, are promising. Her Griseldas have begun to develop a style of their own, one which begins with the self and which explores "issues of great personal importance" (908); now, rather than "sensing and bending to the demands of the outer world," they have begun "to listen to the demands of the inner world as well" (908). No longer do they write as "good girls" but rather as females.

In this way, Bolker has helped us to see the inadequacy of male models for female writers, but what is still needed is greater discussion, clarification, and exemplification of female writing. Helene Cixous, one of the leading French feminists, has addressed many of these needs. In her landmark essay, "The Laugh of the Medusa," Cixous suggests that female writing resembles female speech, that convoluted mode which we have so often been encouraged to hide, change, abandon. Through demonstration and definition. Cixous suggests that female writing is emotional, digressive, imaginative, open; female writing favors questions over answers and seeks response rather than agreement; female writing also recognizes and endorses the inherent subjectivity of language. Cixous suggests, further, that women's "language does not contain, it carries; it does not hold back, it makes possible" (293).

Cixous also argues that it is precisely in the maleness of written language that women have been most oppressed, and it is here that they must begin to create a space through which they may escape. Cixous writes,

> I mean it when I speak of male writing. I maintain unequivocally that there is such a thing as marked writing; that, until now, far more extensively and repressively than is ever suspected or admitted, writing has been run by a libidinal and cultural—hence political, typically masculine—economy; that this is a locus where the repression of women has been perpetuated, over and over, more or less consciously, and in a manner that's frightening since it's often hidden or adorned with the mystifying charms of fiction; that this locus has grossly exaggerated all the signs of sexual opposition (and not sexual difference), where woman has never her turn to speak—this being all the more serious and unpardonable in that writing is precisely the very possibility of change, the space that can serve as a springboard for subversive thought, the precursory movement of a transformation of social and cultural structures. (283)

In her discussion of Cixous and the influence of the French feminists on composition instruction, Clara Juncker notes particularly this emphasis on language and its transformative powers. Juncker claims that the "site of resistance is writing (428) and that like Cixous and her cohorts, Julia Kristeva and Luce Irigaray, we must "never allow language to become invisible, a mere medium for transmitting 'truth' and 'meaning' from master to students" (424). Juncker then demonstrates, by citing extensively from Cixous's writings, the Cixousian penchant for linguistic sport which prohibits just the invisibility against which Juncker warns. In fact, throughout Cixous's writings, she refuses to allow language even the appearance of being a passive receptacle. Her essays are often a never ending spiral of linguistic self-reflectiveness which refers back to and calls attention to itself, and through which she manages to simultaneously embrace the manipulation of language and yet use that manipulation as the focus of her discussion. And by thus repeatedly forcing language into the forefront, Cixous creates a new voice which is distinctly female and which might indeed be seen as Juncker's "site of resistance."

Yet, and here is the catch, women have long been taught to suppress precisely those feminine forms which Cixous advises us to embrace and which she herself does embrace; we have been taught rather to adopt a masculine, and therefore, an inherently unsuitable, mode. And that is precisely what we must not do. We must not follow the advice of the ever-proliferating number of self-help books, groups, and magazines, many of which recommend that women should dress like men, talk like men, write like men, that women should try to beat men at their own game. Such tactics rarely work; they more often backfire. Spender, in fact, specifically warns against this danger:

> It is vitally important for women to start to talk, it is not necessary that we emulate the habits of men. Research has indicated that although men talk more, they exert more control over talk, and that they interrupt more. Women listen more, are more supportive when they do talk and have greater expertise in terms of sustaining conversations. It is precisely these qualities which have been neither valued nor acknowledged. Rather than women learning to talk like men it would seem to be preferable if men were to learn to listen more and to be more supportive of the conversation of others. This could revolutionize our patterns of talk in society, and a vast range of oppressive images, of both men and women will be undermined. ("Talking" 154)

One of these images which I would like to begin undermining is the traditional structure which characterizes both the freshman composition class and the writing center tutorial. To do so invites the "possibility of change" of which Cixous speaks so hopefully because it is here that the definition of good writing is established, and it is here that we should begin to rethink and to recreate. A possible first step in this direction might be the contemplation of the field of composition and its own power structure.

I once heard James Sledd argue that "standard English is the dialect of the elite." It seems to me equally plausible that persuasive essays based on the notion of thesis-and-support are the written dialect of the elite. And who holds the primary power in that elite? Certainly not female composition specialists. For years, as both an instructor and a tutor, I have told students who speak non-standard English that standard English is a language which they must master *if,* and I place heavy emphasis on the *if,* they want to enter the world of the powerful and the elite. But I never before considered the parallels between learning to speak as do those in power and learning to write as do those in power. Yet those parallels suddenly seem striking. Annette Kolodny, in her now famous essay, "Dancing Through the Minefields," asserts that "reading is a highly socialized—or learned—activity" (11). This statement seems to me to be equally applicable to writing, which like reading, "is a highly socialized—or learned—activity." I wonder though, what are the implications of this socialized skill which we have been teaching.

I know for myself, that while in graduate school, I learned both to speak as an academic and to write as an academic, and, undoubtedly, there have been benefits to my learning that language. Nonetheless, I now suspect that the price of that knowledge was a denial of self and my natural mode of expression. I now see the ways in which I was gradually taught to accept that good writing was indeed male writing. And this seems to me, in retrospect, to be the ultimate patriarchal victory, to be able to convince those who are excluded that centrality is quality. In this way, there is constant movement from without to within, a constant swallowing by the center of the margins, a constant reenforcement of the status quo. And in many ways, our composition courses and our writing centers, as they are currently structured, contribute to this movement. Once having uncovered the male orientation of our methods, I find myself in search of the margins, of that space where the opposition exists. I am interested in exploring female writing and the Cixousian "possibilities for change" which that writing may create.

In previous composition courses, I've emphasized facts, logic, organization, closure, answers. I've encouraged objectivity and strong assertions, with objections and refutations for the support of those assertions. I've told my students, each semester, that objections and refutations are essential to any well-constructed argument, that by covering all possible questions and by providing self-contained answers, a writer can eliminate opposition. I have, in effect, instructed my students in ways to effectively discourage debate, response, and dialogue.

And as a tutor, I have been guilty of many of the same offenses. Indeed, I suspect that writing centers are particularly prone to these dangers. Because the tutor-tutee relationship tends to lack continuity and longevity, it is easy for tutors to fall into the trap of teaching writing as a fixed set of rules and procedures: "the thesis statement should be the last sentence in the first paragraph"; "a thesis needs at least three supporting ideas"; "a well developed paragraph should have at least three sentences." And even though the time and budget restraints of most of our writing centers render this type of instruction understandable, the results are nonetheless deadly: the inevitably insipid five paragraph theme, with perhaps a few attempts at variation, often twisted to suit the needs of a pre-selected rhetorical mode.

This formula approach to writing has two advantages: (1) it is imminently teachable, for both the instructor and the tutor, because, to whatever extent possible, it objectifies and quantifies a subject matter that is inherently resistant to both; (2) it appeals to students because it simplifies the writing process by eliminating numerable decisions. Olivia Frey, who taught in a writing program which insisted upon the formula approach to freshman composition and who describes herself as "one of those untrained assistants [who] taught writing according to the traditional paradigm" (95), outlines the limitations to such a program, noting that one of the greatest is the safety appeal:

> It is easier to write by following a formula, abiding by prescribed, predetermined standards. Not writing according to formula requires that the writer make choices at

> every stage. The writer, not the teacher, chooses a topic; the writer decides the shape and length of her essay as well as the rhetorical strategies that she will use; the writer decides whether or not her words are appropriate. Students are unaccustomed to and uncomfortable with having to make so many choices, and composition writers especially cling to any maxim that the writing teacher offers, grasping at rules as the key to good writing. (101)

Frey points out, in addition, that the essays written by students in such a program generally ''lacked style, grace, substance''; indeed, she says ''most were boring'' (96).

And not surprisingly, tutoring sessions based on such an approach lead to similar results. Having been given ''the formula,'' tutees then draw a disastrous, yet sadly commonplace, conclusion: they assume that they now ''know'' how to write an essay in the same way that they ''know'' an algebra equation—they think that they need only fill in the missing information in the proper sequence. Such a method of composition discourages experimentation and is, by definition, supportive of traditional modes of writing; in fact, what results most often is just the kind of writing against which feminists must strain. In this way, writing centers often tend to serve, albeit unintentionally (at least one would hope) as a means of reinforcement; students become ever more convinced that good writing depends upon the joining of several ''objective'' facts, a narrow thesis, and a conclusion which eliminates the need for response.

Feminist theory, on the other hand, is based on the belief that objectivity is an illusion, an unattainable and undesirable goal. Feminist theory values questions, openness, dialogue, and above all, subjectivity, and I would like to see our writing and our teaching of writing uphold these feminist values.

The question becomes, though, that which repeatedly confronts those interested in feminist theory: where do we locate ourselves—inside or out. What direction is most advantageous for feminist endeavors? Is it to our benefit to be isolated—a position which has the advantage of self-containment and independence, a position which avoids the danger of dilution and assimilation? Or is it more productive to work within, to assert, change, fight that which already is, to make a place for ourselves in the existing structure in the hope of changing that structure? This position has the potential for greater contact and wider scope, but runs the risk of adaption and reduction. This question is crucial to all feminist activities, and I have heard it nowhere better addressed than by Annette Kolodny, whom I first heard speak at a conference on the revision of the American canon. When asked that day whether she thought the goals of women's studies are best served by the creation of a separate department or by an influx into all departments, Kolodny replied quickly and emphatically—''as all good feminists know, *we must have both.*''

And so it goes with freshman composition. We must devise separate assignments, readings, and class activities which will encourage female modes of writing, but at the same time, we must try to redirect those traditional assignments, readings, and class activities which encourage and endorse male modes of writing. And in our writing centers, we must create an environment which encourages more experimental, more exploratory, more female styles of writing while simultaneously resisting the formulaic but ubiquitous five-paragraph theme which our students so dearly want to write. In this way, we might, as Juncker suggests, encourage our ''student writers to play with language, to stretch it, form it, caress it'' and we might then ''give (back) to them the pleasure of the text and thus dispense a whiff of a libidinal economy within the classroom'' (432).

In my own classroom, I have long noticed that my students arrive each semester with the sense that I am going to force them to learn a foreign language, and perhaps that is exactly what I have been doing. I used to think that I was teaching my native tongue; now I wonder. A new vision or re-vision of my own history tells me that I have merely, through years of training, become a master of a certain type of rhetoric to which I was not naturally inclined. I taught myself to write logically, clearly, factually, even objectively, to whatever extent that is possible, and I learned to supress impassioned commentary, instinctual insight, and biased observation. In turn, I have taught my students to do likewise. And to be a student is necessarily to be an *other,* a being without power, a creature with limited choices, a person lacking the skill to articulate. Our students are indeed learning a foreign language, but we have the choice as to which language we are going to teach. And while I would still like to empower them in the language of the elite, I would like for them to know it as such, and I would like for them, at the same time, to become skilled in their own discourse. Our students might well become masters of two languages: that of the oppressor and that of the oppressed. They might be taught to develop a keen sense of audience which would enable them to utilize that discourse most effective, most appropriate to their purpose and their readers. This would be true empowerment.

So, suddenly, I find myself contemplating a new composition course, one which is bilingual: English 101—Male and Female Discourse; this course would require students to examine their own language, to recognize the extent to which that language has been influenced by gender, and to develop the ability to identify and imitate both male and female discourse. Robin Lakoff has argued, however, that women are already bilingual because our patriarchal culture requires that they be so. She also contends that women are, consequently, at a linguistic disadvantage. According to Lakoff, a woman

> may never really be master of either language, though her command of both is adequate enough for most purposes. . . . Shifting from one language to another requires special awareness to the nuances of social situations, special alertness to possible disapproval. It may be that the extra energy that must be (subconsciously or otherwise) expended in this game is energy sapped from more creative work, and hinders women from expressing themselves as well, as fully, or as freely as they might otherwise. (7)

And no doubt, as Lakoff suggests, bilingualism can lead to difficulties and can impose limits on the speed and ease of one's speech, but it seems to me that there are also potentially offsetting benefits: there is the possibility of increased linguistic awareness, of heightened sensitivity to the variations of language, and of personal broadening as a result of one's being able to communicate within different linguistic communities. Also, if composition studies were to become bilingual, were men and women both forced to learn the dialect of both sexes, bilingualism would become less gender bound, less of an enforced female skill, and women would no longer be at a disadvantage.

So despite the potential drawbacks which Lakoff has outlined, I continue to speculate on the ways in which I might teach two languages in one semester. This thought might at first seem overwhelming; we have so little time to teach even standard composition skills, but overall, I find the vision freeing. Consider our literature courses, where so many of us have begun to teach multiple approaches to literary texts; these courses have often proven more stimulating for us as teachers and for our students. A similar approach to composition—multiple writings of a text—might well revitalize our writing classes. I envision my new writing assignments to invite a twofold response, one female and the other male. I anticipate having my students write collaboratively, rather than competitively. I have begun to develop what I'm calling "dialogue essays," papers which students will write to one another, papers which will actively encourage response, conversations on paper.[6] I hope to create assignments which will allow for the development of instinct, passion, repetition, ambiguity, abundance—papers that will pause rather than end. I want my students to discover the possibilities of the language of the "other," a role which some of them will long retain and which some of them will shed upon graduation and a white-collar job.

I envision myself, next September, entering my classroom and informing my students that I am indeed going to teach them a foreign language, one which may prove useful, but foreign nonetheless. But at the same time, I hope to be able to encourage them in the exploration of their own discourse, the mastery of their own language, that of the *other*. The contemplation of such a course leaves me, for the first time in years, looking forward to teaching English 101.

Works Cited

Annas, Pamela J. "Style as Politics: A Feminist Approach to the Teaching of Writing." *College English* 47 (1985): 360-371.

Baym, Nina. "Melodramas of Best Manhood: How Theories of American Fiction Exclude Women Authors." *American Quarterly* 33 (1981): 123-139.

Benstock, Shari, ed. *The Private Self: Theory and Practice of Women's Autobiographical Writings*. Chapel Hill: U of North Carolina P, 1988.

Bolker, Joan. "Teaching Griselda to Write." *College English* 40 (1979): 906-908.

Caywood, Cynthia L., and Gillian R. Overing, eds. *Teaching Writing: Pedagogy, Gender, and Equity*. Albany: State U of New York P, 1987.

Cixous, Helene. "The Laugh of the Medusa." Trans. Keith Cohen and Paula Cohen. *The Signs Reader: Women, Gender, and Scholarship*. Ed. Elizabeth Abel. Chicago: U of Chicago P, 1982.

Daumer, Elisabeth, and Sandra Runzo. "Transforming the Composition Classroom." Caywood and Overing 45-62.

Farrell, Thomas J. "The Female and Male Modes of Rhetoric." *College English* 40 (1979): 909-921.

Frey, Olivia. "Equity and Peace in the New Writing Class." Caywood and Overing 93-105.

Gilligan, Carol. *In a Different Voice: Psychological Theory and Women's Development*. Cambridge: Harvard UP, 1982.

---

6. Clara Juncker mentions a similar assignment from her intermediate exposition courses where she assigns a "dialogue paper in which students debate with(in) themselves possible research paper topics" (432).

Goulston, Wendy. "Women Writing." Caywood and Overing 19-29.
Harrison, G. B., ed. *Major British Writers.* 2 vols. New York: Harcourt, Brace, 1959.
Juncker, Clara. "Writing (with) Cixous." *College English* 50 (1988): 424-436.
Kolondy, Annette. "Dancing Through the Minefield: Some Observations on the Theory, Practice and Politics of a Feminist Literary Criticism." *Feminist Studies* 6.1 (1980): 1-25.
———. Address to the Conference on the Revision of the American Canon. University of Maryland. College Park, April 1982.
Lakoff, Robin. *Language and Woman's Place.* New York: Harper, 1973.
McMichael, George, ed. *American Literature.* New York: Macmillan, 1985.
Pigott, Margaret B. "Sexist Roadblocks in Inventing, Focusing, and Writing." *College English* 40 (1979): 922-927.
Sledd, James. "The Successful Failure of E. D. Hirsch." Penn State Conference on Composition and Rhetoric. State College, 8 July 1988.
Smith, Sidonie. *A Poetics of Women's Autobiography: Marginality and the Fictions of Self-Representation.* Bloomington: Indiana UP, 1987.
Spender, Dale. "Talking in Class." *Learning to Lose: Sexism and Education.* Ed. Dale Spender and Elizabeth Sarah. London: The Women's P, 1980. 148-154.
———. "Disappearing Tricks." *Learning to Lose: Sexism and Education.* Ed. Dale Spender and Elizabeth Sarah. London: The Women's P, 1980. 165-173.
Strunk, William, and E. B. White. *The Elements of Style.* New York: Macmillan, 1979.
Warhol, Robyn R. "Toward a Theory of the Engaging Narrator: Earnest Interventions in Gaskell, Stowe, and Eliot." *PMLA* 101 (1986): 811-818.

Harvey S. Wiener

Collaborative Learning in the Classroom: A Guide to Evaluation

Over the last decade collaborative learning has become an important method for college English teachers, who now realize that their own education rarely taught them how colleagues work together to learn and to make meaning in a discipline, and who have rejected philosophically the kinds of approaches to teaching that isolate learners instead of drawing them together. In addition, the problems for education in the seventies and eighties—the changes in student populations, the growth in the number of nontraditional learners in the collegiate body, the alienating nature of learning in large classrooms with too many students, the acknowledged decline of freshmen entry-level skills in reading, writing, speaking, listening, and thinking—these and other challenges to an earlier educational paradigm have shaken our faith in conventional teaching strategies and have called to question our obsession with the major metaphor for learning over the last three hundred years, "the human mind as the Mirror of Nature."

As Ken Bruffee has put it, this old metaphor insists that teachers give students as much information as they can "to insure that their mental mirrors reflect reality as completely as possible" and also insists that we help our students "through the exercise of intellect or development of sensibility, to sharpen and sensitize their inner eyesight" ("Liberal Education" 98). In this ground-breaking essay, Bruffee, drawing upon the works of Thomas Kuhn, L. S. Vygotsky, Jean Piaget, M. L. J. Abercrombie, and Richard Rorty, advances an alternate concept of knowledge as *socially justified belief*. According to this concept, knowledge depends on social relations, not on reflections of reality. Knowledge is "a collaborative artifact" (103) that results from "intellectual negotiations" (107). Bruffee explores the curricular implications of knowledge collaboratively generated, always with one eye on the classroom and the other on the philosophical underpinnings of the new paradigm.

But Bruffee's model, built on the delicate and necessary tension between theory and practice, may not, I suspect, have guided much of what teachers are calling collaborative learning today. I mention this suspicion out of my recent investigations into the issue of assessment generally as a force in postsecondary education and also out of my recent frustration as formal observer of classroom teaching performances in a university-mandated system of evaluation for promotion, retention, and tenure. I realized as I

From *College English* 48 (January 1986): 52-61.

watched these attempts at instruction through collaboration that to apply to the new paradigm the standards we had in place for the old was inappropriate. Our elaborate student evaluation forms and classroom observation checklists had little relation to the classroom activities I observed. What was worse, I realized that we had not established either as an institution or as a profession any standards for judging our attempts to implement the evolving concept of teaching and learning as a social act. Hence the question I intend to address in this essay: How do we assess the effectiveness of collaborative teaching models in the classroom?

Asking this question on evaluation now, as collaborative learning grows more and more popular, is to seize an advantage we have missed many times before. Formal assessment has always been the stepchild of the profession. In the past we have given up important evaluation activities for certifying the success of our students as learners and of ourselves as teachers. Professional testing agencies, for example, not classroom teachers, develop, and oversee college entrance tests for graduates and undergraduates. Despite the obligatory committees of teachers and researchers who are invited to establish standards in general terms and to highlight areas of learning, professional test writers are the ones who produce specifications on most commercially prepared large-scale examinations. Worse still, legislatures, seeing a void, have leaped in to define competencies we have not. In many states, legislatures, not teachers, have mandated and overseen the development of tests for college writers. The Florida Department of Education, for example, has created the College Level Academic Skills Test (an essay and an objective test) for all students in the state, and has prescribed the number of pages to be written each week in writing classes. Georgia has a similar test in progress. Even current measures for judging a teacher's classroom effectiveness have been influenced insufficiently by the teachers themselves who are being judged. Administrative committees, education school faculty, and evaluation specialists often develop the standards for classroom observations and create atomistic, overly-generalized checklists for use in assessing teaching. Or, institutions develop no standards whatsoever, and classroom observation is an exercise in a senior professor's effort to characterize someone else's teaching by means of some vague, unarticulated, and as yet socially unjustified vision of perfection. Even useful efforts by the profession are often too late to do as much good as they might have done had they flowered earlier. The evaluation instruments developed by the Conference on College Composition and Communication's Committee on the Evaluation of the Teaching of Writing, for example, reached English teachers ten years after The City University of New York's faculty negotiating unit, the Professional Staff Congress, wrote an evaluation system into the University's faculty contract, long after precedent set most of the institutional evaluation procedures in cement.

By advancing collaborative learning as a productive instructional mode for teaching literature and writing, however, English teachers have a rare opportunity to evolve a set of standards by which to judge classroom performance in the new paradigm. Our first obligation is to define for ourselves what we see as efficient classroom models for collaborative learning. Our next obligation is to pass on to beginners the standards by which we measure our own performances so that new teachers seeking membership in this intellectual community have a clear paradigm to study. And, finally, we are obliged to lay out for classroom observers what to look for as hallmarks of collaboration so that any judgments evaluators make about teaching performance are judgments our community has justified through thoughtful, disciplined discussion.

In an effort to move forward this evolution of standards for appropriate collaborative teaching models and to provide a temporary set of guidelines for the classroom observer of collaborative learning, I will look at the teacher's role in a collaborative session sequentially. I will confine my remarks to one of the most common kinds of collaborative learning, collaborative group work. Here, students perform some common task in small study and discussion groups. The class is divided into clusters of three to seven students each. Each group chooses a recorder to take notes on the conversation and, when the discussion ends, to report the group's deliberations to the whole class. The time required for a collaborative effort depends on the task, but fifteen or twenty minutes is a bare minimum. The teacher helps the class compare results, resolve differences, and understand features of the task that students did not work out on their own.

The Teacher as Task Setter

The success of the collaborative model depends primarily upon the quality of the initial task students must perform in groups. Hence, the instructor's role as task-setter is one that any observer must view with great attention. "What is essential," Bruffee writes, "is that the task lead to an answer or solution that can represent as nearly as possible the collaborative judgment and labor of the group as a whole" (*Short Course* 45).

The group's effort to reach consensus by their own authority is the major factor that distinguishes collaborative learning from mere work in groups. What is consensus? Unfortunately the word is widely misunderstood as a dimension of collaborative learning. It is not an activity that stifles differences or intends to make conformists out of divergent thinkers. John Trimbur asserts that those new to collaborative learning often miss

> the process of intellectual negotiation that underwrites the consensus. The demand for consensus that's made by the task promotes a kind of social pressure. Sometimes, to be sure, this pressure causes the process of negotiation to short circuit when students rush to an answer. When it works, however, the pressure leads students to take their ideas seriously, to fight for them, and to modify or revise them in light of others' ideas. It can also cause students to agree to disagree—to recognize and tolerate differences and at best to see the value system, set of beliefs, etc. that underlie these differences.

Consensus, he points out, "is intellectual negotiation which leads to an outcome (consensus) through a process of taking responsibility and investing collective judgment with authority."

Certainly methodology in education for many years has depended upon group work, but it is generally not an activity that demands collective judgment. In elementary and secondary schools, for example, teachers of reading, spelling, and mathematics divide students into groups for skills instruction, each group at a different level. Such groupings permit those with like abilities to investigate topics at the same rate and with the same intensity as their peers. But this kind of group work is by no means collaborative learning. It merely subdivides the traditional hierarchical classroom into several smaller versions of the same model. Despite the groups, the teacher remains the central authority figure in the students' attempts to acquire knowledge. Other popular yet perhaps more imaginative types of group activity—clusters of students working on a common project or experiment, say—also rarely build upon the idea of a learning community that leads to joint decisions. Much group work on projects and experiments of this sort is only the sum of its parts, each student contributing his or her piece without the vital "intellectual negotiation" that "places the authority of knowledge in the assent of a community of knowledgeable peers" (Bruffee, "Liberal Education" 107). Students put into groups are only students grouped and are not collaborators, unless a task that demands consensual learning unifies the group activity.

To assure that the teacher in a collaborative learning classroom is guiding students to collective judgments in groups, evaluators are right to insist that the task be written down. A written task provides the language that helps to shape students' conversations. An observer asked to judge a class session in collaborative learning must first scrutinize the task and then comment on it in the evaluation report in the same way he or she would comment on the teacher's preparation for any lesson. To look only at the outward manifestations of the collaborative classroom—the fact that students group together and talk within their groups—is to look at the activity with one eye closed.

Peter Hawkes points out important differences between collaboration and group work and these differences inhere in the nature of the task:

> Sometimes in mere group work the teacher sets a task or poses a question that has an answer that the teacher has already decided on. Groups take on the role of the smart kid in class who guesses what's on the teacher's mind. The evaluator should examine the task assigned and the way the teacher responds to the student reports in the plenary session to see whether the authority of knowledge has been shifted temporarily in the classroom. In CL, the teacher should ask questions that have more than one answer or set problems that are capable of more than one solution. In other words, sincere questions rather than pedagogical ones. The CL teacher is interested in the way the students come up with their consensual answer, the rationale for that answer, the opportunities for debate among groups, the suggestion of how knowledge in a discipline is arrived at rather than in leading students toward an already acknowledged 'right answer.' CL changes the student-teacher relationship; mere group work appears to but does not.

A good written statement of task will probably have a number of components: general instructions about how to collaborate in this particular activity; a copy of the text, if a single text is the focus of the collaboration; and questions appropriately limited in number and scope and offered in sequence from easier to more complex, questions requiring the kind of critical thinking that leads to sustained responses from students at work in their groups. Since collaborative group work normally should move toward consensus, instructions almost always should require a member of the group to record this consensus in writing. But although one member writes the report, the group as a whole shapes it. Some experienced collaborative learning teachers insist that the recorder do something more like a performance after the work in the group ends—a formal presentation to the class, participation in a debate with recorders from other groups, or some other responsible social activity that may be subjected to group judgment. When recorders must perform, these teachers argue, the recorders keep the groups functioning smoothly and efficiently.

The teacher's role as task-setter often must go beyond simply writing the assignment down and distributing it. This is especially true when students consider varied texts collaboratively (their own papers, for example). The instructor may have to guide the manner in which students attack the task by reviewing some of the principles that need attention if activity is to move forward before the group work begins. For example, in a typical collaborative session, dividing students into small groups to read and provide commentary on the coherence of a practice essay, an instructor might explain to the class at the beginning of the hour some of the principles of coherence in expository writing. Or, if students are to comment on drafts of each other's essays, the teacher could begin by asking student groups to generate a *Reader Response Guide*. Asking the class "Which two or three vital questions do you wish to have answered about your draft so that you can take it to the next stage" and then collecting the questions for everyone to see is effective because it reviews whatever was taught in an earlier class or in advance of the assignment; it highlights for the whole class the major issues to be addressed in this writing task; it calls attention immediately to the students' own most pressing concerns; and it gives the class an opportunity to buy into the collaborative process as shapers of their own learning.

For evaluators the key issue here once again is that the task and the teacher's role in setting it must stimulate active learning that leads to an important outcome: consensus (either agreement or agreement to disagree) on the issue at hand. Many collaborative settings I've witnessed do not pay much attention to consensus. Students divided into groups to examine drafts and to "discuss" their papers, but who lack specific guidelines, will flounder. I saw one class session like this where students told to discuss their drafts discussed only their errors in spelling and sentence structure, probably the least valuable things to talk about in the early stages of composing. Perhaps even more troublesome than activities inappropriate to the task is no collaboration at all. The risk is great that, without clear guidelines, students will just pat each other on the back, attack each other counterproductively, or fall silent.

An observer in a collaborative setting, then, must consider the task set by the teacher as the first essential element in any evaluation. The task must figure very prominently in judgments about the class. Questions an observer might ask about the task are: is it clearly worded and unambiguous? does it split the exercise into workable segments? do students know what to do and how to do it? is the task pertinent to the students' needs, goals, and abilities? does the exercise move toward consensus? do the questions students deal with stimulate critical thinking? and, perhaps most important of all, does it call on what students can be expected to know in a way that will lead them together beyond what they already know—is the task difficult enough to challenge but not too difficult to stonewall conversations?

The Teacher as Classroom Manager

The second aspect of collaborative learning for evaluators to consider is the teacher as classroom manager. With the task laid out, how does the teacher implement the actual act of collaboration? How does the teacher organize the social relations in which learning will occur? Have students learned to form groups easily and with relative speed? Are chairs organized in well-spaced clusters so that group conversations do not drown each other out? Do group members demonstrate an ability to work together, one person talking at a time, others listening? Are time limits clear and generally adhered to, and yet flexible? Does the teacher check on how much more time the groups may need as the prescribed end point draws near, and perhaps urge the groups to move on to complete their tasks? If a recorder or reporter is required—the member of each group who acts as synthesizer of the discussion—are his or her functions clear? Does the recorder or reporter take down statements carefully and check with group members for accuracy?

The Teacher's Role During Group Work

The third aspect of collaborative learning that evaluators should examine carefully is the teacher's behavior while the groups are working. Most teachers I have observed travel from group to group answering questions from students, participating in discussions, probing with further questions, guiding responses, and focusing students' attention on the task. Although some of these steps may be necessary from time to time, the teacher's presence as a group member challenges one of the basic tenets of collaboration in the classroom. "The purpose of collaborative learning . . ." Bruffee points out "is to help students gain authority over their knowledge and gain independence in using it" (*Short Course* 49). In the classroom "teachers create social structures in which students can learn to take over the authority for learning as they gain the ability and confidence to do so" (49). A teacher joining a group can easily undermine the development of that authority and that confidence. All attention will turn to the teacher as the central figure in the learning process. Usually, collaboration advances best when groups are left pretty much to the students themselves. At this point in the process, in most cases the best teacher is usually the seemingly most idle teacher, busy with other tasks or even going out of the room from time to time as the groups conduct their business. Evaluators, then, should not judge harshly a practitioner of collaborative learning who reads papers or who leaves the class during small group discussions.

An observer can learn a great deal about prior instruction by watching how students engage in the group task. The noise level in the room, the arrangement of furniture, the ease with which the groups are formed, the tone of conversation among students, the nature of reports emerging from groups all indicate how much the class has practiced efficient collaborative schemes in the past. Evaluators, therefore, should note very carefully how students behave in their groups as a signal of the teacher's advance preparation. Group management is the teacher's responsibility and the collaborative learning teacher pays careful attention to dynamics and composition. Are there too many monopolizers in one group? too many withdrawn students? too many unprepared students? If a group is not working at the task or if a group delivers a weak report, how does the teacher respond? Evaluators should pay particularly close attention to the reporter's role after group activity ends. If selected students make thoughtful, responsible, well-planned presentations to the whole class, the evaluator knows that the teacher has built collaboration theory into the structure of the course prior to the evaluation sessions. Student behavior in groups and at the reporting stage is an important signal for the teacher's skill in the uses of collaborative learning.

The Teacher as Synthesizer

The fourth aspect of collaborative learning that the classroom observer must consider is how the teacher performs in the role of synthesizer after the activity in groups is complete. Once the groups finish their work, it is important for each recorder to share the group's consensus with the rest of the class. With this done, the teacher must help the class as a whole to make sense and order out of the sometimes conflicting and contradictory reports. Writing the points raised by each group on the chalkboard or on a transparency for the overhead projector (or asking recorders themselves to do this) allows everyone to discuss and evaluate the conclusion arrived at by the groups. Even when a consensus report does not follow inevitably from the task, when, for example, students read their drafts aloud to each other for revision, a report on the process itself or on what people think they learned from it may be useful. Questions from the teacher like "What were the general recommendations made to members of the group?" or "What did readers of your paper suggest that you do to take it to the next stage?" help to reinforce what has been learned as well as to establish the value of learning communities and of peer review in any intellectual process.

How the teacher conducts this plenary discussion is very important to the success of collaborative learning. First, the teacher helps students synthesize each group's results with the results produced by other groups. The teacher should lead the class to consider the similarities and contradictions in the recorded points of view and should unite them all, if possible, into a larger vision. The instructor must help students see their differences and to reconcile them. Here "the teacher acts as a referee, directing the energies of the groups on two sides of a divided issue to debate the matter until the parties either arrive at a position that satisfies the whole class or until they agree to disagree" (Bruffee, "Liberal Education" 52).

With agreement, then, the teacher's role once again changes. The teacher now must help the class move further toward joining another community of knowledgeable peers, the community outside the classroom, the scholars who do research in the discipline, who establish the conventions of thinking and writing in those disciplines, who write books and articles and read papers on the problem at hand. "What happens when we learn something," Bruffee writes, "is that we leave a community that justifies certain beliefs in a certain way and join another community that justifies other beliefs in other ways. We leave one community of knowledgeable peers and join another" ("Liberal Education" 105). By synthesizing results of the individual groups, and comparing that synthesis with the consensus of the larger community of knowledgeable peers—the teacher's own community—the teacher helps complete the movement into this larger community.

An observer considering these last two features of the teacher's role—as synthesizer and as representative of the academic community—must be prepared to evaluate the teacher's knowledge of content as well as the teacher's ability to bring the class to perceive differences and similarities in the conclusions of the groups. The teacher must guide students to classify the ideas presented by the various groups without judging one idea right and the other wrong, but by helping the class to investigate the reasoning used to develop and shape the ideas. The teacher also must lead the class to consider how their consensus differs from the consensus of the larger community, and must lead the class to speculate about how that larger community might have arrived at its decision. The skill with which the teacher manages the stages of collaboration is directly related to the teacher's knowledge of and commitment to the philosophical principles upon which collaborative learning is based (see Bruffee, "Collaborative Learning"). An instructor who understands and believes in knowledge as a social construct will see group reporting as an important means of advancing knowledge in the classroom. On the other hand, an instructor willing to experiment with group work but clinging to the Mirror-of-Nature-metaphor will find it hard to avoid using the group setting as anything other than a microcosm of the lecture hall. Many teachers who attempt collaborative learning but abandon it are frequently trying to achieve the same ends in groups that they tried to achieve in the more familiar lecture-recitation session or Socratic dialogue. Thus, an appropriate evaluation should consider the teacher's understanding of collaboration as a means to generate knowledge as a social construct and not simply as the use of a new configuration of students in the classroom.

Yet a one-hour class does not always easily reveal a teacher's knowledge of the rationale for collaboration. Evaluators, therefore, may find it useful to consult with teachers either before or after the class in order to uncover the roots of the particular program of learning for the session. Furthermore, the evaluator's interests must extend to the whole course of study and should not be confined exclusively to a single hour's instruction. Too often collaborative activities are a chain of exercises, unrelated to each other. Thus, in a conference with the teacher an evaluator should aim to discover the goals for the course as a whole and the relation of those goals to the collaborative task just observed.

Summary

I am not unaware of the problems that inhere in the kind of evaluation that this essay is advocating. Collaborative learning is messier in practice than in theory; no one can "live" the theory as clearly as the model suggests. As Harvey Kail points out:

> One doesn't simply eradicate the 'mirror-of-nature metaphor' from one's life as if one were changing from Crest to Colgate. Sometimes I find myself back in the old world, the one where knowledge is 'out there' and my job is to find it and my students' job is to model my search. Other times, more frequently now, I see conversation, it gives and take, as the central manufacturing process of knowledge and appropriate ways of talking (and writing) as the goal. At the same time, I also believe that the lecture is a perfectly legitimate mode of teaching, even within the boundaries established by CL theory. So . . . I contradict myself . . . very well. . . .

Certainly, a commitment to collaborative learning is based on a desire to confront the traditional view of knowledge in our own lives. Like all confrontations, this too is anything but smooth and simple.

Yet my purpose here is to move the practitioner of collaborative learning to an ideal model that will help students achieve knowledge in the classroom. Toward that end, I wish to summarize the features of the collaborative session that an outside evaluator should consider:

1. the nature and quality of the task statement.
2. the social setting of the collaborative activity and the behavior of students during the execution of the task.
3. the teacher's behavior during the execution of the task.

4. the teacher's role in group composition and management.
5. the nature and quality of the reports made by each group.
6. the teacher's performance as synthesizer and as representative of the academic learning community.
7. the relation of the collaborative activity to the design of the course.
8. the teacher's knowledge of and commitment to the rationale of collaborative learning.

The critical underlying principle for evaluators is that in the collaborative learning classroom the instructor is in no sense a passive figure. Collaborative learning is not unstructured learning; it replaces one structure, the traditional one, with another, a collaborative structure. The roles I have attempted to outline here define some of the elements to consider in evaluating a teacher's effectiveness as a leader of collaborative learning within this new structure. Expecting students to engage in productive conversation simply by reshuffling chairs, by telling them to work together in groups, or by requiring, without further guidance, that they read each other's papers, can easily stymie collaboration and not stimulate it. I have seen reflected in the attitude of teachers inexperienced with collaboration and inattentive to its complexities as a mode of learning an often unfulfilled plea to students: "Don't just sit there—collaborate!" Neither inactive nor nondirective, the teacher in the collaborative classroom must plan and organize the session so that students know that the end is not simply to work in groups but to work in groups in an effort to reach consensus for an important task. The effective collaborative learning teacher is one who understands the basis and structure of collaborative learning and who knows how to lead students to work productively within it.[1]

Works Cited

Bruffee, Kenneth A. "Collaborative Learning and the 'Conversation of Mankind.' " *CE* 46 (1984): 635-652.

———. "Liberal Education and the Social Justification of Belief." *Liberal Education* 68 (1982): 95-114.

———. *A Short Course in Writing*. 3rd ed. Boston: Little, 1985.

Hawkes, Peter. Letter to the author. n.d.

Kail, Harvey. Letter to the author. 25 Feb. 1985.

Trimbur, John. Letter to the author. 10 Feb. 1985.

1. I have based my comments in this essay upon many years' experience in observing college English teachers as part of a required program of classroom observation as well as upon my work in supervising teachers across the curriculum in LaGuardia's ongoing faculty development effort, the Integrated Skills Reinforcement Project. But I have shared this paper with a number of colleagues who have long been at the forefront of collaborative learning—including my mentor in all this, Kenneth Bruffee (Brooklyn College), and Marian Arkin (LaGuardia Community College), John Bean (Montana State University), Peter Hawkes (Dutchess Community College), Harvey Kail (University of Maine at Orono), Carol Stanger (John Jay College), and John Trimbur (Boston University). Of course, I assume all responsibility for the points made here, but I acknowledge with gratitude the thoughtful comments and suggestions of my colleagues as this paper evolved from draft to draft.

Hephzibah Roskelly

The Risky Business of Group Work

All of us compositionists believe in group work. In this post-Vygotskian, post-Freirean age it's impossible not to. The terms that dominate our collective conversation in conferences and in our journals—collaboration, peer response, discourse community, shared knowledge—have become symbols for a pedagogical agenda that values talk and activity as learning tools. And Vygotsky, Freire and others provide strong underpinning for our conversation, with the clear link they make between social interaction and learning. A person learns in a group as he listens and speaks, and he learns about himself as well as the culture he inhabits. He may act to change that culture; he most certainly will be changed by it.

Three anecdotes:

Last semester I ask students to write about their past experiences with group work. "One person did all the work," a young woman writes. "It was me. Seems like the teacher counts on the 'smart kid' to keep the others in line. But most of them read the paper in my group."

A few months ago I conduct a workshop for teachers where we talk about the value of group work. They believe in group work they tell me. But they don't use it much. I ask why. "I feel so pressed for time. And group work takes it up." "I feel guilty when I'm not actually teaching." "I'm afraid they'll start socializing."

This year I come up for tenure. I have to write—and ask my co-authors to write—a defense of the collaborative work we've accomplished. How much did you do? is the question. Some of my colleagues are nonplussed by my own and my collaborators' inability to divide up our contributions.

Whatever our belief in group work, and in the collaboration that ensues from it, we haven't translated that belief very effectively to our classrooms, to other educators, to administrations. The gap between talk about groups and talk in groups looms large. I want to argue here that such a gap exists because the purposes of group work are deeply in conflict, and that our refusal to acknowledge and mediate those conflicts has constrained the methods of group work and blunted its effectiveness.

In his moving account of education and literacy, *Lives on the Boundary,* Mike Rose describes students, including himself as a boy, who are sidelined in the academic game by the way their lives separate them from mainstream educational values. The answer? "My students needed to be immersed in talking, reading and writing, they needed to further develop their ability to think critically, and they needed to gain confidence in themselves as systematic inquirers. They had to be let into the academic club" (141).

From *ATAC Forum* 4 (Spring 1992): 1-5. Reprinted by permission of the publisher.

Paulo Freire's students live farther out on the boundary than the disadvantaged in Southern California, in isolated villages and farms in Brazil or Guinnea-Bissau or Cape Verde, but they're outsiders for the same reasons as Rose's students. Freire's plan for their education; that is, their involvement in the discourse of academic literacy, includes more than immersion, inquiry and admission. It it is to "make visible the language, dreams, values and encounters that constitute lives and transforms the structure in which those contexts are placed" (105). The conflict in aims is clear. Does the group work toward socialization—being let into the "academic club"—or conscientization—transforming structures by asserting the value of those without membership?

These two aims—socializing and criticizing—are almost always separate and opposed, even for teachers and theorists who claim both as goals of group work. And many recognize it. Peter Elbow's may be the most honest account of the conflict. In "Pedagogy of the Bamboozled," Elbow wrestles with the problem of how a teacher can both challenge institutional goals through encouraging students to question them and work in an institution. He admits that liberation—Freire's aim—and preservation—the institution's aim—usually match poorly. "We have very little pedagogy truly designed to liberate, and we need more. But in all truth we must admit to ourselves that few of us, because of our temperaments and because of our institutional setting, are in a position to offer it (93). Taking a programmatic rather than pedagogical angle, Daniel Mahala describes the same anomaly as he assesses the Writing Across the Curriculum movement, puzzling over the paradox of WAC goals that promote the valuing of student knowledge while assuring that "writing to learn does not mean changing course content." Mahala is not concerned with group work primarily, but his conclusions about the WAC movement coincide precisely with our concern here. "The question we beg, of course," he says, "is how can we call the knowledge students find their own, if we are assuring our colleagues that writing to learn does mean changing course content?" (778). How can students liberate themselves and indoctrinate themselves at the same time?

Kenneth Bruffee, whose work for years has advocated the methods and goals of group work, is well aware of those double and conflicting aims. In "Collaborative Learning and the Conversation of Mankind," Bruffee begins his argument for the value of group approaches with a short history of the growth of collaboration. The London Schools Project, an early collaborative attempt "sprang from a desire to democratize education by loosening destructive authoritarian forms" (636). Students decided on their own work. Teachers helped them. The agenda for the London students is clearly political one would say. And clearly aimed toward a Freirean conscientization, challenging structures by asserting the power of groups within them. But Bruffee is quick to add that for American college teachers "the roots of collaborative learning lie neither in radical politics nor in research." They come in desperate response to a "pressing educational need."

Support for educational need as the agenda for group work appears to come from Bruffee's description of the research of M. L. J. Abercrombie, a British physician training medical students, who discovered that diagnoses improved when students talked together rather than listened to the physician in charge. Abercrombie's work establishes the educational value of group problem solving. But Bruffee ignores the political implications of that group activity, and his carefully argued study ends by giving the win to "educational need" over politics in the conflict he admits to but downplays. "We should contrive to ensure that students' conversations about what they read and write is similar in as many ways as possible to the way we would like them eventually to read and write" (650). In spite of the way students' own discourse gets valued through the group (as they use their "abnormal discourse," a term Bruffee alters from Richard Rorty), in spite of the way teachers strive to meet both the need to maintain knowledge and to challenge it, the institution gets the final say: students become re-acculturated as they loosen ties to their own communities and form other ties, academic and institutional. If we recognize that students operate with a discourse unlike the talk of academics, the group functions as a way of allowing "abnormal discourse" while eventually privileging the "normal discourse" of the field.

Yet picture the scene around the patient's bed at University Hospital, London the day before Dr. Abercrombie tries her new approach. The lead physician stands at the head of the bed, explaining the case, giving the terms. She then questions, waiting for hands to go up to suggest diagnosis. She calls on one, then another and another, and then she assesses the conclusions each individual has offered, nodding to the most correct responder, and maybe giving an encouraging pat on the shoulder to the patient. See the same scene a week later. Students look at the material on the patient, then look up, begin to talk hesitantly at first perhaps, explaining the status of the patient and offering suggestions about causes. Maybe one or two disagree and a third mediates. They conclude and nod to one another.

What has happened is more than a shift in educational practice that values talk, interaction and consensus. Nobody raises a hand to speak. Even more important, the teacher/physician is not at the head of the bed. Grouped around the patient, students do more than decide cause and treatment, presumably the teacher's agenda. They decide matters of greater potential educational significance. What terms do we use to describe causes? What counts as relevant information? How are we to make decisions about the problem? All of those decisions create new responsibilities for students, accustomed to having the answers to them already ordained. The dynamic has altered radically: institutional tradition that assigns sources of knowledge and social code to one authority has suddenly disappeared within the group.

It's this physical difference in the classroom that makes the small group revolutionary in the institutional setting, for the way in which it implies a criticism of institutional structure. There is suddenly no front of the room. The teacher in this small group setting is, as Bruffee asserts, both "agent for change and conservator of institutional value," but that position is far from comfortable. She doesn't know where to stand. As Elbow indicates, it leads to "bamboozlement." We use the group but distrust it, as Bruffee finds. We say we value talk but we don't allow much, as studies from Shirley Brice Heath to the report from the Dartmouth Conference at NCTE in 1991 indicate: students talk far less than teachers, and teachers talk more than 75% of the time. Classrooms then remain locked into the question/response/lecture/silence dynamic that limits talk or makes it adversarial.

It's not surprising, as Elbow suggests, that Freire's success with the group occurs primarily outside the institutional setting because the conflict of aims is palpable within it. Anne Gere's study of writing groups echoes Elbow's insight, noting that the reason groups have worked is because they've worked primarily outside academic institutions.

Groups have flourished outside institutions because they provide potential power for those traditionally powerless within institutions. Once the group moves inside the institution, they retain their potential power for change, but such change often runs exactly counter to institutional goals.

Freire points out that the "political character of our method is independent of our consciousness of it" (102). Our stance on group work is political whether or not we acknowledge conflicts, political whether we believe we're giving students access to preset educational goals or creating potential dissenters to institutional values. It's this political recognition that power relationships get fuzzy in the small group that help explain institutional distrust of group methods and collaboration in research and writing. It helps explain teachers' fears about loss of control, often masked by questions about loss of time. It helps explain students' own discomfort with work that will be difficult to count for any one person. Teachers and students fear the question of who sets the values once the group takes over, once they're fifteen voices raised rather than one. It's fairly typical among teachers I've observed who continue to opt for group work to assuage their own and their students' suspicion by implementing tight controls on the work the groups do. You set the content: *here's the topic for discussion*. You decide how much work will be done: *I want three questions*. You insist that each group report out and you make decisions about how effective the group has been in doing the work assigned to them. When group work is used as "peer editing," where students assess one another's writing, the controls are often even more stringent.

But I believe both aims, pulling against one another as they most certainly do, can work together once they're rethought and mediated in classroom approaches. Bruffee, Rose, David Bartholomae and others who note and in some ways applaud the double aim of group work, all move too quickly to the socializing institutional end of groups, implying in this move that academic discourse is a product, long ago completed, and that it's only students who operate in process. That's why students, and not the discourse, must change. Bartholomae even encourages teachers to "turn again to the requirements of academic writing, or any writing that moves a writer into a privileged and closed discourse" (88). When you're initiated, they might say to students, perhaps then you can criticize; it's only then that you know what you're criticizing. But because criticism as a group goal is left till later or suppressed within the normal discourse of the subject field, the conclusions of group work are often foregone. House rules are obeyed, and the teacher as willing or unwilling servant in the academic house sees that it happens.

Freire's work suggests that in order to achieve both aims of group work the second, criticizing, aim must precede and then accompany the first as students become literate. I could show how the double aim gets accomplished in this way by giving examples from Freire himself, but Freire has argued, usually to little avail, that it's wrong to take one cultural context and apply it to another, wrong to see Brazilian farmers emerging from a feudal economy analogous to ESL students in Boston. A culture circle needs to be made and remade, according to each political, social and individual reality. So rather than take the farmers and workers in Principe, let me take instead one of Rose's "culture circles" in *Lives on the Boundary*.

Rose's book tells the story of many groups of students who stand outside the normal discourse of fields they study, whose "socially justifying belief," as Rorty puts it, places them as outsiders. As with Bruffee, for Rose the group, and the reading and writing the group accomplishes, become avenues to enter the academic community. Rose compares his students' situations to the new arrival in a foreign land, who needs to learn cultural codes and signals to operate successfully within that culture.

But in these stories of outsider or marginal groups something else is going on, though Rose may not notice it. There's Willie, the Vietnam veteran who writes of his frustration with life in rude comments on the classics, there are the fourth graders who complete assigned narratives with jokes from TV sitcoms and from family tales. And there is the group of homebound students linked together by telephone in a program Rose taught called the Learning Line. When Rose asked students to send along examples of poems they liked, they took him at his word. And he was suddenly caught between the normal discourse of the academic environment he had learned and the abnormal discourse of the one his students valued. "These [poems] threw me," Rose says. "They were sentimental as could be. The rhymes were strained, and the diction archaic. They were poems all my schooling had trained me to dismiss." But he can't dismiss them; the group won't let him. Their enjoyment in sharing bad verse is so real, so much a part of their learning, that it can't be denied. "I realized," Rose says, "that they wanted to participate in some fuller way." The poems they provide become part of the conversation of the class, passed along together with Rose's own more traditionally acceptable ones. The Learning Line group in a way takes Rose into their conversation rather than the other way around. "Here," one woman who continues to write poems her own way says at one point, "here's a poem like one of the ones Mike sends us."

These people needed to get into the academic conversation, but they needed to do something more important first. They needed to name and recognize their alienation from it. *These are poems I like.* They needed to line that up against academic ways. *What makes the poems I like and the ones Mike likes different?* Only then could they participate in the academic way, renaming it, and using their own difference as a tool for renaming. "They found things to value in both kinds of poems, mine and theirs." I don't think they would have found value in Mike's poems had they not insisted on the inclusion of their own.

It's easy enough perhaps to see how alienation might be a tool for learning in a group such as the isolated homebound, or the exploited peasantry. It may be harder to see how students in a freshman composition class or in an advanced writing class or in a twelfth grade Advanced Placement English class would experience alienation of the kind Freire and Rose portray. But the truth is that school is cultural, promoting social and institutional values, and students are alienated by their lack of consciousness of the role they play in that culture. The first work of students in a small group is to understand their own place in that culture, their terms and beliefs, and then to juxtapose that difference to the social knowledge of the classroom, which would involve groups in studying the content and forms of such knowledge. Through reading and talking and writing, groups begin to come up with terms and ideas that merit exploration. The group then learns to remake knowledge, pulling together academic or institutional knowledge within the framework of their own. This is the method for individual learning that Vygotsky describes, and it's also the method groups use in order to achieve power. Its the method students need to learn to use to enter the academic house, a method to add rooms of their own, even tear parts down.

The teacher is the catalyst that helps students follow this process. But to be a catalyst is to allow, or even encourage, change. As Mahala's critique of WAC argues, "To admit that whatever methods or boundaries we or others espouse for making knowledge are in fact inside the dialectic of knowledge-making is to open ourselves radically to dialogue and critique of the institution we inhabit" (785). The risk of dissent, and even the greater risk of having to question oneself, prevents many teachers from using group work effectively. It's like the movie *Risky Business.* In that movie Tom Cruise and friends engage in many risky activities, but the single most frightening one is that Tom's character Joel allows people into his upper middle class house who don't belong. They're strangers from the city, rowdies and gamblers, prostitutes and pimps. They are threatening precisely because they're outsiders, don't operate with the same codes. Joel spends a lot of time once things get out of control wondering what's lost or broken, wondering how much he has to accommodate himself to the group's insistent presence. Of course what happens is that he does have to accommodate, learns to want to: the movie ends with his exuberantly challenging his own codes in front of the symbolic defender of them—the admissions officer from Princeton.

As teachers, we have to take on the risky business of looking at the academic house we live in, and the ways we invite students into it. We have to be willing to look at how we ourselves entered it, how much we brought with us, how much we were forced to leave at the door. We have to make ourselves brave enough to risk the dissent that inevitably comes when democracy is in action. Once teachers do that, we'll see the work of the small groups in our classes become the real work in the class, with students negotiating their own ideas against and around the ideas they're offered. When students find a real voice, their own and not some mimicked institutional voice, both students and teachers acknowledge the possibility of the real change that might ensue. As they find that groups can transform and be transformed, teachers and students learn not only to risk that change but welcome it.

Works Cited

Bartholomae, David. "Released into Language: Errors, Expectations and the Legacy of Mina Shaughnessy." *The Territory of Language: Linguistics, Stylistics and the Teaching of Composition.* Ed. Donald McQuade. Carbondale: Southern Illinois UP, 1986. 65-88.

Bruffee, Kenneth. "Collaboration and the Conversation of Mankind." *College English* 46 (1984): 635-652.

Elbow, Peter. *Embracing Contraries.* NY: Oxford UP, 1986.

Freire, Paulo. *Pedagogy in Process.* NY: Seabury Press, 1978.

Mahala, Daniel. "Writing Utopias." *College English* 53 (1991): 773-789.

Rose, Mike. *Lives on the Boundary.* NY: Penguin Press, 1989.

COMPOSING AND REVISING

Sondra Perl

Understanding Composing

Any psychological process, whether the development of thought or voluntary behavior, is a process undergoing changes right before one's eyes. . . . Under certain conditions it becomes possible to trace this development.[1]

L. S. Vygotsky

It's hard to begin this case study of myself as a writer because even as I'm searching for a beginning, a pattern of organization, I'm watching myself, trying to understand my behavior. As I sit here in silence, I can see lots of things happening that never made it onto my tapes. My mind leaps from the task at hand to what I need at the vegetable stand for tonight's soup to the threatening rain outside to ideas voiced in my writing group this morning, but in between "distractions" I hear myself trying out words I might use. It's as if the extraneous thoughts are a counterpoint to the more steady attention I'm giving to composing. This is all to point out that the process is more complex than I'm aware of, but I think my tapes reveal certain basic patterns that I tend to follow.

Anne
New York City Teacher

Anne is a teacher of writing. In 1979, she was among a group of twenty teachers who were taking a course in research and basic writing at New York University.[2] One of the assignments in the course was for the teachers to tape their thoughts while composing aloud on the topic, "My Most Anxious Moment as a Writer." Everyone in the group was given the topic in the morning during class and told to compose later on that day in a place where they would be comfortable and relatively free from distractions. The result was a tape of composing aloud and a written product that formed the basis for class discussion over the next few days.

One of the purposes of this assignment was to provide teachers with an opportunity to see their own composing processes at work. From the start of the course, we recognized that we were controlling the situation by assigning a topic and that we might be altering the process by asking writers to compose aloud. Nonetheless we viewed the task as a way of capturing some of the flow of composing and, as Anne later observed

From *College Composition and Communication* 31 (December 1980): 363-69.

1. L. S. Vygotsky, *Mind in Society,* trans. M. Cole, V. John-Steiner, S. Scribner, and E. Souberman (Cambridge, Mass: Harvard University Press, 1978), p. 61.
2. This course was team-taught by myself and Gordon Pradl, Associate Professor of English Education at New York University.

in her analysis of her tape, she was able to detect certain basic patterns. This observation, made not only by Anne, then leads me to ask "What basic patterns seem to occur during composing?" and "What does this type of research have to tell us about the nature of the composing process?"

Perhaps the most challenging part of the answer is the recognition of recursiveness in writing. In recent years, many researchers including myself have questioned the traditional notion that writing is a linear process with a strict plan-write-revise sequence.[3] In its stead, we have advocated the idea that writing is a recursive process, that throughout the process of writing, writers return to substrands of the overall process, or subroutines (short successions of steps that yield results on which the writer draws in taking the next set of steps); writers use these to keep the process moving forward. In other words, recursiveness in writing implies that there is a forward-moving action that exists by virtue of a backward-moving action. The questions that then need to be answered are, "To what do writers move back?" "What exactly is being repeated?" "What recurs?"

To answer these questions, it is important to look at what writers do while writing and what an analysis of their processes reveals. The descriptions that follow are based on my own observations of the composing processes of many types of writers including college students, graduate students, and English teachers like Anne.

Writing does appear to be recursive, yet the parts that recur seem to vary from writer to writer and from topic to topic. Furthermore, some recursive elements are easy to spot while others are not.

1. The most visible recurring feature or backward movement involves rereading little bits of discourse. Few writers I have seen write for long periods of time without returning briefly to what is already down on the page.

For some, like Anne, rereading occurs after every few phrases; for others, it occurs after every sentence; more frequently, it occurs after a "chunk" of information has been written. Thus, the unit that is reread is not necessarily a syntactic one, but rather a semantic one as defined by the writer.

2. The second recurring feature is some key word or item called up by the topic. Writers consistently return to their notion of the topic throughout the process of writing. Particularly when they are stuck, writers seem to use the topic or a key word in it as a way to get going again. Thus many times it is possible to see writers "going back," rereading the topic they were given, changing it to suit what they have been writing or changing what they have written to suit their notion of the topic.

3. There is also a third backward movement in writing, one that is not so easy to document. It is not easy because the move, itself, cannot immediately be identified with words. In fact, the move is not to any words on the page nor to the topic but to feelings or non-verbalized perceptions that *surround* the words, or to what the words already present *evoke* in the writer. The move draws on sense experience, and it can be observed if one pays close attention to what happens when writers pause and seem to listen or otherwise react to what is inside of them. The move occurs inside the writer, to what is

3. See Janet Emig, *The Composing Processes of Twelfth-Graders,* NCTE Research Report No. 13 (Urbana, Ill: National Council of Teachers of English, 1971); Linda Flower and J. R. Hayes, "The Cognition of Discovery," *CCC,* 31 (February, 1980), 21-32; Nancy Sommers, "The Need for Theory in Composition Research," *CCC,* 30 (February, 1979), 46-49.

physically felt. The term used to describe this focus of writers' attention is *felt sense.* The term "felt sense" has been coined and described by Eugene Gendlin, a philosopher at the University of Chicago. In his words, felt sense is

> the soft underbelly of thought . . . a kind of bodily awareness that . . . can be used as a tool . . . a bodily awareness that . . . encompasses everything you feel and know about a given subject at a given time. . . . It is felt in the body, yet it has meanings. It is body *and* mind before they are split apart.[4]

This self sense is always there, within us. It is unifying, and yet, when we bring words to it, it can break apart, shift, unravel, and become something else. Gendlin has spent many years showing people how to work with their felt sense. Here I am making connections between what he has done and what I have seen happen as people write.

When writers are given a topic, the topic itself evokes a felt sense in them. This topic calls forth images, words, ideas, and vague fuzzy feelings that are anchored in the writer's body. What is elicited, then, is not solely the product of a mind but of a mind alive in a living, sensing body.

When writers pause, when they go back and repeat key words, what they seem to be doing is waiting, paying attention to what is still vague and unclear. They are looking to their felt experience, and waiting for an image, a word, or a phrase to emerge that captures the sense they embody.

Usually, when they make the decision to write, it is after they have a dawning awareness that something has clicked, that they have enough of a sense that if they begin with a few words heading in a certain direction, words will continue to come which will allow them to flesh out the sense they have.

The process of using what is sensed directly about a topic is a natural one. Many writers do it without any conscious awareness that that is what they are doing. For example, Anne repeats the words "anxious moments," using these key words as a way of allowing her sense of the topic to deepen. She asks herself, "Why are exams so anxiety provoking?" and waits until she has enough of a sense within her that she can go in a certain direction. She does not yet have the words, only the sense that she is able to begin. Once she writes, she stops to see what is there. She maintains a highly recursive composing style throughout and she seems unable to go forward without first going back to see and to listen to what she has already created. In her own words, she says:

> My disjointed style of composing is very striking to me. I almost never move from the writing of one sentence directly to the next. After each sentence I pause to read what I've written, assess, sometimes edit and think about what will come next. I often have to read the several preceding sentences a few times as if to gain momentum to carry me to the next sentence. I seem to depend a lot on the sound of my words and . . . while I'm hanging in the middle of this uncompleted thought, I may also start editing a previous sentence or get an inspiration for something which I want to include later in the paper.

---

4. Eugene Gendlin, *Focusing* (New York: Everest House, 1978), pp. 35, 165.

What tells Anne that she is ready to write? What is the feeling of ''momentum'' like for her? What is she hearing as she listens to the ''sound'' of her words? When she experiences ''inspiration,'' how does she recognize it?

In the approach I am presenting, the ability to recognize what one needs to do or where one needs to go is informed by calling on felt sense. This is the internal criterion writers seem to use to guide them when they are planning, drafting, and revising.

The recursive move, then, that is hardest to document but is probably the most important to be aware of is the move to felt sense, to what is not yet *in words* but out of which images, words, and concepts emerge.

The continuing presence of this felt sense, waiting for us to discover it and see where it leads, raises a number of questions.

Is ''felt sense'' another term for what professional writers call their ''inner voice'' or their felling of ''inspiration''?

Do skilled writers call on their capacity to sense more readily than unskilled writers?

Rather than merely reducing the complex act of writing to a neat formulation, can the term ''felt sense'' point us to an area of our experience from which we can evolve even richer and more accurate descriptions of composing?

Can learning how to work with felt sense teach us about creativity and release us from stultifyingly repetitive patterns?

My observations lead me to answer ''yes'' to all four questions. There seems to be a basic step in the process of composing that skilled writers rely on even when they are unaware of it and that less skilled writers can be taught. This process seems to rely on very careful attention to one's inner reflections and is often accompanied with bodily sensations.

When it's working, this process allows us to say or write what we've never said before, to create something new and fresh, and occasionally it provides us with the experience of ''newness'' or ''freshness,'' even when ''old words'' or images are used.

The basic process begins with paying attention. If we are given a topic, it begins with taking the topic in and attending to what it evokes in us. There is less ''figuring out'' an answer and more ''waiting'' to see what forms. Even without a predetermined topic, the process remains the same. We can ask ourselves, ''What's on my mind?'' or ''Of all the things I know about, what would I most like to write about now?'' and wait to see what comes. What we pay attention to is the part of our bodies where we experience ourselves directly. For many people, it's the area of their stomachs; for others, there is a more generalized response and they maintain a hovering attention to what they experience throughout their bodies.

Once a felt sense forms, we match words to it. As we begin to describe it, we get to see what is there for us. We get to see what we think, what we know. If we are writing about something that truly interests us, the felt sense deepens. We know that we are writing out of a ''centered'' place.

If the process is working, we begin to move along, sometimes quickly. Other times, we need to return to the beginning, to reread, to see if we captured what we meant to say. Sometimes after rereading we move on again, picking up speed. Other times by rereading we realize we've gone off the track, that what we've written doesn't quite ''say it,'' and we need to reassess. Sometimes the words are wrong and we need to change them. Other times we need to go back to the topic, to call up the sense it initially evoked to see where and how our words led us astray. Sometimes in rereading we discover that the topic is "wrong," that the direction we discovered in writing is where we really want to go. It is important here to clarify that the terms "right" and "wrong" are not necessarily meant to refer to grammatical structures or to correctness.

What is "right" or "wrong" corresponds to our sense of our intention. We intend to write something, words come, and now we assess if those words adequately capture our intended meaning. Thus, the first question we ask ourselves is "Are these words right for me?" "Do they capture what I'm trying to say?" "If not, what's missing?"

Once we ask "what's missing?" we need once again to wait, to let a felt sense of what is missing form, and then to write out of that sense.

I have labeled this process of attending, of calling up a felt sense, and of writing out of that place, the process of *retrospective structuring*. It is retrospective in that it begins with what is already there, inchoately, and brings whatever is there forward by using language in structured form.

It seems as though a felt sense has within it many possible structures or forms. As we shape what we intend to say, we are further structuring our sense while correspondingly shaping our piece of writing.

It is also important to note that what is there implicitly, without words, is not equivalent to what finally emerges. In the process of writing, we begin with what is inchoate and end with something that is tangible. In order to do so, we both discover and construct what we mean. Yet the term "discovery" ought not lead us to think that meaning exists fully formed inside of us and that all we need do is dig deep enough to release it. In writing, meaning cannot be discovered the way we discover an object on an archeological dig. In writing, meaning is crafted and constructed. It involves us in a process of coming-into-being. Once we have worked at shaping, through language, what is there inchoately, we can look at what we have written to see if it adequately captures what we intended. Often at this moment discovery occurs. We see something new in our writing that comes upon us as a surprise. We see in our words a further structuring of the sense we began with and we recognize that in those words we have discovered something new about ourselves and our topic. Thus when we are successful at this process, we end up with a product that teaches us something, that clarifies what we know (or what we knew at one point only implicitly), and that lifts out or explicates or enlarges our experience. In this way, writing leads to discovery.

All the writers I have observed, skilled and unskilled alike, use the process of retrospective structuring while writing. Yet the degree to which they do so varies and seems, in fact, to depend upon the model of the writing process that they have internalized. Those who realize that writing can be a recursive process have an easier time with waiting, looking, and discovering. Those who subscribe to the linear model find themselves easily frustrated when what they write does not immediately correspond to what they planned or when what they produce leaves them with little sense of accomplishment. Since they have relied on a formulaic approach, they often produce writing that is formulaic as well, thereby cutting themselves off from the possibility of discovering something new.

Such a result seems linked to another feature of the composing process, to what I call *projective structuring*, or the ability to craft what one intends to say so that it is intelligible to others.

A number of concerns arise in regard to projective structuring; I will mention only a few that have been raised for me as I have watched different writers at work.

1. Although projective structuring is only one important part of the composing process, many writers act as if it is the whole process. These writers focus on what they think others want them to write rather than looking to see what it is they want to write. As a result, they often ignore their felt sense and they do not establish a living connection between themselves and their topic.

2. Many writers reduce projective structuring to a series of rules or criteria for evaluating finished discourse. These writers ask, ''Is what I'm writing correct?'' and ''Does it conform to the rules I've been taught?'' While these concerns are important, they often overshadow all others and lock the writer in the position of writing solely or primarily for the approval of readers.

Projective structuring, as I see it, involves much more than imagining a strict audience and maintaining a strict focus on correctness. It is true that to handle this part of the process well, writers need to know certain grammatical rules and evaluative criteria, but they also need to know how to call up a sense of their reader's needs and expectations.

For projective structuring to function fully, writers need to draw on their capacity to move away from their own words, to decenter from the page, and to project themselves into the role of the reader. In other words, projective structuring asks writers to attempt to become readers and to imagine what someone other than themselves will need before the writer's particular piece of writing can become intelligible and compelling. To do so, writers must have the experience of being readers. They cannot call up a felt sense of a reader unless they themselves have experienced what it means to be lost in a piece of writing or to be excited by it. When writers do not have such experiences, it is easy for them to accept that readers merely require correctness.

In closing, I would like to suggest that retrospective and projective structuring are two parts of the same basic process. Together they form the alternating mental postures writers assume as they move through the act of composing. The former relies on the ability to go inside, to attend to what is there, from that attending to place words upon a page, and then to assess if those words adequately capture one's meaning. The latter relies on the ability to assess how the words on that page will affect someone other than the writer, the reader. We rarely do one without the other entering in; in fact, again in these postures we can see the shuttling back-and-forth movements of the composing process, the move from sense to words and from words to sense, from inner experience to outer judgment and from judgment back to experience. As we move through this cycle, we are continually composing and recomposing our meanings and what we mean. And in doing so, we display some of the basic recursive patterns that writers who observe themselves closely seem to see in their own work. After observing the process for a long time we may, like Anne, conclude that at any given moment the process is more complex than anything we are aware of; yet such insights, I believe, are important. They show us the fallacy of reducing the composing process to a simple linear scheme and they leave us with the potential for creating even more powerful ways of understanding composing.

Nancy Sommers

Between the Drafts

I cannot think of my childhood without hearing voices, deep, heavily-accented, instructive German voices.

I hear the voice of my father reading to me from *Struvelpater,* the German children's tale about a messy boy who refuses to cut his hair or his fingernails. Struvelpater's hair grows so long that birds nest in it, and his fingernails grow so long that his hands become useless. He fares better, though, than the other characters in the book who don't listen to their parents. Augustus, for instance, refuses to eat his soup for four days, becomes as thin as a thread, and on the fifth day he is dead. Fidgety Philip tilts his dinner chair like a rocking horse until his chair falls backwards; the hot food falls on top of him, and suffocates him under the weight of the table cloth. The worst story by far for me is that of Conrad, an incorrigible thumb-sucker, who couldn't stop sucking his thumb and whose mother warned him that a great, long, red-legged scissor-man would—and, yes, did—snip both his thumbs off.

As a child, I hated these horrid stories with their clear moral lessons, exhorting me to listen to my parents: do the right thing, they said; obey authority, or else catastrophic things—dissipation, suffocation, loss of thumbs—will follow. It never occurred to me as a child to wonder why my parents, who had escaped Nazi Germany in 1939, were so deferential to authority, so beholden to sanctioned sources of power. I guess it never occurred to them to reflect or to make any connections between generations of German children reading *Struvelpater,* being instructed from early childhood to honor and defer to the parental authority of the state, and the Nazis' easy rise to power.

When I hear my mother's voice, it is usually reading to me from some kind of guide book showing me how different *They,* the Americans, were from us, the German Jews of Terre Haute. My parents never left home without their passports; we had roots somewhere else. When we traveled westward every summer from our home in Indiana, our bible was the AAA tour guide, giving us the officially sanctioned version of America. We attempted to "see" America from the windows of our 1958 two-tone green Oldsmobile. We were literally the tourists from Terre Haute, those whom Walker Percy describes in "The Loss of the Creature," people who could never experience the Grand Canyon because it had already been formulated for us by picture postcards, tourist folders, guide books, and the words *Grand Canyon.*

Percy suggests that tourists never see the progressive movement of depths, patterns, colors, and shadows of the Grand Canyon, but rather measure their satisfaction by the degree to which the canyon conforms to the expectations in their minds. My mother's

From *College Composition and Communication* 43 (February 1992): 23-31.

AAA guide book directed us, told us what to see, how to see it, and how long it should take us to see it. We never stopped anywhere serendipitously, never lingered, never attempted to know a place.

As I look now at the black-and-white photographs of our trips, seeing myself in ponytail and pedal pushers, I am struck by how many of the photos were taken against the car or, at least, with the car close enough to be included in the photograph. I am not sure we really saw the Grand Canyon or the Painted Desert or the Petrified Forest except from the security of a parking lot. We were travelling on a self-imposed visa that kept us close to our parked car; we lacked the freedom of our own authority and stuck close to each other and to the book itself.

My parents' belief that there was a right and a wrong way to do everything extended to the way they decided to teach us German. Wanting us to learn the correct way, not trusting their own native voices, they bought language-learning records with an officially sanctioned voice of an expert language teacher; never mind that they spoke fluent German.

It is 1959; I am 8 years old. We sit in the olive-drab living room with the drapes closed so the neighbors won't see in. What the neighbors would have seen strikes me now as a scene out of a *Saturday Night Live* skit about the Coneheads. The children and their parental-unit sitting in stiff, good-for-your-posture chairs that my brother and I call the electric chairs. Those chairs are at odd angles to each other so we all face the fireplace; we don't look at each other. I guess my parents never considered pulling the chairs around, facing each other, so we could just talk in German. My father's investment was in the best 1959 technology he could find; he was proud of the time and money he had spent, so that we could be instructed in the right way. I can still see him there in that room removing the record from its purple package, placing it on the hi-fi:

Guten Tag.
Wie geht es Dir?
Wie geht es Werner/Helmut/Dieter?
Werner ist heute krank.
Oh, das tut mir Leid.
Gute Besserung.

We are disconnected voices worrying over the health of Werner, Dieter, and Helmut, foreign characters, names, who have no place in our own family. We go on and on for an eternity with that dialogue until my brother passes gas, or commits some other unspeakable offense, something that sets my father's German sensibility on edge, and he finally says, "We will continue another time." He releases us back into another life, where we speak English, forgetting for yet another week about the health of Werner, Helmut, or Dieter.

I thought I had the issue of authority all settled in my mind when I was in college. My favorite T-shirt, the one I took the greatest pleasure in wearing, was one with the bold words *Question Authority* inscribed across my chest. It seemed that easy. As we said then, either you were part of the problem or you were part of the solution; either you deferred to authority or you resisted it by questioning. Twenty years later, it doesn't seem that simple. I am beginning to get a better sense of my legacy, beginning to see just how complicated and how far-reaching is this business of authority. It extends into my life and touches my students' lives, reminding me again and again of the delicate relationship between language and authority.

In 1989, 30 years after my German lessons at home, I'm having dinner with my daughters in an Italian restaurant. The waiter is flirting with 8-year-old Rachel, telling her she has the most beautiful name, that she is *una ragazza bellissima.* Intoxicated with this affectionate attention, she turns to me passionately and says, "Oh, Momma, Momma, can't we learn Italian?" I, too, for the moment am caught up in the brio of my daughter's passion. I say, "Yes, yes, we must learn Italian." We rush to our favorite bookstore where we find Italian language-learning tapes packaged in 30-, 60-, and 90-day lessons, and in our modesty buy the promise of fluent Italian in 30 lessons. Driving home together, we put the tape in our car tape player, and begin lesson number 1:

Buon giorno.
Come stai?
Come stai Monica?

As we wind our way home, our Italian lessons quickly move beyond preliminaries. We stop worrying over the health of Monica, and suddenly we are in the midst of a dialogue about Signor Fellini who lives at 21 Broadway Street. We cannot follow the dialogue. Rachel, in great despair, betrayed by the promise of being a beautiful girl with a beautiful name speaking Italian in 30 lessons, begins to scream at me: "This isn't the way to learn a language. This isn't language at all. These are just words and sentences; this isn't about us; we don't live at 21 Broadway Street."

And I am back home in Indiana, hearing the disembodied voices of my family, teaching a language out of the context of life.

In 1987, I gave a talk at CCCC entitled "New Directions for Researching Revision." At the time, I liked the talk very much because it gave me an opportunity to illustrate how revision, once a subject as interesting to our profession as an autopsy, had received new body and soul, almost celebrity status, in our time. Yet as interesting as revision had become, it seemed to me that our pedagogies and research methods were resting on some shaky, unquestioned assumptions.

I had begun to see how students often sabotage their own best interests when they revise, searching for errors and assuming, like the eighteenth-century theory of words parodied in *Gulliver's Travels,* that words are a load of things to be carried around and exchanged. It seemed to me that despite all those multiple drafts, all the peer workshops that we were encouraging, we had left unexamined the most important fact of all: revision does not always guarantee improvement; successive drafts do not always lead to a clearer vision. You can't just change the words around and get the ideas right.

Here I am four years later, looking back on that abandoned talk, thinking of myself as a student writer, and seeing that successive drafts have not led me to a clearer vision. I have been under the influence of a voice other than my own.

I live by the lyrical dream of change, of being made anew, always believing that a new vision is possible. I have been gripped, probably obsessed, with the subject of revision since graduate school. I have spent hundreds of hours studying manuscripts, looking for clues in the drafts of professional and student writers, looking for the figure in the carpet. The pleasures of this kind of literary detective work, this literary voyeurism, are the peeps behind the scenes, the glimpses of the process revealed in all its nakedness, of what Edgar Allan Poe called "the elaborate and vacillating crudities of thought, the true purposes seized only at the last moment, the cautious selections and rejections, the painful erasures."

My decision to study revision was not an innocent choice. It is deeply satisfying to believe that we are not locked into our original statements, that we might start and stop, erase, use the delete key in life, and be saved from the roughness of our early drafts. Words can be retracted; souls can be reincarnated. Such beliefs have informed my study of revision, and yet, in my own writing, I have always treated revising as an academic subject, not a personal one. Every time I have written about revision, I have set out to argue a thesis, present my research, accumulate my footnotes. By treating revision as an academic subject, by suggesting that I could learn something only by studying the drafts of other experienced writers, I kept myself clean and distant from any kind of scrutiny. No Struvelpater was I; no birds could nest in my hair; I kept my thumbs intact. I have been the bloodless academic creating taxonomies, creating a hierarchy of student writers and experienced writers, and never asking myself how I was being displaced from my own work. I never asked, "What does my absence *signify?*"

In that unrevised talk from CCCC, I had let Wayne Booth replace my father. Here are my words:

> Revision presents a unique opportunity to study what writers know. By studying writers' revisions we can learn how writers locate themselves within a discourse tradition by developing a persona—a fictionalized self. Creating a persona involves placing the self in a textual community, seeing oneself within a discourse, and positing a self that shares or antagonizes the beliefs that a community of readers shares. As Wayne Booth has written, 'Every speaker makes a self with every word uttered. Even the most sincere statement implies a self that is at best a radical selection from many possible roles. No one comes on in exactly the same way with parents, teachers, classmates, lovers, and IRS inspectors.'

What strikes me now, in this paragraph from my own talk, is that fictionalized self I invented, that anemic researcher, who set herself apart from her most passionate convictions. In that paragraph, I am a distant, imponderable, impersonal voice—inaccessible, humorless, and disguised like the packaged voice of Signor Fellini giving lessons as if absolutely nothing depends on my work. I speak in an inherited academic voice; it isn't mine.

I simply wasn't there for my own talk. Just as my father hid behind his language-learning records and my mother behind her guide books, I disguised myself behind the authority of "the researcher," attempting to bring in the weighty authority of Wayne Booth to justify my own statements, never gazing inward, never trusting my own authority as a writer.

Looking back on that talk, I know how deeply I was under the influence of a way of seeing: Foucault's "Discourse on Language," Barthes' *S/Z*, Scholes' *Textual Power*, and Bartholomae's "Inventing the University" had become my tourist guides. I was so much under their influence that I remember standing in a parking lot of a supermarket,

holding two heavy bags of groceries, talking with a colleague who was tellin
his teaching. Without any reference, except to locate my own authority somewhere else, I felt compelled to suggest to him that he read Foucault. My daughter Alexandra, waiting impatiently for me, eating chocolate while pounding on the hood of the car with her new black patent-leather party shoes, spoke with her own authority. She reminded me that I, too, had bumped on cars, eaten Hershey Bars, worn party shoes without straps, never read Foucault, and knew, nevertheless, what to say on most occasions.

One of my colleague's put a telling cartoon on the wall of our Xerox room. It reads "Breakfast Theory: A morning methodology." The cartoon describes two new cereals: Foucault Flakes and Post-Modern Toasties. The slogan for Foucault Flakes reads: "It's French so it must be good for you. A breakfast commodity so complex that you need a theoretical apparatus to digest it. You don't want to eat it; you'll just want to read it. Breakfast as text." And Post-Modern Toasties: "More than just a cereal, it's a commentary on the nature of cereal-ness, cerealism, and the theory of cerealtivity. Free decoding ring inside."

I had swallowed the whole flake, undigested, as my morning methodology, but, alas, I never found the decoding ring. I was lost in the box. Or, to use the metaphor of revision, I was stuck in a way of seeing: reproducing the thoughts of others, using them as my guides, letting the post-structuralist vocabulary give authority to my text.

Successive drafts of my own talk did not lead to a clearer vision because it simply was not my vision. I, like so many of my students, was reproducing acceptable truths, imitating the gestures and rituals of the academy, not having confidence enough in my own ideas, nor trusting the native language I had learned. I had surrendered my own authority to someone else, to those other authorial voices.

Three years later, I am still wondering: Where does revision come from? Or, as I think about it now, what happens between the drafts? Something has to happen or else we are struck doing mop and broom work, the janitorial work of polishing, cleaning, and fixing what is and always has been. What happens between drafts seems to be one of the great secrets of our profession.

Between drafts, I take lots of showers, hot showers, talking to myself as I watch the water play against the gestures of my hands. In the shower, I get lost in the steam. There I stand without my badges of authority. I begin an imagined conversation with my colleague, the one whom I told in the parking lot of the grocery store, "Oh, but you must read Foucault." I revise our conversation. This time I listen.

I understand why he showed so much disdain when I began to pay homage to Foucault. He had his own sources aplenty that nourished him. Yet he hadn't felt the need to speak through his sources or interject their names into our conversation. His teaching stories and experiences are his own; they give him the authority to speak.

As I get lost in the stream, I listen to his stories, and I begin to tell him mine. I tell him about my father not trusting his native voice to teach me German, about my mother not trusting her own eyes and reading to us from guide books, about my own claustrophobia in not being able to revise a talk about revision, about being drowned out by a chorus of authorial voices. And I surprise myself. I say, Yes, these stories of mine provide powerful evidence; they belong to me; I can use them to say what I must about revision.

I begin at last to have a conversation with all the voices I embody, and I wonder why so many issues are posed as either/or propositions. Either I suck my thumb *or* the great long-legged scissor-man will cut it off. Either I cook two chickens *or* my guests will go away hungry. Either I accept authority *or* I question it. Either I have babies and be in service of the species *or* I write books and be in service of the academy. Either I be personal *or* I be academic.

These either/or ways of seeing exclude life and real revision by pushing us to safe positions, to what is known. They are safe positions that exclude each other and don't allow for any ambiguity, uncertainty. Only when I suspend myself between either *and* or can I move away from conventional boundaries and begin to see shapes and shadows and contours—ambiguity, uncertainty, and discontinuity, moments when the seams of life just don't want to hold; days when I wake up to find, once again, that I don't have enough bread for the children's sandwiches or that there are no shoelaces for their gym shoes. My life is full of uncertainty; negotiating that uncertainty day to day gives me authority.

Maybe this is a woman's journey, maybe not. Maybe it is just my own, but the journey between home and work, between being personal and being authoritative, between the drafts of my life, is a journey of learning how to be both personal and authoritative, both scholarly *and* reflective. It is a journey that leads me to embrace the experiences of my life, and gives me the insight to transform these experiences into evidence. I begin to see discontinuous moments as sources of strength and knowledge. When my writing and my life actually come together, the safe positions of either/or will no longer pacify me, no longer contain me and hem me in.

In that unrevised talk, I had actually misused my sources. What they were saying to me, if I had listened, was pretty simple: don't follow us, don't reproduce what we have produced, don't live life from secondary sources like us, don't disappear. I hear Bob Scholes' and David Bartholomae's voices telling me to answer them, to speak back to them, to use them and make them anew. In a word, they say: revise me. The language lesson starts to make sense, finally: by confronting these authorial voices, I find the power to understand and gain access to my own ideas. Against all the voices I embody—the voices heard, read, whispered to me from off-stage—I must bring a voice of my own. I must enter the dialogue on my own authority, knowing that other voices have enabled mine, but no longer can I subordinate mine to theirs.

The voices I embody encourage me to show up as a writer and to bring the courage of my own authority into my classroom. I have also learned about the dangers of submission from observing the struggles of my own students. When they write about their lives, they write with confidence. As soon as they begin to turn their attention toward outside sources, they too lose confidence, defer to the voice of the academy, and write in the voice of Everystudent to an audience they think of as Everyteacher. They disguise themselves in the weighty, imponderable voice of acquired authority: "In today's society," for instance, or "Since the beginning of civilization mankind has. . . ." Or, as one student wrote about authority itself, "In attempting to investigate the origins of authority of the group, we must first decide exactly what we mean by authority."

In my workshops with teachers, the issue of authority, or deciding exactly what we mean by authority, always seems to be at the center of many heated conversations. Some colleagues are convinced that our writing programs should be about teaching academic writing. They see such programs as the welcome wagon of the academy, the

Holiday Inn where students lodge as they take holy orders. Some colleagues fear that if we don't control what students learn, don't teach them to write as scholars write, we aren't doing our job and some great red-legged scissor-man will cut off our thumbs. Again it is one of those either/or propositions: either we teach students to write academic essays or we teach them to write personal essays—and then who knows what might happen? The world might become uncontrollable: Students might start writing about their grandmother's death in an essay for a sociology course. Or even worse, something more uncontrollable, they might just write essays and publish them in professional journals claiming the authority to tell stories about their families and their colleagues. The uncontrollable world of ambiguity and uncertainty opens up, my colleagues imagine, as soon as the academic embraces the personal.

But, of course, our students are not empty vessels waiting to be filled with authorial intent. Given the opportunity to speak their own authority as writers, given a turn in the conversation, students can claim their stories as primary source material and transform their experiences into evidence. They might, if given enough encouragement, be empowered not to serve the academy and accommodate it, not to write in the persona of Everystudent, but rather to write essays that will change the academy. When we create opportunities for something to happen between the drafts, when we create writing exercises that allow students to work with sources of their own that can complicate and enrich their primary sources, they will find new ways to write scholarly essays that are exploratory, thoughtful, and reflective.

I want my students to know what writers know—to know something no researchers could ever find out no matter how many times they pin my students to the table, no matter how many protocols they tape. I want my students to know how to bring their life and their writing together.

Sometimes when I cook a chicken and my children scuffle over the one wishbone, I wish I had listened to my grandmother and cooked two. Usually, the child who gets the short end of the wishbone dissolves into tears of frustration and failure. Interjecting my own authority as the earth mother from central casting, I try to make their life better by asking: On whose authority is it that the short end can't get her wish? Why can't both of you, the long and the short ends, get your wishes?

My children, on cue, as if they too were brought in from central casting, roll their eyes as children are supposed to do when their mothers attempt to impose a way of seeing. They won't let me control the situation by interpreting it for them. My interpretation serves my needs, temporarily, for sibling compromise and resolution. They don't buy my story because they know something about the sheer thrill of the pull that they are not going to let *me* deny *them.* They will have to revise my self-serving story about compromise, just as they will have to revise the other stories I tell them. Between the drafts, as they get outside my authority, they too will have to question, and begin to see for themselves their own complicated legacy, their own trail of authority.

It *is* in the thrill of the pull between someone else's authority and our own, between submission and independence that we must discover how to define ourselves. In the uncertainty of that struggle, we have a chance of finding the voice of our own authority. Finding it, we can speak convincingly . . . at long last.

James A. Reither

Writing and Knowing: Toward Redefining the Writing Process

Who is this that darkeneth counsel by words without knowledge?

Job, 38:2

Composition Studies was transformed when theorists, researchers, and teachers of writing began trying to find out what actually happens when people write. Over the last decade or so, members of the discipline have striven primarily to discover and teach the special kinds of thinking, the processes, that occur during composing.[1] The goal has been to replace a prescriptive pedagogy (select a subject, formulate a thesis, outline, write, proofread) with a descriptive discipline whose members study and teach "process not product." Although the methodologies of process research have been challenged, its contributions to our understanding of composing have been applauded by theorists and practitioners alike. The consensus has generally been that process researchers have done a good job of answering the questions they have asked. Still, some are beginning to point to questions that, if they've been raised at all, have certainly not been answered.

Richard Larson, for example, has asked, "How does the impulse to write arise?" And, "How does the writer identify the elements needed for a solution [to a rhetorical problem], retrieve from memory or find in some other source(s) the items needed in the solution, and then test the trial solution to see whether it answers the problem?" (250-251).

Lee Odell, in a Four Cs paper entitled "Reading and Writing in the Workplace," observed that our questions about composing and inquiry processes have tended to stay "too close to the text." Odell's own research has led him to conclude that writing and inquiry are often (if not always) "socially collaborative" and that invention, discovery, and inquiry are closely tied to institutional relationships and strategies. Interpersonal and institutional contexts are, according to Odell, far more important than our literature has

From *College English* 47 (October 1985): 620-28.

1. Some well-known examples: Emig combined composing-aloud sessions, observation, and interviews to examine the composing processes of twelfth-grade writers. Perl used thinking-aloud protocols to uncover patterns or subroutines that occur and recur during composing. Flower and Hayes also tape and analyze thinking-aloud protocols, created by skilled and unskilled writers; their special concern has been to construct an accurate model of what happens as writers manage such subprocesses as planning, translating, and reviewing. Matsuhashi video-taped writers in the act of writing, paying special attention to planning and decision-making processes during pauses in composing. Sommers interviewed skilled and unskilled writers after they had revised pieces of writing, and then analyzed the pieces to determine the kinds of writer-concerns that motivated changes made from draft to draft. And, just as important, Murray has written to watch himself writing to learn what was happening as he wrote.

acknowledged, and he urges us to study more closely these contexts and strategies as necessary components of writing and inquiry processes.

Taking a different tack, Patricia Bizzell has divided composition theorists and researchers into two theoretical camps—those "interested in the structure of language-learning and thinking processes in their earliest state, prior to social influence"; and those "more interested in the social processes whereby language-learning and thinking capabilities are shaped and used in particular communities" (215). Bizzell laments the dominance of the "inner-directed" camp, arguing that Flower and Hayes, for example, pay too little attention to the role of knowledge in composing (229), and that "what looks like a cognitive difference [between unskilled and skilled writers often] turns out to have a large social component" (233). She thus argues that student writing difficulties often stem not from faulty or inefficient composing processes but, rather, from unfamiliarity with academic discourse conventions. "What is underdeveloped," she suggests, "is their knowledge of the ways experience is constituted and interpreted in the academic discourse community . . ." (230).

One result, as John Gage notes, is that the classical concept of *stasis* has all but vanished from the textbooks. The typical writing situation, according to Gage, is one in which reader and writer already share knowledge, "and it is the difference between what they know that motivates the need for communication—in both directions—and which therefore compels the act of writing" (2). Our practice, however, is to "send students in search of something to intend, . . . as if intention itself were subject to free choice. Students do not begin writing in order to fulfill an intention; rather, they are assumed to begin intentionless to search for something to want to say" (2).

What Larson, Odell, Bizzell, and Gage all point to is the tendency in composition studies to think of writing as a process which begins with an impulse to put words on paper; and the issues they raise should lead us to wonder if our thinking is not being severely limited by a concept of process that explains only the cognitive processes that occur as people write. Their questions and observations remind us that writing is not merely a process that occurs within contexts. That is, writing and what writers do during writing cannot be artificially separated from the social-rhetorical situations in which writing gets done, from the conditions that enable writers to do what they do, and from the motives writers have for doing what they do. Writing is not to context what a fried egg is to its pan.[2] Writing is, in fact, one of those processes which, in its use, *creates* and *constitutes* its own contexts.

Assisted, however, by the notion that writing is itself a mode of learning and knowing, and by the popularity of such developments as the attacks on "English" (with the concomitant emphasis on the values of expressive writing), process research—precisely because it has taught us so much—has bewitched and beguiled us into thinking of writing as a self-contained process that evolves essentially out of a relationship between writers and their emerging texts. That is, we conceptualize and teach writing on the "model of the individual writer shaping thought through language" (Bazerman, "Relationship" 657), as if the process began in the writer (perhaps with an experience of cognitive dissonance) and not in the writer's relationship to the world. In this truncated view, all writing—whether the writer is a seasoned veteran or a "placidly inexperienced

2. I owe the metaphor to my colleague Alan W. Mason (personal communication).

nineteen-year-old'' (Schor 72)—begins naturally and properly with probing the contents of the memory and the mind to discover the information, ideas, and language that are the substance of writing. This model of what happens when people write does not include, at least not centrally, any substantive coming to know beyond that which occurs as writers probe their own present experience and knowledge. Composition studies does not seriously attend to the ways writers know what other people know or to the ways mutual knowing motivates writing—does not seriously attend, that is, to the knowing without which cognitive dissonance is impossible.

The upshot is that we proceed as if students come to us already widely-experienced, widely-read, well-informed beings who need only learn how to do the kinds of thinking that will enable them to probe their experience and knowledge to discover what Rohman calls the ''writing ideas'' (106) for their compositions. We teach them to look heuristically into their own hearts, experiences, long-term memories, information- and idea-banks to discover what they have to say on the assigned or chosen subject. In so doing, we send several obviously problematic messages. One, identified by Bizzell, is that ''once students are capable of cognitively sophisticated thinking and writing, they are ready to tackle the problems of a particular writing situation'' (217). Another is that composing can be learned and done outside of full participation in the knowledge/discourse communities that motivate writing. Another is that other kinds of learning which can and do impel and give substance to writing—those, for example, that result from deliberate, purposeful learning through observation, reading, research, inquiry—are not really part of writing.[3] Yet another is that those kinds of learning have already occurred sufficiently to impel and ''authorize'' writing. That is, writers do not need to know what they are talking about: they can learn what they are talking about as they compose; they can write their way out of their ignorance.

We need to broaden our concept of what happens when people write. Writing is clearly a more multi-dimensioned process than current theory and practice would have us believe, and one that begins long before it is appropriate to commence working with strategies for invention. If we are going to teach our students to *need* to write, we will have to know much more than we do about the kinds of contexts that conduce—sometimes even force, certainly enable—the impulse to write. The ''micro-theory'' of process now current in composition studies needs to be expanded into a ''macro-theory'' encompassing activities, processes, and kinds of knowing that come into play long before the impulse to write is even possible.

To bring about that expansion, we need to press some new questions; and we need to know more than we now know, not only about cognitive processes during composing, but also about processes involved in coming to know generally. The focus of composition studies is presently on the first three of the five parts of classical rhetoric—on invention, arrangement, and style. It is time to look for ways to bring *stasis* back into the process and to learn more about its role in writing. We should use case studies,

3. This reductive notion of writing allows one widely-adopted composition textbook, Cowan and Cowan's *Writing*, to advise students that in writing a research paper ''you have to have a large number of skills—some writing skills, some nonwriting'' (428). Students learning the shape and scope of the writing process from this textbook are advised that using the library, taking notes, incorporating notes into an essay, documenting sources, and using appropriate research paper forms are ''nonwriting skills'' (428). Inquiry outside the mind and memory of the writer, and the knowing required for conducting such inquiry, are not necessarily related and therefore readily separable from ''writing.''

ethnographic studies, longitudinal studies, textual analysis, thinking-aloud protocol analysis, to answer such questions as these: What is the precise role in composing of substantive knowing—of concentrated participation in a knowledge/discourse community; of, simply, a fund of information on and ideas about the subject at hand? What, in this regard, is the precise relation between writing and reading? Where do we get our language for talking about things? What exactly *are* discourse conventions,[4] where do they come from, and how do we learn them? Are writers who know a great deal—who have engaged in direct and indirect sorts of inquiry within specific knowledge/discourse communities—likely to be better or different writers? Are writers who *know how* to find out likely to be better or different? What happens when people conduct inquiry and research? How *do* writers acquire the authority that impels writing? What *kinds* of knowing, and what kinds of knowing how, enable and assist writing?

Bizzell (238-239), Elaine Maimon, and Kenneth Bruffee (''Peer Tutoring'') all argue that we must analyze and teach the conventions of academic discourse. It seems clear, however, that that's not enough. To do that is to continue to confine students to the ''impoverished'' ''meanings carried by the conventional rules of language'' (Cooper 108). Bruffee, citing Richard Rorty, notes that ''In normal discourse . . . everyone agrees on the 'set of conventions about what counts as a relevant contribution, what counts as a question, what counts as having a good argument for that answer or a good criticism of it' '' (8). He goes on to say, rightly, that ''Not to have mastered the normal discourse of a discipline, no matter how many 'facts' or data one may know, is not to be knowledgeable in that discipline'' (9). But the obverse is equally true: What counts as a relevant contribution, question, answer, or criticism is determined not only by adherence to a set of discourse conventions, but also by such concerns as whether or not the contribution, question, answer, or criticism has already appeared in ''the literature''—whether or not it is to the point, relevant, or timely. A writer addressing dead issues, posing questions already answered, or voicing irrelevant criticisms is judged ignorant and viewed as, at best, an initiate—not yet an insider, not yet a full member of the discipline. Rather more basically, what counts as relevant is a contribution in which the writer's version of ''the facts of the matter'' accords with the version held in general by the community addressed by the writing.

To belong to a discourse community is to belong to a knowledge community—an ''inquiry community''; and the ways things are talked or written about are no more vital than the content of what's talked and written about. As Bruffee says, ''Ordinary people write to inform and convince other people within the writer's own community . . .'' (8). Because that's true, we must think not merely in terms of analysis and explanation; we must also think in terms of the other kinds of knowing required to belong to a community. We need to extend our understanding of the process of writing so that it will include not only experience- and memory-probing activities, but also inquiry strategies and techniques that will enable students to search beyond their own limited present experience and knowledge. We need to help students learn how to do the kinds of learning that will allow them, in their writing, to use what they *can* know, through effective inquiry, rather than suffer the limits of what they already know. We need to

4. In this regard, see Bazerman, ''What Written Knowledge Does.'' See also, on a different level, the two textbooks that have come out of the Beaver College writing-across-the-curriculum program: Maimon, et al., *Writing in the Arts and Sciences* and *Readings in the Arts and Sciences*.

bring curiosity, the ability to conduct productive inquiry, and an obligation for substantive knowing into our model of the process of writing. To do that, we need to find ways to immerse writing students in academic knowledge/discourse communities so they can write from within those communities.

The writing-across-the-curriculum movement, when it's done well, seems to have a chance of doing that. So also does Bruffee's own collaborative learning, if it can be untied from the notion of peer tutoring. As matters now stand, however, neither of these adequately addresses the problem of teaching students how to come to know so they can write literally as "knowledgeable peers" (Bruffee, "Peer Tutoring" 6) in academic communities. Neither gives students opportunities to "indwell" (Polanyi) an actual academic knowledge/discourse community, to learn, from the inside, its major questions, its governing assumptions, its language, its research methods, its evidential contexts, its forms, its discourse conventions, its major authors and its major texts—that is, its knowledge and its modes of knowing. Only this kind of immersion has a real chance of giving substance to their coming to know through composing.

The title of a course in which this immersion is to occur does not really matter. Neither does the name of the discipline or department in which the course is taught. It need not be a writing course. (In fact, obviously, this immersion need not occur in the context of a course at all. Most of us learned to do what we do on our own—perhaps in spite of the courses we took—and some students continue to do the same.) What does matter is that the course should be "organized as a collaborative investigation of a scholarly field rather than the delivery of a body of knowledge."[5]

As I have claimed above, discourse communities are also knowledge communities. The business of knowledge communities is inquiry—coming to know. In academia, inquiry necessarily begins with reading in the literature of a "scholarly field" (which may be almost anything: rhetoric or evolution, for instance; or deviant behavior, the literature of eighteenth-century England, the comedies of Shakespeare, Islamic religions, literacy, and so on). Because, in an essential way, the literature of a scholarly field *is* the scholarly field, reading in that literature is elemental to all other kinds and levels of investigation, including writing; and for all of us, but particularly for students, reading in the literature normally means library research. Furthermore, academic writing, reading, and inquiry are collaborative, social acts, social processes,[6] which not only result in, but also—and this is crucial—result *from,* social products: writing processes and written products are both elements of the *same social process*. Hence, academic writing, reading, and inquiry are inseparably linked; and all three are learned not by doing any one alone, but by doing them all at the same time. To "teach writing" is thus necessarily to ground writing in reading and inquiry.

In general terms, then, this immersion—this initiation—should image in important

---

5. Russell A. Hunt, my colleague at St. Thomas University, phrased it this way in a course description.

6. Bizzell's article and Bruffee's "Peer Tutoring" (or his recontextualization of that article, "Collaborative Writing and the 'Conversation of Mankind,' ") are important here, not only for their discussions of the social grounding of writing, but also for their references to much of the important literature in this particular scholarly field. See (for example) the following theoretical works: Fleck, *Genesis and Development of a Scientific Fact;* Rorty, *Philosophy and the Mirror of Nature,* esp. Part III; Fish, *Is There a Text in This Class?;* and, most important, Kuhn, *The Structure of Scientific Revolutions.* Finally, for a sampling of various kinds of research in this area, see Bazerman's "What Written Knowledge Does"; the work of Odell and Goswami—for example, "Writing in a Nonacademic Setting"; and Myers, "Texts as Knowledge Claims."

ways the "real world" of active, workaday academic inquirers. The course most effectively operates as a workshop[7] in which students read and write not merely for their teacher, but for themselves and for each other. In fact, students and teachers function best as co-investigators, with reading and writing being used collaboratively to conduct the inquiry. Organizing a course in this way allows an incredible range of reading activities—in everything from bibliographies to books; and a similar range of writing activities—from jotting down call numbers to writing formal articles of the sorts they are reading. What matters is that this should be language in use. In such a context, writing, reading, and inquiry are evaluated according to their pragmatic utility: the important question is not "How good is it?" but, instead, "To what extent and how effectively does it contribute to and further the investigation?" The inquiry is made manageable in the same way all such inquiries are made manageable, not by "choosing" and "focusing" a topic, but by seeking answers to the questions which impel the investigation.

Out of this immersion in academic inquiry and out of the ways they see themselves and others (both their immediate peers and those who have authored the literature of the field) using reading and writing to conduct the inquiry, students can construct appropriate models. That is, they can see effective and ineffective writing, reading, and inquiry conventions, strategies, and behaviors at work—not just as those conventions and behaviors can be inferred by reading in the literature, but also as they are evolved and used by their teachers and *each other*. Student and teacher roles in the workshop evolve out of their own participation in the investigation: reading and writing; exchanging and using each others' information, ideas, notes, annotations, sources; defining goals and making plans; applying "truth-seeking procedures" (Bach and Harnish 43); bringing to bear topic and world knowledge to conduct what Bereiter and Scardamalia call "reflective inquiry" (5-6).

At the core of composition studies is the virtually unchallenged conviction that what we have to study and what we have to teach is "process not product." By process, however, we presently mean something that encourages in our students the notion that through writing they can, like Plato's Gorgias, "answer any question that is put to [them]" (20). Because we routinely put our students in arhetorical situations in which they can only write out of ignorance, they have little choice but to "hunt more after words than matter" (Bacon 29), and we stand open to the charge that we advocate "mere rhetoric" over writing informed by a profound relationship between writers and their worlds. It is time to redefine the writing process so that substantive social knowing is given due prominence in both our thinking and our teaching.

Works Cited

Bach, K., and R. M. Harnish. *Linguistic Communication and Speech Acts*. Cambridge: MIT P, 1979.

---

7. For a model of the kind of workshop this might be, see Knoblauch and Brannon. A major difference between their ideal workshop and mine is that I would embed the discourse community of the workshop in the socially-constructed knowing available in the record of the larger conversation going on in the literature of the scholarly field being investigated.

Bacon, Francis. *The Advancement of Learning*. Oxford: Oxford UP, 1951.

Bazerman, Charles. "A Relationship Between Reading and Writing: The Conversational Model." *College English* 41 (1980): 656-661.

———. "What Written Knowledge Does: Three Examples of Academic Discourse." *Philosophy of the Social Sciences* 11 (1981): 361-387.

Bereiter, Carl, and Marlene Scardamalia. "Levels of Inquiry in Writing Research." *Research on Writing: Principles and Methods*. Eds. Peter Mosenthal, Lynne Tamor, and Sean A. Walmsley. New York: Longman, 1983. 3-25.

Bizzell, Patricia. "Cognition, Convention, and Certainty: What We Need to Know About Writing." *PRE/TEXT* 3 (1982): 213-243.

Bruffee, Kenneth A. "Collaborative Learning and the 'Conversation of Mankind.' " *College English* 46 (1984): 635-652.

———. "Peer Tutoring and the 'Conversation of Mankind.' " *Writing Centers: Theory and Administration*. Ed. Gary A. Olson. Urbana: NCTE, 1984. 3-15.

Cooper, Marilyn M. "Context as Vehicle: Implicatures in Writing." *What Writers Know: The Language, Process, and Structure of Written Discourse*. Ed. Martin Nystrand. New York: Academic, 1982. 105-128.

Cowan, Gregory, and Elizabeth Cowan. *Writing*. New York: Wiley, 1980.

Emig, Janet. *The Composing Processes of Twelfth Graders*. Urbana: NCTE, 1971.

Fish, Stanley. *Is There a Text in This Class?: The Authority of Interpretive Communities*. Cambridge: Harvard UP, 1980.

Fleck, Ludwik. *Genesis and Development of a Scientific Fact*. 1935. Chicago: U of Chicago P, 1979.

Flower, Linda, and John R. Hayes. "A Cognitive Process Theory of Writing." *College Composition and Communication* 32 (1981): 365-387.

Gage, John. "Towards an Epistemology of Composition." *Journal of Advanced Composition* 2 (1981): 1-9.

Knoblauch, C. H., and Lil Brannon. "Modern Rhetoric in the Classroom: Making Meaning Matter." *Rhetorical Traditions and the Teaching of Writing*. Upper Montclair, NJ: Boynton/Cook, 1984. 98-117.

Kuhn, Thomas. *The Structure of Scientific Revolutions*. 2nd ed., enlarged. Chicago: U of Chicago P, 1970.

Larson, Richard L. "The Writer's Mind: Recent Research and Unanswered Questions." *The Writer's Mind: Writing as a Mode of Thinking*. Ed. Janice M. Hays, et al. Urbana: NCTE, 1983. 239-251.

Maimon, Elaine P. "Maps and Genres: Exploring Connections in the Arts and Sciences." *Composition and Literature: Bridging the Gap*. Ed. Winifred Bryan Horner. Chicago: U of Chicago P, 1983. 110-125.

Maimon, Elaine P., et al. *Readings in the Arts and Sciences*. Boston: Little, 1984.

———. *Writing in the Arts and Sciences*. Cambridge, MA: Winthrop, 1981.

Matsuhashi, Ann. "Pausing and Planning: The Tempo of Written Discourse Production." *Research in the Teaching of English* 15 (1981): 113-134.

Murray, Donald M. "Write Before Writing." *College Composition and Communication* 29 (1978): 375-381.

———. "Writing as Process: How Writing Finds Its Own Meaning." *Eight Approaches to Teaching Composition*. Ed. Timothy R. Donovan and Ben W. McClelland. Urbana: NCTE, 1980. 3-20.

Myers, Greg. ''Texts as Knowledge Claims: The Social Construction of Two Biology Articles.'' *Social Studies of Science* 15 (1985): 593-630.

Odell, Lee. ''Reading and Writing in the Workplace.'' Conference on College Composition and Communication. New York City, 31 March 1984.

Odell, Lee, and Dixie Goswami. ''Writing in a Nonacademic Setting.'' *New Directions in Composition Research.* Ed. Richard Beach and Lillian S. Bridwell. New York: Guilford, 1984. 233-258.

Perl, Sondra. ''Understanding Composing.'' *College Composition and Communication* 31 (1980): 363-369.

Plato. *Gorgias.* Trans. Walter Hamilton. Harmondsworth: Penguin, 1971.

Polanyi, Michael. *Personal Knowledge: Towards a Post-Critical Philosophy.* Chicago: U of Chicago P, 1962.

Rohman, D. Gordon. ''Pre-writing: The Stage of Discovery in the Writing Process.'' *College Composition and Communication* 16 (1965): 106-112.

Rorty, Richard. *Philosophy and the Mirror of Nature.* Princeton: Princeton UP, 1979.

Schor, Sandra. ''Style Through Control: The Pleasures of the Beginning Writer.'' *Linguistics, Stylistics, and the Teaching of Composition.* Ed. Donald McQuade. Akron, OH: L and S, 1979. 72-80.

Sommers, Nancy. ''Revision Strategies of Student Writers and Experienced Adult Writers.'' *College Composition and Communication* 31 (1980): 378-388.

ASSIGNING AND RESPONDING

Christopher C. Burnham

Crumbling Metaphors: Integrating Heart and Brain Through Structured Journals

Beginning college writers use a wide range of cognitive skills; however, many do not use them systematically or rigorously. My purpose here is to present two journal exercises teachers can use to help their students develop and apply their cognitive skills. These exercises help students integrate what many consider to be dichotomous and frequently contradictory activities—feeling and thinking. George Lakoff and Mark Johnson's *Metaphors We Live By* (Chicago: University of Chicago Press, 1980) will provide the theoretical framework to explain the integrating process.

Structured Personal and Academic Journals

Teachers can introduce students to system and rigor through structured journals. These journals should attend to both personal and academic development. In *Writing from the Inside Out* (Harcourt, 1989), I present both kinds of journals as they could be used in a beginning writing course. The "Personal Development Journal" uses a sequence of expressive writing exercises to help students to examine the relation between past and present and to speculate upon their futures. A personal, ungraded journal, it offers them a private, safe place to take risks and examine their lives honestly. Working through the journal, students formulate and analyze statements about themselves and their experience in order to discover and subsequently evaluate the more or less conscious value systems that, according to William Perry in *Forms of Intellectual and Ethical Development in the College Years: A Scheme* (Holt, 1968), direct and limit their abilities to act and learn.

Each exercise in the 14-day sequence repeats a pattern. Students first generate material, then structure it as a metaphor, dialogue, or narrative, and, finally, complete an evaluation ending with an action statement. The structure employs the full psychological and linguistic power of expressive writing as students search the self in order to discover and formulate meaning (Christopher Burnham, "Expressive Writing," *Focuses* 2 [Spring 1989]: 5-18). The entire journal works on the principle that awareness precedes evaluation and that evaluation encourages change. Writing the journal enables students

From *College Composition and Communication* 43 (December 1992): 508-15.

to gain some distance and objectivity toward their experience, to articulate their values and beliefs, and to gain control of their lives and begin making informed choices.

The second journal, "The Learning Log," a structured academic journal, causes students to become aware of and to practice and apply the various cognitive skills needed for the higher-level learning required in the university. These activities include abstracting and summarizing, classifying and categorizing, formulating questions and developing applications, and using critical thinking and creative problem-solving strategies. The goal is to make students aware of the processes used in critical thinking and creative problem solving and to provide them a repertoire of specific strategies, a bag of tricks, to use when they are confronted with a problem or assignment. Both journals encourage students to integrate—to make connections between feeling and thinking, writing and acting, and going to school and living their lives.

The "Purer Mind"

Integration is the point. The journal writing connects what in the lives of many students have become seemingly unrelated activities—thinking and feeling—and unconnected realms—the public and the private. Structured journals can begin to repair the fragmentation so many experience in contemporary life. Contemporary theories of language, cognition, and moral development inform the sequence and structure of the journals, but the journals also tap an older source, the theories of creativity and imagination developed by Coleridge and Wordsworth. Throughout their work they concentrate on both the need for and process of integration, accomplished through the imagination. In the *Biographia Literaria,* Coleridge defines imagination as the "esemplastic power" through which we create new concepts and metaphors from sensation and experience and lead ourselves through the mundane to the divine. This is the analog of the process of integration. For Wordsworth, poetry is emotion recollected in tranquillity. He describes the poetic process in "Tintern Abbey": "In hours of weariness, sensations sweet/Felt in the blood, and felt in the heart;/And passing even into my purer mind,/ With tranquil restoration" (11.27-30).

The intellect, fired by the imagination, transforms sensation from immediate experience, "in blood" and "heart," into concept and category in the intellect, "the purer mind," resolving the dichotomies. Integration. Each student, each single separate mind in its particular context, can use the journal to complete the same process—identify the parts, establish their connections, and transform them into a whole. Structured journal exercises can demonstrate to even the greatest skeptic, in some instances the students themselves, the power of the mind at work.

Student responses to one of the personal-journal exercises will demonstrate integration at work. "Centering" exercises involve students in a simple though highly structured exercise in which metaphor-making provides a means for examining and assessing the students' current personal context. First, students complete a relaxation exercise to foster concentration, then they brainstorm a list of the sensations they have been experiencing recently, noting briefly the sources of these sensations. After listing a number of these sensations, students select one and use it to develop a metaphor that describes their current life. Finally, students evaluate the metaphors, analyzing them to determine what lessons the metaphors offer about their lives.

They have little difficulty creating metaphors that capture their reality for the moment, freezing it so they can investigate, understand, and, after evaluation, use their analysis of the metaphors to guide subsequent behavior. For example, a sophomore engineering major reviewed her list of emotions, a list characterized by diffuseness rooted in seemingly unrelated activities and course work—studying Spanish, writing essays, factoring equations—and portrayed herself as "a bowl of A B C vegetable soup." She mentions in her evaluation that she feels overloaded by all the information she is getting, "but the vegetables are calming down the feelings—I'm beginning to think more clearly." From the confusion comes a sense of calm, a direction that accepts temporary disorientation for long-term gain. Sometimes these metaphors contain conceits. A first-year male, previously a big man on campus who now feels like just another face in the thousands roaming the campus, portrayed himself as "the lead dog of a racing team sold to a business as a watchdog." Though egocentric, these metaphors demonstrate the power of minds at work in our students. Sharp minds are at work shaping and structuring sensation in even problematic metaphors. Whether successful from a literary viewpoint or not, these metaphors accomplish their purpose, providing students material for fruitful reflection. A first-year female portrayed herself as a "mole who has burrowed deeper and deeper into the ground, only to be finally thrust out onto cold concrete." The structured journal exercises can free the purer mind in all our students, leading them to integrate sensation and intellect, helping them make sense of themselves.

"Conventional" and "Unconventional" Metaphors

According to Lakoff and Johnson, exercises such as the "Centering" exercise tap one of the basic functions of metaphor. Metaphor is not an extraordinary language activity; rather, it is an everyday use of language through which we structure reality. Everyday discourse brims with words that function as metaphors, that are icebergs of meaning. These common metaphors expose some facet of reality but hide other facets beneath the waterline. When examined, they allow us to make sense of our lives. Lakoff and Johnson use many examples through their book, but the one used most frequently, to demonstrate how metaphors structure reality is "argument is war." They demonstrate how our language about argument exposes the metaphorical organization of the concept and defines our behavior. For example, we *attack* each other's weak points. We *demolish* arguments; sometimes we even *shoot them down.* If successful, we *win* the argument (4).

As teachers of argument, we can appreciate Lakoff and Johnson's analysis; we use the same metaphorical terms when we teach our students. But as teachers of writing who understand the collaborative nature of learning and the need for cooperation as well as competition, we see the limits of the metaphor. As is the case with all conceptual thinking, metaphors, as Lakoff and Johnson demonstrate, highlight only parts of our experience of argument. They mask other parts. The focusing function is the root of coherence. We fit ourselves and our experience into the relevant aspects of certain conventional metaphors. Metaphors of this sort are culture bound. In this lies a lesson for teachers: we must remain aware of the metaphorical nature of our language and be sure we share sufficient context and culture with our students so we can communicate.

This is surely not a new idea, but one I have registered lately while examining the metaphors my students make. The same young woman who saw herself as a mole later portrayed herself as "Dorothy in the Wizard of Oz—my dream world and reality are constantly interwoven." I have long carried the prejudice that an Oz fan was necessarily soft-brained. I would entertain analogies drawn from the movie only skeptically. But encountering that metaphor when I did, while living in London during a sabbatical, I made a vital connection. There, despite daily adventures and access to culture and scholarly resources unmatched for my purposes in all the world, I understood "there's no place like home, there's no place like home" as I never had before.

For my student the metaphor functioned as a "conventional" metaphor expressing her reality, at least in that time and place. In my instance, the metaphor functioned as what Lakoff and Johnson call a novel or "unconventional" metaphor. It became strikingly meaningful to me only because I was living for a time in an Oz in which my imagination and reality comingled. My brain was swollen from eight hours of *Henry IV, I & II,* all day Saturday at the Old Vic; my behind ached from the marathon sit. Overstimulated, I longed for lazy days of warming myself under a hot New Mexico desert sun. At least I did as I read my student's metaphor.

Lakoff and Johnson analyze the dynamic by which novel or "unconventional" metaphors work (174-175). The example they use is "Life's . . . a tale told by an idiot, full of sound and fury, signifying nothing." We understand this metaphor because it evokes the conventional metaphor: life is a story, invoking the concepts of significance and coherence, characteristics our culture values. We make sense of our lives through reflection, remembering incidents according to a chronology as if one event led logically to another. But we necessarily remember selectively, revising our lives to emphasize significance and coherence. This is the case for us, to greater or lesser degrees. But not for Macbeth. The power of the metaphor, "life is a story told by an idiot without purpose," lies in our *trying* to understand, and through this developing a sympathy for Macbeth. Many of us know opportunism and ambition, and their result. Our conventional expectation of life is coherent narrative; Macbeth's experience is an idiot's mumblings. Afterwards we live differently, wondering what fate might bring to obstruct our own sense of coherence. Unconventional metaphors, then, serve a deconstructive function. They help us pierce the fog and haze of our normal language and discover the shore of meaning and a course of action.

Metaphor-making exercises in personal journals can be structured to involve both constructive and deconstructive functions. We discover the double edge only when we are made to examine our language, to assess, to rethink and evaluate the metaphor by explaining it. That's clear when the A B C vegetable-soup student works through the confusion of her life. The symbols she cannot read, the random alphabet letters, finally combine into the wholeness of soup. Her dissonance resolved, she is beginning to see clearly. So is the big man on campus who eventually writes a metaphor expressing a liberating vision: "a tiger, escaped from the zoo, finding my way back to the jungle." And the mole, later Dorothy, finally portrays herself as "a bear who woke up from hibernation too early but who is too busy anticipating spring to go back to sleep." The pattern is constructing, then deconstructing to establish meaning; creating metaphors, then crumbling them to make sense. The key is using both types and functions of metaphor. Moreover, this use of metaphor is normal; it is not the sole province of the very intelligent or highly imaginative. It is the human way.

Reading: Dichotomous Response

With some training and a little practice, students learn to use structured personal-journal exercises to create, then crumble metaphors—to construct then deconstruct meaning—allowing them to integrate blood and heart, sensation and passion, and enter a "purer mind," the intellect. Their work in personal journals can lead to increased self-awareness and openness to change. But what about more formal or public instances whose success depends upon integrating heart and mind, such as understanding literature?

Most of our students have mastered the mechanics of reading. They can isolate main ideas, distinguish detail from generalization, and write adequate if flat summaries. They have learned to read by mastering a set of discrete skills. They do not integrate because they assume reading is no more than the sum of such discrete skills as identifying vocabulary and main ideas and drawing inferences and generalizations. Standardized reading tests, such as the Nelson-Denny, have turned reading into a positivistic science. Reading as science, as manipulation of discrete skills, frustrates integration. Moreover, it perpetuates the heart/brain, feeling/thinking dichotomies. Borrowing Wordsworth's lines from "The Tables Turned," our institutionalized approach to reading, "Misshapes the beauteous forms of things:—/We murder to dissect" (11.26-28).

The dichotomy surfaces when we ask students to respond to literature. Typically, they respond "I like it" or "It means." They do not see that feeling and thinking—sensation and intellect—complete each other. Further, the "I like it" response remains subjective, entirely personal, while the "It means" response pretends to the objective, generally echoing the voice of authority—what the teacher, the "Introduction," or the criticism says. Reliance on external authority reveals another dichotomy—self versus other; generally the uninformed, incapable voice of the student opposed to the voices of authority. Students have learned through experience that teachers often value the objective and authoritative, sometimes to the exclusion of the subjective. Students survive through slavish imitation of the external authoritative voices.

Response Through Metaphor

Our students' responses to literature will reflect their subjective/objective, self/authority dichotomies until they are trained otherwise. Structured academic-journal exercises in which readers create metaphors to respond to literature have helped many students begin to integrate the subjective and the objective, the personal and the public. The exercise is simple. Students, already familiar with metaphor-making from their work with the "Centering" exercises, create metaphors for the work or the writer. If students are having difficulty, providing them a formulaic opening, for example, "*Walden* is like . . ." or "Thoreau is . . . ," can get them started. Having created a metaphor, they then write a paragraph explaining the metaphor.

How does this metaphor exercise work? In the framework provided by Lakoff and Johnson, the metaphor exercise requires shuttling back and forth between constructive and deconstructive activities. In this shuttling, this interaction, lies integration. In fact, the procedure fosters integration on two levels. In order to create the metaphor, students have to have come to understand the literature in a personal sense. The metaphor is initially a constructive personal response tying the experience of reading with previous experience, combining new information with old, linking sensation and intellect. This is integration on one level. Having created the metaphor, they must then explain it. Here is the second level of integration. The explanation requires crumbling the metaphor, penetrating the veil of normal language. Personal meaning must be deconstructed, articulated, and elaborated so others can understand. The private is made public; the subjective bridged to the objective. Integration.

Examples: Metaphors for Understanding

The following examples demonstrate this phenomenon. They were collected in an advanced composition course using a structured academic journal, a learning log, as the source for extended personal and academic essays. The following extracts represent raw writing but admirable thinking about *Walden.* They represent a range of responses. One uses an extended metaphor to capture *Walden*'s substance and personal impact. The other two represent more personal responses to *Walden.* Each, however, demonstrates integration.

The first example comes from a senior male, taking a double major in psychology and communication. In interviews, he discussed how *Walden* gave him a way of thinking about a personal dilemma he was experiencing: a conflict between the life paths his peers were choosing—well-paid jobs and marriage—and his urge to break the mold through writing and working in theatre. He had been aware of this conflict, but until reading *Walden* he could not articulate it. He had encountered *Walden* before but reported that previously, "it always put me to sleep." This time, however, he seems not to have fallen asleep. His metaphor demonstrates an understanding of Thoreau's iconoclasm:

Metaphor: *Walden* is a green thumbprint placed near the top of the Sears Tower—splatter painted among an infinite number of red lines.

Explanation: *Walden* is one man's defiant explanation of the way he feels about himself and his relationship with his fellow man. The thumbprint is green because it contrasts with the popular color of red. It is unique, intricate, unlike any concept another man could compose. The tower is marked with an infinite number of uniform red lines—these represent the way the vast majority of the human race conduct their lives—conventional patterns which conform to preconceived notions and popular traditions. The thumbprint lies near the top of the tower where few others will ever see it, yet it sticks out like a beacon, proclaiming its existence. It is *placed* there artfully and intentionally, unlike the random but uniform splatterings of others.

The themes of self-reliance and the deliberate life sound through this metaphor and explanation, doing justice to Thoreau's work. Placing the thumbprint in Chicago shows some of the bridging going on. This student is not worrying about rural Massachusetts; he lives in contemporary America. His metaphor, statements in interviews, subsequent journal work, and essays all suggest that the student has accepted the challenge of *Walden* and will strive to make his own life a distinctive thumbprint on the tower.

The other examples demonstrate a different approach but nevertheless reflect understanding that signals integration. These personal responses demonstrate how the writers were affected by a close reading of the book.

The following exercise was completed by a female, a sophomore double-majoring in English and journalism. She had never been exposed to *Walden* before. Though she had heard about it, she had no preconceptions, no sense of what authorities said the book meant:

Metaphor: *Walden*—a funny bone.

Explanation: In *Walden,* Thoreau uses humor. But the book also has a painful side. Rereading selected parts of the book, I was able to pick out the more subtle humor and it pained me to realize I do some of the things he despised or used sarcasm against.

The key words in this metaphor are "it pained me." There is a difference between a phrase like "I noticed" or "it dawned on me" and "it pained me." Here is someone responding on a very personal level, trying to deal with her own disappointment in discovering that in laughing with Thoreau at the John Fields of this world, those "born to be poor," who live lives of "quiet desperation" and "resignation," that she was laughing at herself. Here is the bridge between self and other. This book is about me! She has crumbled the distance between her life and the book. Such recognition warms writers and teachers of writing alike; it signals integration.

The final example was written by a male, a junior majoring in government and minoring in professional communication. He had not read *Walden* before though he was told about *Walden* in a high-school literature survey. Sense the urgency in this metaphor:

Metaphor: In writing *Walden* Thoreau made an alarm clock that cannot be shut off.

Explanation: The clock buzzes and buzzes trying to shake us to where we will wake up. Throughout the book Thoreau is telling us to keep awake. If we are asleep we must wake up and notice things. To see things with our own eyes, to maybe rub out the glaze that has settled on our eyes from sleeping, or perhaps to take off the blinders that family or school or society put on us—whatever—but do WAKE UP!

For this student *Walden* is buzzing even now, stirring him from slumber.

These examples show how we can use structured personal- and academic-journal exercises to help our students create and then crumble metaphors, helping them bridge the gap between heart and brain, encouraging integration. In their "Afterword," Lakoff and Johnson assert that when dealing with experience, metaphor functions as "a sense, like seeing or touching or hearing, with metaphors providing the only ways to perceive and experience much of the world" (239). As we teach our students to use metaphor to understand their own lives and literature, we are only making conscious and purposeful a natural human activity. And, in helping them integrate sensation and intellect, we help them enter "the purer mind."

Richard L. Larson

The "Research Paper" in the Writing Course: A Non-Form of Writing

Let me begin by assuring you that I do not oppose the assumption that student writers in academic and professional settings, whether they be freshmen or sophomores or students in secondary school or intend to be journalists or lawyers or scholars or whatever, should engage in research. I think they should engage in research, and that appropriately informed people should help them learn to engage in research in whatever field these writers happen to be studying. Nor do I deny the axiom that writing should incorporate the citation of the writer's sources of information when those sources are not common knowledge. I think that writers must incorporate into their writing the citation of their sources—and they must also incorporate the thoughtful, perceptive evaluation of those sources and of the contribution that those sources might have made to the writer's thinking. Nor do I oppose the assumption that a writer should make the use of appropriate sources a regular activity in the process of composing. I share the assumption that writers should identify, explore, evaluate, and draw upon appropriate sources as an integral step in what today we think of as the composing process.

In fact, let me begin with some positive values. On my campus, the Department of English has just decided to request a change in the description of its second-semester freshman course. The old description read as follows:

> This course emphasizes the writing of formal analytic essays and the basic methods of research common to various academic disciplines. Students will write frequently in and out of class. By the close of the semester, students will demonstrate mastery of the formal expository essay and the research paper. Individual conferences.

The department is asking our curriculum committee to have the description read:

> This course emphasizes the writing of analytical essays and the methods of inquiry common to various academic disciplines. Students will write frequently in and out of class. By the close of the semester, students will demonstrate their ability to write essays incorporating references to suitable sources of information and to use appropriate methods of documentation. Individual conferences.

I applauded the department for requesting that change, and I wrote to the college curriculum committee to say so.

While thinking about this paper—to take another positive example—I received from

From *College English* 44 (December 1982): 811-16.

the University of Michigan Press a copy of the proofs of a forthcoming book titled *Researching American Culture: A Guide for Student Anthropologists,* sent to me because members of the English Composition Board of the University of Michigan had decided that the book might be of use as a supplementary text at Michigan in writing courses that emphasize writing in the academic disciplines. Along with essays by professional anthropologists presenting or discussing research in anthropology, the book includes several essays by students. In these essays the students, who had been instructed and guided by faculty in anthropology, report the results of research they have performed on aspects of American culture, from peer groups in high school to connections between consumption of alcohol and tipping in a restaurant, to mortuary customs, to sports in America. If anyone was in doubt about the point, the collection demonstrates that undergraduate students can conduct and report sensible, orderly, clear, and informative research in the discipline of anthropology. I am here to endorse, indeed to applaud, such work, not to question the wisdom of such collections as that from Michigan or to voice reservations about the capacity of undergraduates for research.

Why, then, an essay whose title makes clear a deep skepticism about "research papers"? First, because I believe that the generic "research paper" as a concept, and as a form of writing taught in a department of English, is not defensible. Second, because I believe that by saying that we teach the "research paper"—that is, by acting as if there is a generic concept defensibly entitled the "research paper"—we mislead students about the activities of both research and writing. I take up these propositions in order.

We would all agree to begin with, I think, that "research" is an activity in which one engages. Probably almost everyone reading this paper has engaged, at one time or another, in research. Most graduate seminars require research; most dissertations rely upon research, though of course many dissertations in English may also include analytical interpretation of texts written by one or more authors. Research can take many forms: systematically observing events, finding out what happens when one performs certain procedures in the laboratory, conducting interviews, tape-recording speakers' comments, asking human beings to utter aloud their thoughts while composing in writing or in another medium and noting what emerges, photographing phenomena (such as the light received in a telescope from planets and stars), watching the activities of people in groups, reading a person's letters and notes: all these are research. So, of course, is looking up information in a library or in newspaper files, or reading documents to which one has gained accesss under the Freedom of Information Act—though reading filed and catalogued documents is in many fields not the most important (it may be the least important) activity in which a "researcher" engages. We could probably define "research" generally as the seeking out of information new to the seeker, for a purpose, and we would probably agree that the researcher usually has to interpret, evaluate, and organize that information before it acquires value. And we would probably agree that the researcher has to present the fruits of his or her research, appropriately ordered and interpreted, in symbols that are intelligible to others, before that research can be evaluated and can have an effect. Most often, outside of mathematics and the sciences (and outside of those branches of philosophy that work with nonverbal symbolic notation), maybe also outside of music, that research is presented to others, orally or in writing, in a verbal language.

But research still is an activity; it furnishes the substance of much discourse and can furnish substance to almost any discourse except, possibly, one's personal reflections on one's own experience. But it is itself the subject—the substance—of no distinctively identifiable kind of writing. Research can inform virtually any writing or speaking if the author wishes it to do so; there is nothing of substance or content that differentiates one paper that draws on data from outside the author's own self from another such paper—nothing that can enable one to say that this paper is a "research paper" and that paper is not. (Indeed even an ordered, interpretive reporting of altogether personal experiences and responses can, if presented purposively, be a reporting of research.) I would assert therefore that the so-called "research paper," as a generic, cross-disciplinary term, has no conceptual or substantive identity. If almost any paper is potentially a paper incorporating the fruits of research, the term "research paper" has virtually no value as an identification of a kind of substance in a paper. Conceptually, the generic term "research paper" is for practical purposes meaningless. We can not usefully distinguish between "research papers" and non-research papers; we can distinguish only between papers that should have incorporated the fruits of research but did not, and those that did incorporate such results, or between those that reflect poor or inadequate research and those that reflect good or sufficient research. I would argue that most undergraduate papers reflect poor or inadequate research, and that our responsibility as instructors should be to assure that each student reflect in each paper the appropriate research, wisely conducted, for his or her subject.

I have already suggested that "research" can take a wide variety of forms, down to and including the ordered presentation of one's personal reflections and the interpretations of one's most direct experiences unmediated by interaction with others or by reference to identifiably external sources. (The form of research on composing known as "protocol analysis," or even the keeping of what some teachers of writing designate as a "process journal," if conducted by the giver of the protocol or by the writer while writing, might be such research.) If research can refer to almost any process by which data outside the immediate and purely personal experiences of the writer are gathered, then I suggest that just as the so-called "research paper" has no conceptual or substantive identity, neither does it have a procedural identity; the term does not necessarily designate any particular kind of data nor any preferred procedure for gathering data. I would argue that the so-called "research paper," as ordinarily taught by the kinds of texts I have reviewed, implicitly equates "research" with looking up books in the library and taking down information from those books. Even if there is going on in some departments of English instruction that gets beyond those narrow boundaries, the customary practices that I have observed for guiding the "research paper" assume a procedural identity for that paper that is, I think, nonexistent.

As the activity of research can take a wide variety of forms, so the presentation and use of research within discourse can take a wide variety of forms. Indeed I cannot imagine any identifiable design that any scholar in rhetoric has identified as a recurrent plan for arranging discourse which cannot incorporate the fruits of research, broadly construed. I am not aware of any kind or form of discourse, or any aim, identified by any student of rhetoric or any theorist of language or any investigator of discourse theory, that is distinguished primarily—or to any extent—by the presence of the fruits of "research" in its typical examples. One currently popular theoretical classification of discourse, that by James Kinneavy (*A Theory of Discourse* [Englewood Cliffs, N.J.: Prentice-Hall, 1971]), identifies some "aims" of discourse that might seem to furnish a home for papers based on research: "referential" and "exploratory" discourse. But, as I understand these aims, a piece of discourse does not require the presence of results of ordered "research" in order to fit into either of these classes, even though discourse incorporating the results of ordered research might fit there—as indeed it might under almost any other of Kinneavy's categories, including the category of "expressive" discourse. (All discourse is to a degree "expressive" anyway.) The other currently dominant categorization of examples of discourse—dominant even over Kinneavy's extensively discussed theory—is really a categorization based upon plans that organize discourse: narration (of completed events, of ongoing processes, of possible scenarios), causal analysis, comparison, analogy, and so on. None of these plans is differentiated from other plans by the presence within it of fruits from research; research can be presented, so far as I can see, according to any of these plans. And if one consults Frank J. D'Angelo's *A Conceptual Theory of Rhetoric* (Cambridge, Mass.: Winthrop, 1975) one will not find, if my memory serves me reliably, any category of rhetorical plan or any fundamental human cognitive processes—D'Angelo connects all rhetorical plans with human cognitive processes—that is defined by the presence of the fruits of research. If there is a particular rhetorical form that is defined by the presence of results from research, then, I have not seen an effort to define that form and to argue that the results of research are what identify it as a form. I conclude that the "research paper," as now taught, has no formal identity, as it has no substantive identity and no procedural identity.

For me, then, very little is gained by speaking about and teaching, as a generic concept, the so-called "research paper." If anything at all is gained, it is only the reminder that responsible writing normally depends on well-planned investigation of data. But much is lost by teaching the research paper in writing courses as a separately designated activity. For by teaching the generic "research paper" as a separate activity, instructors in writing signal to their students that there is a kind of writing that incorporates the results of research, and there are (by implication) many kinds of writing that do not and need not do so. "Research," students are allowed to infer, is a specialized activity that one engages in during a special course, or late in a regular semester or year, but that one does not ordinarily need to be concerned about and can indeed, for the most part, forget about. Designating the "research paper" as a separate project therefore seems to me to work against the purposes for which we allegedly teach the research paper: to help students familiarize themselves with ways of gathering, interpreting, drawing upon, and acknowledging data from outside themselves in their writing. By talking of the "research paper," that is, we undermine some of the very goals of our teaching.

We also meet two other, related difficulties. First, when we tend to present the "research paper" as in effect a paper based upon the use of the library, we misrepresent "research." Granted that a good deal of research in the field of literature is conducted in the library among books, much research that is still entitled to be called humanistic takes place outside the library. It can take place, as I mentioned earlier, wherever "protocol" research or writers' analyses of their composing processes take place; it can take place in the living room or study of an author who is being interviewed about his or her habits of working. It can take place in the home of an old farmer or rancher or weaver or potter who is telling a student about the legends or songs of his or her people, or about the historical process by which the speaker came from roots at home or abroad. Much research relies upon books, but books do not constitute the corpus of research data except possibly in one or two fields of study. If we teach the so-called "research paper" in such a way as to imply that all or almost all research is done in books and in libraries, we show our provincialism and degrade the research of many disciplines.

Second, though we pretend to prepare students to engage in the research appropriate to their chosen disciplines, we do not and cannot do so. Faculty in other fields may wish that we would relieve them of the responsibility of teaching their students to write about the research students do in those other fields, but I don't think that as teachers of English we can relieve them of that responsibility. Looking at the work of the students who contributed to the University of Michigan Press volume on *Researching American Culture,* I can't conceive myself giving useful direction to those students. I can't conceive myself showing them how to do the research they did, how to avoid pitfalls, assure representativeness of data, draw permissible inferences, and reach defensible conclusions. And, frankly, I can't conceive many teachers of English showing these students what they needed to know either. I can't conceive myself, or very many colleagues (other than trained teachers of technical writing), guiding a student toward a report of a scientific laboratory experiment that a teacher of science would find exemplary. I can't conceive myself or many colleagues guiding a student toward a well-designed experiment in psychology, with appropriate safeguards and controls and wise interpretation of quantitative and nonquantitative information. In each of these fields (indeed probably in each academic field) the term "research paper" may have some meaning—quite probably a meaning different from its meaning in other fields. Students in different fields do write papers reporting research. We in English have no business claiming to teach "research" when research in different academic disciplines works from distinctive assumptions and follows distinctive patterns of inquiry. Such distinctions in fact are what differentiate the disciplines. Most of us are trained in one discipline only and should be modest enough to admit it.

But let me repeat what I said when I started: that I don't come before you to urge that students of writing need not engage in "research." I think that they should engage in research. I think they should understand that in order to function as educated, informed men and women they have to engage in research, from the beginning of and throughout their work as writers. I think that they should know what research can embrace, and I think they should be encouraged to view research as broadly, and conduct it as imaginatively, as they can. I think they should be held accountable for their opinions and should be required to say, from evidence, why they believe what they assert. I think that they should be led to recognize that data from "research" will affect their entire lives, and that they should know how to evaluate such data as well as to gather them. And I think they should know their responsibilities for telling their listeners and readers where their data came from.

What I argue is that the profession of the teaching of English should abandon the concept of the generic "research paper"—that form of what a colleague of mine has called "messenger service" in which a student is told that for this one assignment, this one project, he or she has to go somewhere (usually the library), get out some materials, make some notes, and present them to the customer neatly wrapped in footnotes and bibliography tied together according to someone's notion of a style sheet. I argue that the generic ''research paper,'' so far as I am familiar with it, is a concept without an identity, and that to teach it is not only to misrepresent research but also quite often to pander to the wishes of faculty in other disciplines that we spare them a responsibility that they must accept. Teaching the generic ''research paper'' often represents a misguided notion of ''service'' to other departments. The best service we can render to those departments and to the students themselves, I would argue, is to insist that students recognize their continuing responsibility for looking attentively at their experiences; for seeking out, wherever it can be found, the information they need for the development of their ideas; and for putting such data at the service of every piece they write. That is one kind of service we can do to advance students' humanistic and liberal education.

Jeanne Fahnestock and Marie Secor

Teaching Argument: A Theory of Types

The climax of many composition courses is the argumentative essay, the last, longest, and most difficult assignment. An effective written argument requires all the expository skills the students have learned, and, even more, asks for a voice of authority and certainty that is often quite new to them. Aware of the difficulty and importance of argument, many composition programs are devoting more time to it, even an entire second course. At Penn State, for example, the second of our required composition courses is devoted entirely to written argument, out of our conviction that written argument brings together all other writing skills and prepares students for the kinds of writing tasks demanded in college courses and careers.

We know what we want our students to do by the end of our second course: write clear, orderly, convincing arguments which show respect for evidence, build in refutation, and accommodate their audience. The question is, how do we get them to do it? What is the wisest sequence of assignments? What and how much ancillary material should be brought in? The composition teacher setting up a course in argument has three basic approaches to choose from: the logical/analytic, the content/problem-solving, and the rhetorical/generative. All of these approaches teach the student something about argument, but each has problems. Our purpose here is to defend the rhetorical/generative approach as the one which reaches its goal most directly and most reliably.

The teacher who uses the logical/analytic approach in effect takes the logic book and its terminology into the classroom and introduces students to the square of opposition, the syllogisms categorical and hypothetical, the enthymeme, the fallacies, induction and deduction. It has not been demonstrated, however, that formal logic carries over into written argument. Formal logic, as Chaim Perelman and Stephen Toulmin have pointed out, is simply not the same as the logic of discourse;[1] students who become adept at manipulating fact statements in and out of syllogisms and Venn diagrams still may not have any idea how to construct a written argument on their own.

Another supposed borrowing from logic is the distinction between induction and deduction as forms of reasoning and therefore as distinct forms of written argument. Induction and deduction are sometimes seen to be as different as up and down, induction reaching a generalization from particulars and deduction affirming a particular from a

From *College Composition and Communication* 34 (February 1983): 20-30.

1. Chaim Perelman and L. Olbrechts-Tyteca, *The New Rhetoric: A Treatise on Argument* (Notre Dame, IN: University of Notre Dame Press, 1969), pp. 1-4; Stephen Toulmin, *The Uses of Argument* (Cambridge, England: Cambridge University Press, 1958), p. 146.

generalization. Actually the exact distinction between the two is a matter of some controversy. In his *Introduction to Logic* Irving M. Copi defines the two not as complementary forms of reasoning, but as reasoning toward a certain conclusion (deduction) and reasoning toward a probable conclusion (induction).[2] And Karl Popper in *Conjectures and Refutations* obliterates the distinction by showing that induction, as traditionally defined, is not valid.

> But in fact the belief that we can start with pure observations alone, without anything in the nature of a theory, is absurd; as may be illustrated by the story of the man who dedicated his life to natural science, wrote down everything he could observe, and bequeathed his priceless collection of observations to the Royal Society to be used as inductive evidence. This story should show us that though beetles may profitably be collected, observations may not.
>
> Twenty-five years ago I tried to bring home the same point to a group of physics students in Vienna by beginning a lecture with the following instructions: 'Take pencil and paper; carefully observe, and write down what you have observed!' They asked, of course, *what* I wanted them to observe. Clearly the instruction, 'Observe!' is absurd. . . . Observation is always selective. It needs a chosen object, a definite task, an interest, a point of view, a problem. And its description presupposes a descriptive language, with property words; it presupposes similarity of classification, which in its turn presupposes interests, points of view and problems.[3]

Thus according to Popper, the observations that supposedly lead to a conclusion are, in fact, controlled by a prior conclusion. There is no pure form of reasoning which goes "example + example = conclusion," as it is represented in many rhetorics. We cannot reason "x chow is vicious + y chow is vicious = z chow is vicious = most chows are vicious" unless we assume "Chows x, y and z are typical chows." The conclusion of a so-called inductive argument depends not on the number of examples but on their typicality. The reasoning in such an argument does not leap from particular to general but proceeds from an assumption of typicality and particular evidence (in this case three examples of vicious chows) to a conclusion. This process is not essentially different from deduction. Students are misled if they think their minds work in two gears, inductive forward and deductive reverse, or if they believe it is possible to argue purely from evidence without assumptions. But students who recognize the necessity for typical evidence can fruitfully consider whether their audience will accept their evidence as representative or whether they must explicitly argue that it is.

Another continually attractive if indirect way of teaching argument in the composition classroom is the content/problem-solving approach, which assumes that students will absorb the principles and methods of written argument simply by doing it. In such content-based courses, which may use a case book (now rare) or a group of related readings or even the lectures and readings of another course, the instructor may not even define the writing as argument. Instead, students write papers with "theses" which grow naturally out of their readings or are suggested by the instructor. Another variety

2. Irving M. Copi, *Introduction to Logic*, 5th ed. (New York: Macmillan, 1978), pp. 23-26.
3. Karl Popper, *Conjectures and Refutations: The Growth of Scientific Knowledge* (New York: Harper & Row, 1963), pp. 46-47.

of this approach is the problem-solving method, as in *Cases for Composition* (by John P. Field and Robert H. Weiss, Boston: Little, Brown & Co., 1979), which frames assignments not only by specifying topics but also by defining rhetorical situations. Students write their way out of problems, arguing in letters, memos, reports, and brief articles.

The content/problem-solving approach effectively approximates real-life writing situations which supply both purpose and content. Moreover, a course that teaches writing this way is attractive because the instructor can present for discussion a coherent body of material from philosophy, sociology, psychology or even literature; if such a course works well, invention is not a problem because students are directly stimulated by the content, and they do practice writing arguments. And at best students may learn a method of problem-solving which they can apply to other writing situations when their instructor is no longer suggesting topics nor the controlled reading stimulating invention. However, the content in such courses tends to crowd out the writing instruction or, increasingly, it is given away to the real experts in other departments and the composition teacher reduced to an overseer of the revision process, a police officer with a red pencil.

The composition course which does not organize itself around a body of content can take what we will call the "rhetorical/generative" approach and explicitly teach invention. Now the composition course devoted entirely to argument can turn to the classical sources which are still the only scheme of invention purely for argument. These sources (definition, comparison, cause and effect, and authority) do help students find premises for the proposals and evaluations they usually come up with when left on their own to generate theses for arguments. But the sources are less help when we ask students to take one step further back and support the very premises which the sources have generated. If, for example, the student wants to argue that "the federal government should not subsidize the airlines," thinking about definition might yield a premise like "because airline subsidies are a form of socialism," and thinking about cause and effect might yield a premise like "because once an industry is subsidized, the quality of its service deteriorates." But how is either of these premises to be supported? How does one actually argue for a categorization such as the first or a cause and effect relationship such as the second? To tell the student to continually reapply the four sources is rather discouraging advice. Thus while the classical sources are powerful aids to invention in the large-scale arguments that evaluations and proposals require, they do not help students construct smaller-scale supporting arguments.

It is possible to give students more specific aid in inventing arguments if we begin by distinguishing the basic types of arguments and the structures characteristic of each. We derived this approach when confronted by the variety of propositions our students volunteered as subjects for argument. After collecting scores of these, we found they could be sorted into four main groups answering the questions 1. "What is this thing?" 2. "What caused it or what effects does it have?" 3. "Is it good or bad?" and 4. "What should be done about it?" Propositions which answer these questions are, respectively, categorical propositions, causal statements, evaluations, and proposals. The thesis of any argument falls into one of these categories. The first two, which correspond to the classical sources of definition and cause and effect, demand their own forms of argument with distinctive structures. Arguments for the third and fourth, evaluations and proposals, combine the other two. If we take students through these four types of argument, from the simpler categorical proposition to the complex proposal, we have a coherent rationale for organizing a course in argument.

Any statement about the nature of things fixed in some moment of time can be cast as what logicians call a categorical proposition (CP), a sentence which places its subject in the category of its predicate. The pure form of a CP is

Subject	*Linking Verb*	*Predicate*
All art	is	an illusion
Caligula	was	a spoiled brat
Ballet dancers	are	really athletes

Statements about the nature of things do not always come in such neat packages, but even a proposition without a linking verb, like "Some dinosaurs cared for their young," is still a CP which could be recast into pure form, "Some dinosaurs were caring parents."

Whenever a CP is the thesis of an argument, it makes certain structural demands. Since supporting a CP is always a matter of showing that the subject belongs in the category of or has the attributes of the predicate, that predicate must be defined whenever its meaning cannot be assumed, and evidence or examples must be given to link the subject up with that predicate. The arguer for a CP, then, works under two constraints: the definition of the predicate must be acceptable to the audience and the evidence or examples about the subject must be convincing and verifiable. We can see these two constraints operating on the arguer constructing support for a CP like "America is a class society." For most audiences, the definition of "class society" cannot be assumed. If our arguer defined a "class society" as one in which people live in different sized towns, he or she could produce plenty of evidence that Americans do indeed live in towns small, medium, and large, but "class society" has been defined in what speakers of English would intuitively recognize as a completely unacceptable way. It may be a vague term, but there are some meanings it cannot have. On the other extreme, the writer could define "class society" more acceptably as "a society structured into clearly defined ranks, from peasantry to nobility," but where could he or she find the non-metaphoric American duke or serf? Obviously, the arguer must construct a definition which is acceptable to its audience while it fits real evidence.

But suppose a student writes a brief argument supporting a CP like "My roommate is generous." He or she will bypass definition and go straight to examples of the roommate's generosity: the lending of money, clothes, shampoo, and time. The student can go right to such evidence because he or she has a clear definition of "generous" in mind and cannot imagine any audience having a different one. Still a definition of "generous," whether or not articulated in the argument, controls the choice of examples. It was not the roommate's behavior which led to the label "generous," but a definition of "generous" which led to the categorization of the behavior. Because we tend to forget the controlling power of definition, we delude ourselves into thinking that the examples come first and lead inductively to the thesis when in fact the process goes the other way.

Once the student understands that definition and specific evidence are the structural requirements of a CP argument, several organizational options become available. The controlling definition can sit at the beginning of the argument, can emerge at the end, or can have its elements dispersed.[4] In this last option, the definition of the predicate is broken down into components, each supported with appropriate evidence. Take a CP like "Wilkie Collins's *Armadale* is a sensation novel." An arguer for this proposition might specify a multi-part definition of "sensation novel" which would supply the whole structure of the paper: "a sensation novel is characterized by its ominous setting, grotesque characters, suspenseful plot, and concern with the occult." Each of the elements from this definition becomes the predicate of its own CP (again requiring definition where necessary) and the topic sentence of its own paragraph," e.g. "*Armadale* has grotesque characters," "Collins dabbles in the occult in *Armadale*."

Once students have learned the fundamentals of the CP argument, they have the tools to support a comparison or a contrast as well. An arguer for a single comparison, "Kissinger is like Metternich," for instance, finds one or more traits that the two subjects have in common. "Both Kissinger and Metternich had no chauvinistic pride." This is simply a CP with a compound subject which can be divided into two simple CPs ("Kissinger had no chauvinistic pride," "Metternich had no chauvinistic pride"), each supported, as much as the audience requires, by definition and evidence.

A second type of proposition needs quite a different kind of argument. An assertion of cause and effect adds the dimension of time and is therefore not supported with definition but with another kind of ruling assumption, that of *agency,* a basic belief about what can cause what. Just as users of the same language share a set of definitions, so do people in the same culture share many causal assumptions. We have a common-sense understanding, for instance, of such natural agencies as light, heat and gravity, as well as many accepted human agencies whose operation we believe in as readily as we believe in the operation of physical law. Philosophers, psychologists, anthropologists and social scientists debate about what to call these agencies—motives, instincts, or learned patterns of behavior. But we recognize a believable appeal to the way human nature works, just as we recognize an appeal to the way physical nature works; we no more accept happiness as a motive for murder than we would accept the power of rocks to fly.

Definition and agency, then, are the warrants (to use Toulmin's term) behind the two basic kinds of arguments.[5] If we make a claim about the nature of things (a CP), we rely on an assumption about the nature of things, a definition. If we make a claim about causal relations, we rely on an assumption about what can cause what, an agency. And whether or not we articulate agency in a causal argument depends largely on audience. For example, if we argue that a significant cause of teenage vandalism is violence on TV, the agency between these two is imitation. Since most audiences will readily accept imitation as a human motive, we would not have to stop and argue for it. But if we claimed that wearing a mouth plate can improve athletic performance (*Sports Illustrated,* 2 June 1980), we will certainly have to explain agency. (The article did.)

---

4. Here perhaps is the only legitimate use of the terms "inductive" and "deductive" in written argument. They can be used to describe the organization of arguments, the deductive setting out the thesis at the beginning and the inductive disclosing it at the end.

5. Stephen Toulmin, finding the syllogism ambiguous, created a new pattern for analyzing arguments. In his terminology, a "claim" is supported by "data" linked to the claim by a "warrant." Warrants are "inference licences," "the general hypothetical statements which can act as bridges between the data and the claim" (Toulmin, p. 98). Warrants often require backing themselves. They are not always interchangeable with the minor premise of a syllogism, which may be either a warrant or its backing.

Students have two problems with causal argument. First, they need help thinking up the possible causes of an event. Students tend to overlook the complex interaction of factors, conditions, and influences that yield an effect; they will seize on one cause without understanding how that cause works in connection with others. We have to teach them to think backwards along the paths of known causal agencies, and we can help them do this by introducing the existing terminology of causality. Causes can be identified as necessary and sufficient, as remote and proximate, or as conditions and influences acted on by a precipitating cause. Or sometimes any linear model of causality is a falsification and we have to look at causes and effects as reciprocal, as acting on each other; inflation urges pay raises and higher wages fuel inflation. And, oddly enough, students have to consider what was missing when they think about causes, for an event can take place because a blocking cause was absent. Finally, whenever people are involved in a consideration of causes, the question of responsibility arises. We look for whoever acted or failed to act, or at the person in charge, as causes (usually with the ribbon of praise or the stigma of blame in our hands). In the aftermath of the Three Mile Island reactor breakdown, for instance, the operators in the control room, the engineers who designed the reactor, and even the Nuclear Regulatory Commission officials whose safety regulations controlled its operation were all considered in varying degrees as causes of the accident. Students who are familiar with these possible frames or sets of causes, from necessary cause to responsible cause, can put together models of how causes interacted to bring about the effect they are interested in.

Convincing an audience that a particular cause did in fact operate is the second problem students need help with. The writer of a causal argument can choose from several tactics for presenting evidence that two events are connected as cause and effect. A remote cause, for instance, can be linked to an effect by a chain of causes. NASA provides us with a good example of this technique in their argument claiming that sunspots caused Skylab's fall. Sunspots are storms on the sun which hurl streams of electromagnetic particles into space; these streams, the solar wind, heat up the earth's outer atmosphere. The heated atmosphere expands into Skylab's orbit, increasing the drag on the craft; the craft therefore slows down and falls. Identifying such a chain of causes in effect replaces an implausible leap from cause to effect, a leap an audience is not prepared to take, with a series of small steps they are willing to follow.

Although proximity in time is by itself insufficient evidence of a causal relationship (indeed this is the *post hoc* fallacy), nevertheless, in the presence of plausible agency, time sequence is another tactic for supporting a causal assertion. So is causal analogy, a parallel case of cause and effect; we believe for instance that saccharin or red dye no. 2 causes cancer in humans because it causes cancer in animals. And in the case of a causal generalization such as "Jogging increases self-confidence," a series of individual cases, so long as agency is plausible (and in this instance a definition of self-confidence established), will lend support.

John Stuart Mill's four methods for discovering causes are also powerful aids to invention in causal argument. If the student can find at least two significantly parallel cases, one in which an effect occurred and one in which it didn't, the *single difference* between them can be convincingly nominated as a cause. Or if the same effect occurs several times, any *common factor* in the antecedent events is possibly a cause; this was the method of the health officials searching for the cause of Legionnaire's disease.

Another method, that of *concomitant variation,* is the favorite of the social scientist who looks for influences and contributing factors; when two trends vary proportionately, when the hours of TV watching increase over a decade as SAT scores decline, a causal relationship is suggested, especially when a plausible agency can be constructed between the two. Mill's fourth method, *elimination,* is the ruling-out of all but one possible sufficient cause. It is the favorite of Sherlock Holmes and other detectives faced with a limited number of possible causes.

The student who uses one of Mill's methods can construct a convincing causal argument by repeating the process in writing. Take our sample proposition, "Violence on TV encourages teenage vandalism." Support may come from concomitant variation if we can document an increase in TV violence and a corresponding increase in vandalism perpetrated by teenagers. (The propositions that teenage vandalism exists and has increased along with TV violence can be supported with CP arguments, which the student knows by now require careful definition. What, for instance, precisely constitutes "violence" on TV?) And since this causal claim is a generalization, it could also be supported by citing specific acts of violence clearly inspired by similar acts on TV.

Arguments for CPs and causal statements are the two basic types. Once students have learned to construct these simpler arguments they can combine them into the more complex arguments required for evaluations and proposals. An evaluation is a proposition which makes a value judgment: e.g. "The San Diego Padres are a bad team," "*Jane Eyre* is a great novel," "The open classroom is a poor learning environment." We have to encourage our students to see such propositions as genuinely arguable, as claims which an audience can be convinced of and not merely as occasions for the expression of personal taste. The key is, once again, finding and, when necessary, articulating and defending the sharable assumptions or criteria on which the evaluation is made. Just as the CP argument rests on definition and the causal on agency, so do all evaluations rest ultimately on criteria or assumptions of value.

Students can be taught to construct evaluations by first learning to distinguish the various subjects of evaluations. We evaluate objects both natural and man-made, including the practical and the aesthetic. We also judge people, both in roles and as whole human beings, and we evaluate actions, events, policies, decisions and even abstractions such as lifestyles and institutions which are made up of people, things and actions. Constructing a good evaluation argument is a matter of finding acceptable criteria appropriate to the subject. Our students are already familiar with the typical standards behind the "consumer" evaluations of practical objects. They are far less familiar with the formal criteria used or implied in aesthetic judgments. Most challenging of all are the evaluations of people, actions, and events which require the application of ethical criteria. Students must be encouraged to see that arguing about ethics is not the exclusive province of religion or law, but that we all have beliefs about what is right, proper, or of value which an arguer can appeal to in an evaluation.

In form, an evaluation proposition looks exactly like a CP and, overall, the argument is carried on like a CP argument. Our example above places *Jane Eyre* in the class "great novel." An arguer for this proposition must construct a plausible "definition" or set of criteria for "great novel" which fits the evidence from the book. But the criteria or standard of an evaluation can easily include good or bad consequences as well as qualities, and thus evaluations often require causal arguments showing that the subject does indeed produce this or that effect. If we want to argue, for example, that it was right to bring the Shah to the U.S. for medical treatment, we could do so by classifying that decision as an humanitarian one in a CP argument; or we could argue that the decision was wrong by exploring its consequences in a causal argument. (Of course, whether a consequence can be labeled good or bad gets us right back to ethical assumptions which must be either appealed to or defended, depending on one's audience. Evaluations can lead us into an infinite regress unless we stop eventually on an appeal to shared values.)

The fourth and final type of proposition is the proposal, the call to action. The specific proposal which recommends an exact course of action requires a special combination of smaller CP and causal arguments. We can imagine this argument's structure as something like an hour glass, preliminary arguments funneling in from the top, proposal statement at the neck, and supporting arguments expanding to the base.

We can see how that structure works if we imagine ourselves carrying through an argument for a proposal such as, "Wolves should be reestablished in the forests of northern Pennsylvania and a stiff fine levied for killing them." No one will feel a desire to take action on this proposal unless first convinced that some problem exists which needs this solution; that is the work of the preliminary arguments. An opening CP argument establishes the existence of a situation, in this case the absence or extreme rarity of wolves in certain areas. But an audience may agree that a situation exists yet not perceive it as a problem; it may take a further causal argument to trace the bad consequences (i.e. deer herds are out of control) or show the ethical wrongness of the situation (i.e. a species has been removed from its rightful habitat). These opening parts amount to a negative evaluation. Another preliminary step might be a causal argument singling out the dominant reason for the problem, for ideally the proposal should remove or block this cause or causes, rather than simply patch up the effect alone. If wolves have become nearly extinct in northern Pennsylvania because of unrestricted hunting, then a ban on hunting wolves ought to take care of that.

After the specific proposal is disclosed, it can be supported with another series of CP and causal arguments. The proposal will lead to good consequences (the causal: deer population will be controlled), and it will be ethically right (CP: the balance of nature will be as it ought to be). And most important, the proposal is feasible; the time, money, and people are available (CP), and the steps to its achievement are all planned out (causal).

Just how much of this full proposal outline is actually needed for the writer to make a convincing recommendation depends entirely on audience and situation. A problem may be so pressing that preliminary arguments can be dispensed with entirely; after the last flood, the people of Johnstown did not have to be convinced they had a problem. And not every call to action requires a full proposal argument. One which ends with an unspecified plea, "We really ought to do something about this," is actually a negative evaluation. Such a vague call to awareness is really a coda resting on the widely-held assumption that if something is wrong it ought to be corrected.

In addition to dealing with the four types of arguments we have described, any course in argument must treat the two elements common to all arguments: accommodation and refutation. Consideration of audience (accommodation) and consideration of potential or actual opposition (refutation) inform all argument, affecting invention, arrangement, and style. Where do they sensibly come in a course? The only answer is first, last, and all the way through, worked into every discussion of every type of argument.

Once students have learned the necessary structures of CP and causal arguments and learned how these types combine in support of evaluations and proposals, they have not only the help they need for constructing arguments but tools for the critical analysis of argumentative discourse as well. They can recognize what type of proposition an argument is trying to support, identify the necessary structural elements both explicit and implied and, considering the argument's audience, determine whether all was skillfully done. We might illustrate how this process works by taking a brief look at James Madison's *Federalist* No. 10, reprinted in Corbett's *Classical Rhetoric for the Modern Student* and followed by a careful analysis of its logical elements and the arguments from the sources (2nd edition, New York: Oxford University Press, 1971, pp. 239-256). Madison's argument is an all but perfectly symmetrical full proposal with preliminary arguments (pars. 1-13), explicit proposal (par. 14), and supporting arguments (pars. 15-23). Given his audience and purpose, Madison needs no lengthy demonstration of the existence of a problem; he has only to appeal to "the evidence of known facts." He turns quickly, therefore, to a causal analysis of the problem and finds it in "faction," rooted in the corrupt nature of man. The ethical problem facing his audience is that of preserving two self-evident goods, the control of faction and some form of popular government. The solution which will bridle the effects (for the causes, as he cogently argues, are untouchable) is the federal union, which Madison then goes on to support by tracing the good consequences that will flow from it. Its feasibility has of course been argued elsewhere. Such an analysis is possible to the student who recognizes types of arguments and can thus identify the necessary structural elements in a given argument and even come to a satisfying understanding of why they are where they are.

The approach to argument we have outlined, then, makes it possible to teach argument coherently while avoiding some of the pitfalls of existing approaches. We can avoid extensive and unnecessary diversions into formal logic while keeping to the principles of sound reasoning. And the overall method of building from simple, basic types of argument to types requiring a combination of steps gives the student transferable structures which are suitable for any subject but are not so automatic as to preclude the student from doing his or her own thinking.

Catherine E. Lamb

Beyond Argument in Feminist Composition

Current discussion of feminist approaches to teaching composition emphasizes the writer's ability to find her own voice through open-ended, exploratory, often autobiographical, writing in which she assumes a sympathetic audience. These approaches are needed and appropriate: they continue to show us the richness and diversity of women's voices. My intent in this essay is to suggest a means by which one can enlarge the sphere of feminist composition by including in it an approach to argument, ways to proceed if one is in conflict with one's audience—in other words, the beginning of a feminist theory of composition. The place to start is not with particular forms—those close off options too easily—but by understanding the range of power relationships available to a writer and her readers. One then determines which are consistent with the emphasis on cooperation, collaboration, shared leadership, and integration of the cognitive and affective which is characteristic of feminist pedagogy (Schniedewind 170-179). This line of exploration has taken me to the study of negotiation and mediation, and how these well-established forms of oral discourse can be adapted for a feminist composition class. Argument still has a place, although now as a means, not an end. The end—a resolution of conflict that is fair to both sides—is possible even in the apparent one-sidedness of written communication.

Broadening the Scope of Feminist Modes of Discourse

Much has now been written about women writing and feminist modes of discourse. To illustrate representative approaches, I have selected two essays that have appeared recently in composition journals—one by Elizabeth Flynn describing patterns in women's narratives; the other by Clara Juncker playing out some of the implications if one applies French feminists' theories in the classroom, especially those of Hélène Cixous. Neither pretends to be an exhaustive treatment of the subject. However, because both deal in the content and form that we have come to associate with the broad topic of women and writing, one could quite easily get the idea that these are the only areas in which feminist composition has a contribution to make.

In "Composing as a Woman," Elizabeth Flynn uses what we know about gender differences in social and psychological development to interpret the content of narratives her students wrote. The four essays she uses, two by women and two by men, are not

From *College Composition and Communication* 42 (February 1991): 11-24.

meant to be definitive proof that women write one way and men another, but rather to show that the connections between psychological theory and narrative content are there and may illuminate each other. Her findings are not surprising: women write "stories of interaction, of connection, or of frustrated connection"; men write "stories of achievement, of separation, or of frustrated achievement" (428). This essay and one which followed it fourteen months later, "Composing 'Composing as a Woman': A Perspective on Research," emphasize the open-ended, provisional nature of Flynn's thinking—another quality that has come to be associated with (and prized in) feminist composition.

What I have learned from Flynn's essay and others like it helps me when I am working with women students and reading some literature by women. I need something else, though, if I am to develop a comprehensive approach to feminist composition, guidelines that could be used throughout a course, including the emphasis I used to give argument as a mode in which one's goal is to persuade another to one's point of view. I would also like to be as free as possible from the charge of essentialism, to which an essay like Flynn's is vulnerable. A feminist composition class could easily be a place where matriarchal forms are as oppressive as the patriarchal ones once were, even if in different ways.

Clara Juncker's essay "Writing (with) Cixous" is quite a different piece, written in the exuberant manner of the theorists whose work she is describing. Like them, she is much less interested in women as gendered beings possessing certain characteristics (a possible extension of Flynn's argument) than she is in "woman" as a feminine linguistic position from which to critique phallogocentrism, "the fantasy of a central, idealized subject and the phallus as signifier of power and authority" (425). If this order is dislocated, students may be able to find their own voices on the margins. Playing with language, as well as stressing pre-writing and revision, can sensitize students to the open-endedness of writing. And if they read material sufficiently outrageous, they are more likely to empathize with otherness, whether "racially, politically, sexually, herstorically" (433).

With Flynn's essay, in spite of its value, I see its potential for reductiveness. My concerns with Juncker's essay are theoretical and practical in a different way. I admire the energy in her essay and, having heard her read a shortened version of it at CCCC in 1988, I don't doubt she is able to convey the same to her students. But after the disruptions, then what? I can imagine an essay written this way that is every bit as combative as the masculinist discourse we are seeking to supplant. Further, how can students take these forms and use them in other classes or in the world of work? If we are serious about the feminist project of transforming the curriculum and even affecting the way students think, write, and act once they leave us, we need an approach to teaching composition that is more broadly based and accessible to our students.

Without such a framework, we are also left battling the dichotomies that Cynthia Caywood and Gillian Overing identify in the introduction to their anthology *Teaching Writing: Pedagogy, Gender, and Equity*. Noting that "the model of writing as product is inherently authoritarian," they continue, "certain forms of discourse and language are privileged: the expository essay is valued over the exploratory; the argumentative essay set above the autobiographical; the clear evocation of a thesis preferred to a more organic exploration of a topic; the impersonal, rational voice ranked more highly than the intimate, subjective one'' (xii). I don't know anyone who would deny that these dichotomies exist and are evaluated in the manner described. Neither do I deny the value of continuing to emphasize and explore the potentials of the categories in the second half of each of these dichotomies, as do Flynn and Juncker, along with the contributors to this anthology. We need as well, however, to consider a feminist response to conflict, at the very least to recast the terms of the dichotomy so that ''argumentation'' is opposed not to ''autobiography'' but, perhaps, to ''mediation.''

One half of the problem I am addressing is the narrow range of feminist composition as defined so far. The other is its incompatibility with the values of what I am calling here ''monologic argument,'' the way most (all?) of us were taught to conceptualize arguments: what we want comes first, and we use the available means of persuasion to get it, in, one hopes, ethical ways. We may acknowledge the other side's position but only to refute it. We also practice what we were taught. Keith Fort, in a 1975 essay that uses language we have come to expect in feminist critiques, sees stating a thesis as a competitive act, a way to claim mastery over the subject matter. Similar competition may be generated between the reader and the text (179).[1] More recently, Olivia Frey demonstrates that the antagonism in our writing is much more overt than Fort implies. Using a sample that included all the essays published in *PMLA* from 1975 to 1988 as well as articles in a variety of other professional journals, she found some version of the ''adversarial method'' in all but two of the essays she examined (512). We have uncritically assumed there is no other way to write—at least that attitude was present in much of the discussion about ways to respond to conflict at the 1990 Wyoming Conference on English. Even a text like Gregory Clark's *Dialogue, Dialectic, and Conversation,* which does a superb job of laying out the theory for a cooperative, collaborative approach to writing and reading that is consistent with much of what I shall present later in this essay, sees the act of writing as by definition authoritarian in an even broader sense than do Caywood and Overing. Thus, it is the reader, not the writer, who has the primary responsibility for how a text functions in a community (see especially 49-59). Ideally, wouldn't we want the reader *and* the writer to share that responsibility?[2]

If we as teachers pass on without reflection what we have been taught and ourselves practice concerning argument, whether the rest of our pedagogy intends it or not, we are contributing to education as ''banking,'' Paulo Freire's metaphor for education that is an act of ''depositing'' information into students who are only to receive and have no say in what or how something is taught (58-59). We are doing so because we are teaching students to form ''banking'' relationships with their readers, resisting dialogue, which, for Freire, means they are precluded from any possibility of naming the world, the essential element of being human (76). One of my first-year students this past year

---

1. Fort's essay is cited by William Covino in *The Art of Wondering*. Fort's solution for the critical essay is to recommend ''process criticism,'' where one might explore the correctness of a particular thesis rather than begin with it and show only how the work being analyzed fits. Fort notes that, to some extent, such essays are about the *process* of criticism; in the same way, the form I have students use when they are using negotiation as an alternative to argument is about the process of negotiation.

2. S. Michael Halloran works under assumptions similar to Clark's in ''Rhetoric in the American College Curriculum: The Decline of Public Discourse.'' He is arguing for returning the practice of teaching public discourse to our teaching of rhetoric (a goal I support, as does Clark, for the ways it would reinforce the social function of writing). Although the form of the discourse is not his major concern, the examples he gives all assume a debate model of interaction.

knew at some basic level what Freire was talking about when he described himself as a writer at the beginning of the semester: "For myself writing as a whole is not very important. . . . I would much rather interact with someone by voice rather than writing. Writing is one-sided where no argument or opinion from others can be intervened."

In my discussion thus far of monologic argument, I have intentionally avoided associating it with classical rhetoric, especially Aristotle's. While the connections can surely be made and have been for more than two thousand years, recent scholarship is much more likely to explore ways in which both Plato and Aristotle comment on the social, dialogical context in which knowledge is acquired and exchanged. (See, for example, chapter 2 in Clark, "Rhetoric in Dialectic: The Functional Context of Writing.") Here, I wish only to remind readers of some of those connections without discussing them in detail. The feminist alternatives I am advocating do not follow necessarily, but they are clearly consistent with them. With respect to Plato, what is most important is the example of his dialogues themselves illustrating the dialectic he is advocating, even though the goal, immutable truth, may not be one we share. In the *Phaedrus,* Socrates criticizes writing (in writing), seeing it as something static which inhibits dialectic (95-103). However one interprets his condemnation,[3] the dissonance resulting from an attack on writing itself, also made directly by Plato in *Letter VII* (136), contributes to the dialectic. Aristotle is much more explicitly connected to monologic argument, especially if one stops at his definition of rhetoric as no more than dealing with "the available means of persuasion," a set of techniques to be used. Andrea Lunsford and Lisa Ede have refuted the contradictory claims that Aristotelian rhetoric over-emphasizes the logical and is manipulative ("On Distinctions"), making use of William Grimaldi's work on Aristotle. Grimaldi maintains that in Aristotle's *Rhetoric* one person is speaking to another *as a person;* the rhetor's task is to put before the audience the means by which the audience can make up his or her mind, but it is then up to the audience to decide. The enthymeme is most often cited by these writers and others as illustrating Aristotle's recognition of the proper use of both reason and emotion. The speaker, in constructing an enthymeme, must take the audience into account since it is the audience who supplies the unstated premise. As Lloyd Bitzer says, the audience in effect persuades itself (408).

A Feminist Theory of Power

While it is helpful to view Plato and Aristotle in the ways I have just summarized, neither provides ways to get to concrete alternatives to monologic argument. Considering writer/reader relationships in the context of a feminist theory of power allows us to see more clearly the disjuncture between monologic argument and the modes of discourse advocated by Flynn and Juncker. It also provides a framework for evaluating any alternatives to resolving conflict. Because the emphasis is on values available to men as well as women, essentialist aspects of this approach are minimized.

---

3. For examples of the range of interpretations possible, see Ronna Burger, *Plato's Phaedrus,* in which she argues for Plato's developing a "philosophic art of writing"; Jasper Neel, on the other hand, in *Plato, Derrida, and Writing,* argues that Plato is not using writing but trying to "use it up," appropriating both Socrates's voice and then his means of expression (1-29). Walter Ong, in *Orality and Literacy,* is more relaxed. He notes that Plato's criticisms of writing are the same that were made with the advent of printing and now of computers (979).

In an earlier essay, I note that we understand power in a common-sense way as "the ability to affect what happens to someone else" (100). Monologic argument fits in here easily. There are, however, a number of feminist theorists who view power not as a quality to exercise on others, but as something which can energize, enabling competence and thus reducing hierarchy.[4] More than thirty years ago, Hannah Arendt, in her discussion of "action" in *The Human Condition,* showed us what this use of power might look like. She wrote about the *polis* in classical Greece, in which rhetoric as a spoken art, and therefore argument, would have functioned to maintain the *polis* as she describes it. Its essential character is not its physical boundaries, but "the organization of the people as it arises out of acting and speaking together, and its true space lies between people living together for this purpose" (198). Power maintains this space in which people act and speak: no single person can possess it (as an individual can possess strength). It "springs up" when people act together and disappears when they separate. This sort of power is limitless; it can, therefore, "be divided without decreasing it, and the interplay of powers with their checks and balances is even liable to generate more power" (200-201). I am reminded here of Bakhtin's familiar image of the carnival as the place where hierarchy is suspended and with it the distance between people (e.g., *Problems of Dostoyevsky's Poetics* 122-126). The image is much less dignified than Arendt's idealized description of the *polis,* but the impulse that drives and sustains them is, I believe, the same.

In discussing Dostoyevsky's world view, Bakhtin says its governing principle is "To affirm someone else's 'I' not as an object but as another subject" (10). Some feminist theorists contribute to articulating how such a relationship might develop through insights gained from studying women's experience. They are arguing from what has come to be called a "feminist standpoint," defined by Sara Ruddick as "a superior vision produced by the political conditions and distinctive work of women" (129). The superiority of the standpoint derives from the manner in which it is acquired. An oppressed group, in this case, women, gains knowledge only through its struggle with the oppressor, men, who have no need to learn about the group they are oppressing. With this experience, women's knowledge has at least the potential to be more complete than men's (Harding, "Conclusion" 184-185). There are, admittedly, dangers in using standpoint theory: It can imply the moral superiority of women, easily become essentialist,[5] and ignore the reality that many of the qualities we ascribe to women can just as accurately be called non-Western—possibilities that my anthropologist friends have pointed out to me and that Harding notes in a later essay ("Instability" 29-30). I continue to use this approach, however, because of the teaching power of concrete experience reflected on, to which the success of a book like Belenky et al.'s *Women's Ways of Knowing* is eloquent testimony.

---

4. In addition to the theorists discussed in this essay, I also refer to Jean Baker Miller, *Toward a New Psychology of Women,* and Elizabeth Janeway, *Powers of the Weak.* Another important source is Nancy Hartsock, *Money, Sex, and Power.*

5. I invite readers to consider whether Ruddick's approach as I go on to summarize it is essentialist. Perhaps in the final analysis it is, although in her book she goes to considerable lengths to discuss varieties of mothering experiences. The potential oppression of any essentialist features is also reduced because the process she describes is available to men as well as women and has as its hallmark a deep respect for the other as person.

The most complete feminist discussion of the thought and action which makes possible the use of power described above, in individual relationships as well as those between nations, is Sara Ruddick's *Maternal Thinking*. Ruddick deliberately uses "maternal" because women still have most of the responsibility for raising children; mothering work, she says, can be done as well by men as by women (xi). One need not be a biological parent either. I want to summarize the main features of maternal thinking and then apply them to writer/reader relationships. (They are also readily applicable to teacher/student relationships—but that is another essay. One of the pleasures of teaching this approach to conflict resolution is that it invites attention to the congruence between what and how one teaches.) Central to the idea and experience of maternal thinking is "attentive love, or loving attention" (120). Loving attention is much like empathy, the ability to think or feel as the other. In connecting with the other, it is critical that one already has and retains a sense of one's self. The process requires, ultimately, more recognition and honoring of difference than it does searching for common ground. The vulnerability of the child, combined with the necessity for it and the mother to grow and change, place apparently contradictory demands on the relationship. On the one hand, maternal work requires an attitude of "holding," in which the mother does what is necessary to protect the child without unduly controlling it. On the other hand, she must continually welcome change if she is to foster growth (78-79, 89-93, 121-123).

In the second half of her book, Ruddick shows how maternal thinking can be applied to conflict resolution more generally. One begins by recognizing that equality often does not exist in relationships; even with this reality, individuals or groups in unequal relationships do not have to resort to violence to resolve conflicts. Making peace in this context requires both "giving and receiving while remaining in connection" (180-181). In *Composing a Life,* a discussion of the shaping of five women's lives, Mary Catherine Bateson reflects on these asymmetrical, interdependent relationships and how ill-prepared we are to function in them. Typically, we value symmetrical relationships—buddies and colleagues—which happen also to promote competition. Instead of honoring difference, which makes interdependence possible (both are qualities which "loving attention" cultivates), we want to reduce difference to inequality (102-106).

Monologic argument, even at its best, inevitably separates itself very quickly from the qualities I have just described because of its subject/object, I/it orientation. As I shall demonstrate later, where we still need this kind of argument is at the early stages of resolving a conflict, where both parties need to be as clear as possible about what they think and feel. Our students need to learn it for their survival in other contexts, and, more fundamentally, as part of the process of becoming adults. It promotes differentiation, the sense of self that Ruddick says must precede maternal thinking or integration more generally. This essay is itself a kind of monologic argument because I am asking readers to consider a different (and better, I believe) approach to resolving conflicts in writing. For any change to occur, however, readers first need to know what it is I am proposing.

At this point, readers might be thinking of Rogerian argument as an alternative to monologic argument. In it, the writer goes to great lengths to show the audience that he understands their point of view and the values behind it. The hope is that the audience, feeling less threatened, will do the same. My experience using Rogerian argument and teaching it to my students is that it is feminine rather than feminist. It has always been women's work to understand others (at Albion, it is women, not men, who sign up for The Psychology of Men); often that has been at the expense of understanding self. Rogerian argument has always felt too much like giving in. (In "Feminist Responses to Rogerian Argument," Phyllis Lassner makes these points and others about the difficulties of using Rogerian argument, and the hostilities it may arouse in users, especially if they do not yet have a clear sense of self.)

Mediation and Negotiation as Alternatives

What we need as an alternative to the self-assertiveness of monologic argument is not self-denial but an approach which cultivates the sense of spaciousness Arendt describes in the working of the *polis*. My very brief comments on Plato and Aristotle were intended as another way of saying they are concerned with knowledge as something that people do together rather than something anyone possesses (Gage 156). In a reversal of Bacon's dictum, we could say that Arendt's notion of power makes possible knowledge realized this way. We are ready now to apply this relationship of knowledge and power more specifically to a conflict situation. In it, both parties can retain the interdependence that permits connectedness while also going through the giving and receiving necessary if they are to resolve their conflict. The result is a paradoxical situation where the distance between writer and audience is lessened (as they explore the dimensions of the conflict together) while the "space" in which they are operating has enlarged because they see more possibilities (Lamb 102-103). Jim Corder, in "Argument as Emergence, Rhetoric as Love," also asks us to visualize the writer/audience relationship in terms of physical space. Argument, he says, is too often a matter of "presentation" and "display." Instead, it should just "be." Rather than objectifying the other, we need to "emerge" toward it. In a corollary to the idea of creating more space in which writer and audience can operate, he says we should expect to have to "pile time" into our arguments: we can do so by relying less on closed, packaged forms and more on narratives that show who we are and what our values are (26-31).

When I read Fisher and Ury's *Getting to Yes,* a layperson's version of how the process of negotiation works, I saw that here were some new (to me, as a composition teacher) ways of thinking about argument and conflict resolution. I later attended a seminar on mediation and have mediated cases of sexual harassment at Albion, as one of the people designated by the College to hear these complaints. What quickly became apparent, in both negotiation and mediation, is that the goal has changed: it is no longer to win but to arrive at a solution in a just way that is acceptable to both sides. Necessarily, the conception of power has changed as well: from something that can be possessed and used on somebody to something that is available to both and has at least the potential of being used for the benefit of both. When negotiation and mediation are adapted for a writing class, talk is still central for either process. Writing marks critical stages but cannot occur without conversation that matters, before and after. With all of the currently fashionable and often obscure discourse about writing as dialogue, here is a simple, concrete way of actually doing it.

Central to understanding this broadened and re-focused "practice" of power—how it creates more space and the possibility for loving attention—is articulating the place of conflict in it. As a culture, we learn much more about how to repress or ignore conflict than how to live with and transform it. When we practice and teach monologic argument as an end, we are teaching students that conflict can be removed by an effort that is fundamentally one-sided. Morton Deutsch, in *The Resolution of Conflict,* reminds us that conflicts arise in order that tensions between antagonists might be resolved. They can be healthy ways of finding a new stability and of clarifying values and priorities (9), especially if both parties participate in the resolution in ways that are mutually satisfactory. Negotiation and mediation are *cooperative* approaches to resolving conflicts that increase the chances of these goals occurring. They focus on the future, not the past (as does the law), and seek to restore trust between the two parties. A win-lose orientation encourages narrowness and a wish to use resources only for the goal one has already identified. Deutsch notes that the outcomes of a cooperative approach are those which encourage creative problem-solving: "openness, lack of defensiveness, and full utilization of available resources" (363). Negotiation and mediation are also *collaborative,* with both parties using the process to identify interests and outcomes they share. (See Clark, xvi, for distinctions between cooperation and dialogue on the one hand and collaboration and dialectic on the other.) Finally, both cooperation and collaboration are facilitated by negotiation and mediation as *structured* forms of conflict resolution. The point is important, for the guidelines which provide the structure are the mechanism whereby space between the two parties can be increased, making it possible for the distance between them to lessen as they move toward each other.

Negotiation as it is described in *Getting to Yes* begins with a recognition that focusing on the particularities of the *positions* of both parties will get them nowhere. Instead, identifying underlying *interests or issues* is a way to get at root causes of the problem as well as seeing where there might be common ground. The parties brainstorm a number of possible solutions, evaluating them using criteria both sides can accept. For Fisher and Ury, the ideal outcome is to reach a solution to which both sides can unequivocally answer "yes." Mediation extends and elaborates the process of negotiation with the introduction of an impartial third party. The nature of the outcome is still the responsibility of the disputants, as is carrying out the settlement. The parties in a dispute often appeal to a mediator when they believe they cannot resolve the conflict themselves. The presence of a mediator is also extremely valuable if there is a power imbalance between the two parties, as with, in my experience, cases of sexual harassment involving a student and professor. One of the mediator's main functions, especially at the beginning, is collecting information: What are the problems for each side? What are the interests these problems reflect? Where are the areas of agreement? What are the outcomes each side wants? The mediator's goal is a written agreement, which all parties sign, consisting of concrete statements describing actions both parties will take to resolve the conflict.[6]

I am not yet prepared to recommend one process of conflict resolution over the other in a composition classroom. Mediation may be somewhat more accessible because the roles of negotiator and disputant are separate. I have also taught both only in upper-

6. I have taken this very abbreviated description of the mediation process from Christopher Moore, *The Mediation Process,* and from the *Mediator Training Manual for Face-to-Face Mediation* (Boston: Department of Attorney General, 1988), used at the Mediation Institute taught each spring at the University of Massachusetts at Amherst by staff of the Mediation Project.

level writing courses, negotiation in Advanced Expository Writing and mediation in Technical Writing. (I originally used mediation in Technical Writing because a good mediation agreement is also a model of good technical writing: its function is instrumental, and it must be straightforward, concrete, and unambiguous.) In both courses, the pedagogy is feminist, but only in the expository writing class do I use the theoretical orientation I describe in this paper as a guiding principle for the entire course. Here, I shall describe my use of mediation first to show how the roles operate separately. It also illustrates how a traditional-looking, writing-as-product piece of discourse actually functions quite differently because of the context out of which it comes.

Students work in groups of three, deciding what problem they will work with and who will take what role. Projects come from their reading or current college issues. Last semester, they were as disparate as mediating a property settlement between Donald and Ivana Trump and a dispute between the Inter-Fraternity Council and the administration at Albion College over the social function fraternities serve on campus. Much of the training for being a mediator (or a disputant) goes on in role plays. Of the many skills a mediator needs, I concentrate on just a few: getting as complete a picture as possible from both sides, separating the facts of the situation from the issues, and getting the parties they are working with to come up with as many options as possible in the process of arriving at a solution.

The first piece of writing is one they do individually after they have met several times as a group. If they are one of the disputants, they write a memo to the mediator in which they explain the problem as they see it, including an attempt to separate the immediate ways in which the problem has exhibited itself from the underlying issues or interests. They gain more from the experience if they are willing to take on a role opposite from their own actual position: a fraternity member representing the administration; or a woman playing a man whose spouse has just been offered a high-paying position hundreds of miles away—accepting it would mean serious disruptions in the family and in his career. If a student is the mediator, he or she writes a memo to a supervisor, summarizing the issues for both parties as they appear at that point. Here, all three are using the analytical skills we associate with monologic argument, although not with the goal of persuasion. The memos are part of what will give the mediator a sense of the dimensions of the conflict. For the disputants, they act to "pile time." All of these actions encourage maternal thinking, which is especially desirable between the mediator and both disputants; one hopes it also occurs between the disputants by the end of the process. The second piece of writing is the mediation agreement itself, which all three prepare together. Here are two of ten clauses in an agreement the Inter-Fraternity/administration group reached to resolve their differences:

1. Fraternities agree to restrict the number of house parties to two per semester for the spring 1990 and fall 1990 terms.

2. The administration agrees to begin free shuttle services to cities (Ann Arbor and Lansing) to widen the available social possibilities.

All these pieces of writing in the mediation process are products and not, as will be seen in the discussion of negotiation, a record of a process. Because of the interaction that must occur, particularly when the agreement is being developed, and because everyone involved is both writer and audience, I am not willing to accept Caywood and Overing's judgment that "writing as product is inherently authoritarian." The group's inventing has quite literally been a collaborative, social act, as Karen LeFevre has urged us to see invention more generally (see especially 35-40). Developing and carrying out a mediation agreement is clearly an illustration of what Arendt is getting at when she describes how power works, a point LeFevre also makes. The mediator and disputants, acting together in good faith, can move beyond the conflict that divides them. They are likely to have the experience described by one of the professionals Andrea Lunsford and Lisa Ede interviewed for their book on collaborative writing, *Singular Texts/Plural Authors:* "Working with someone else gives you another point of view. There is an extra voice inside your head; that can make a lot of difference" (29). If, however, one disputant pulls out, or the mediator gives up her neutrality, the energizing power is gone.

When I teach negotiation, it, like mediation, comes in the second half of the course when students trust me and one another and are accustomed to working in groups on various projects. Many of the features of teaching mediation (sources for topics, how to do the training, using writing in different ways at various stages of the process) apply as well to negotiation. Students work in pairs, selecting an issue of some substance in which they are both interested and which will require outside research. Individually, they each write a paper in which they take a contrasting position on the issue. I expect a monologic argument in the best sense of that term. Students see they cannot hope to negotiate a solution with integrity unless they are first clear about the characteristics and values of the viewpoint they are presenting, especially critical if it is one with which they do not agree. When the students have finished the first paper, I meet with the pairs to discuss their arguments. Sometimes, students on their own will take the initiative to begin negotiating a resolution during the conference, ignoring me. We can all then see the process occurring; their next essay, which they write together, is a record of it. They have little trouble differentiating the effect of reading it, its greater sense of spaciousness, from the much more linear effect of reading a monologic argument. The most common form of resolution is some kind of compromise, for example, merit pay for teachers, with the conditions limiting its application making it acceptable to its opponents. (The dynamics of power between the students working together are something I have not yet tried to identify in any systematic way. My impression, from anecdotal evidence, is that most pairs function in a fairly egalitarian way. Of course, they also know that's what I want to hear them say.)

Taking together my discussion of mediation and negotiation, these several features of a feminist alternative to monologic argument are apparent: (1) Knowledge is seen as cooperatively and collaboratively constructed (what the groups have created has come out of the relationships among their members). (2) The "attentive love" of maternal thinking is present at least to some degree (or they would not have been able to come up with a solution acceptable to both of them). (3) The writing which results is likely to emphasize process. (4) Finally, overall, power is experienced as mutually enabling.

These forms, along with the contexts in which they are produced, may also be ways to respond to Lunsford and Ede's call for written discourse which reflects dialogic collaboration in the texts themselves (*Singular* 136). They will not necessarily be of interest to all feminists. Sandra Gilbert and Susan Gubar, among the best known feminist collaborators, have said in a public discussion that they do not see any particular value to writing in a way that would reflect their collaboration and, by extension, more overtly invite the reader into the text. For those of us who *are* interested, these forms show how the writing of a text need not be "an inherently unethical act" (Clark 61), saved only by its readers and their responses. The forms are expressions of writer/reader relationships which reflect an understanding of power consistent with feminist values. As we use them, the forms themselves will change to mirror our evolving understanding of what we are constructing. We *can* move beyond argument. It may not even be foolish to hope for a time when wanting to do so is beyond argument.

Works Cited

Arendt, Hannah. *The Human Condition.* Chicago: U of Chicago P, 1958.

Bakhtin, Mikhail. *Problems of Dostoyevsky's Poetics.* Trans. and ed. Caryl Emerson. Theory and History of Literature 8. Minneapolis: U of Minnesota P, 1984.

Bateson, Mary Catherine. *Composing a Life.* New York: Atlantic Monthly, 1989.

Belenky, Mary Field, et al. *Women's Ways of Knowing.* New York: Basic Books, 1986.

Bitzer, Lloyd F. "Aristotle's Enthymeme Revisited." *Quarterly Journal of Speech* 45 (1959): 399-408.

Burger, Ronna. *Plato's Phaedrus.* University, AL: U of Alabama P, 1980.

Caywood, Cynthia L., and Gillian R. Overing. Introduction. *Teaching Writing: Pedagogy, Gender, and Equity.* Albany: State U of New York P, 1987, xi-xvi.

Clark, Gregory. *Dialogue, Dialectic, and Conversation.* Carbondale: Southern Illinois UP, 1990.

Corder, Jim W. "Argument as Emergence, Rhetoric as Love." *Rhetoric Review* 4 (1985): 16-32.

Covino, William. *The Art of Wondering.* Portsmouth: Heinemann, 1988.

Department of Attorney General. *Mediator Training Manual for Face-to-Face Mediation.* Boston: Department of Attorney General, 1988.

Deutsch, Morton. *The Resolution of Conflict.* New Haven: Yale UP, 1973.

Fisher, Roger, and William Ury. *Getting To Yes: Negotiating Agreement Without Giving In.* New York: Penguin, 1983.

Flynn, Elizabeth A. "Composing as a Woman." *College Composition and Communication* 39 (1988): 423-435.

———. "Composing 'Composing as a Woman': A Perspective on Research." *College Composition and Communication* 41 (1990): 83-91.

Fort, Keith. "Form, Authority, and the Critical Essay." *Contemporary Rhetoric.* Ed. W. Ross Winterowd. New York: Harcourt, 1975. 171-183.

Freire, Paulo. *Pedagogy of the Oppressed.* Trans. Myra Bergman Ramos. New York: Seabury, 1970.

Frey, Olivia. "Beyond Literary Darwinism: Women's Voices and Critical Discourse." *College English* 52 (1990): 507-526.

Gage, John. "An Adequate Epistemology for Composition: Classical and Modern Perspectives." *Essays on Classical Rhetoric and Modern Discourse.* Ed. Robert J.

Connors, Lisa S. Ede, and Andrea A. Lunsford. Carbondale: Southern Illinois UP, 1984. 152-169, 281-284.
Grimaldi, William M. A. *Aristotle, Rhetoric I: A Commentary*. New York: Fordham UP, 1980.
Halloran, S. Michael. "Rhetoric in the American College Curriculum: The Decline of Public Discourse." PRE/TEXT 3 (1982): 245-269.
Harding, Sandra. "Conclusion: Epistemological Questions." *Feminism and Methodology*. Ed. Sandra Harding. Bloomington: Indiana UP, 1987. 181-190.
———. "The Instability of the Analytical Categories of Feminist Theory." *Signs* 11 (1986). Rpt. in *Feminist Theory in Practice and Process*. Ed. Micheline R. Malson et al. Chicago: U of Chicago P, 1989. 15-34.
Hartsock, Nancy. *Money, Sex, and Power*. Boston: Northeastern UP, 1983.
Janeway, Elizabeth. *Powers of the Weak*. New York: Knopf, 1980.
Juncker, Clara. "Writing (with) Cixous." *College English* 50 (1988): 424-436.
Lamb, Catherine E. "Less Distance, More Space: A Feminist Theory of Power and Writer/Audience Relationships." *Rhetoric and Ideology: Compositions and Criticisms of Power*. Ed. Charles W. Kneupper. Arlington: Rhetoric Society of America, 1989, 99-104.
Lassner, Phyllis. "Feminist Responses to Rogerian Argument." *Rhetoric Review* 8 (Spring 1990): 220-232.
LeFevre, Karen Burke. *Invention as a Social Act*. Carbondale: Southern Illinois UP, 1987.
Lunsford, Andrea, and Lisa Ede. "On Distinctions Between Classical and Modern Discourse." *Essays on Classical Rhetoric and Modern Discourse*. Ed. Robert J. Connors, Lisa S. Ede, and Andrea A. Lunsford. Carbondale: Southern Illinois UP, 1984. 37-49, 265-267.
———. *Singular Texts/Plural Authors: Perspectives on Collaborative Writing*. Carbondale: Southern Illinois UP, 1990.
Miller, Jean Baker. *Toward a New Psychology of Women*. Boston: Beacon, 1975.
Moore, Christopher. *The Mediation Process*. San Francisco: Jossey-Bass, 1987.
Neel, Jasper. *Plato, Derrida, and Writing*. Carbondale: Southern Illinois UP, 1988.
Ong, Walter. *Orality and Literacy*. New York: Methuen, 1982.
Plato. *Phaedrus and Letters VII and VIII*. Trans. Walter Hamilton. London: Penguin, 1973.
Ruddick, Sara. *Maternal Thinking*. Boston: Beacon, 1989.
Schniedewind, Nancy. "Feminist Values: Guidelines for Teaching Methodology in Women's Studies." *Freire for the Classroom*. Ed. Ira Shor. Portsmouth: Heinemann, 1987. 170-179.

Brooke K. Horvath

The Components of Written Response: A Practical Synthesis of Current Views

Written responses to student writing continue to play an important part in most composition classes despite increased employment of peers and tutors as sources of informed opinion and despite increasing emphasis on the importance of oral response. How best to respond to students' essays therefore concerns us all, at whatever level we teach. Yet valuable as we believe our penciled comments to be, this time-consuming, difficult task proves too frequently a confused, unsatisfying experience for us; worse, our efforts prove too often apparently unhelpful to students who, if uninstructed, are alienated, antagonized, by our thought-heavy marginalia and terminal remarks. I suspect many of us, seated before a stack of papers, wonder over late-night coffee if we are doing this job well, if the results are worth the effort.

Much of the research done on response remains buried in unpublished dissertations (for accounts of such research, see Jarabek and Dieterich, Knoblauch and Brannon 1981, Lamberg*), and the published material, scattered throughout the professional literature, is not readily available for comprehensive review. Compared to the growing body of literature devoted to other compositional and rhetorical topics, the amount of accessible advice on how to respond productively to student writing is scant. Nevertheless, enough such material of a practical nature exists to warrant attention. What follows is an attempt to summarize and to synthesize some of the guidelines for writing effective comments that this literature suggests, thereby supplementing C. W. Griffin's recent review-essay, which deals exclusively with the components of a theory of evaluation. To bring together and to group under general rubrics the eighty-one items here reviewed may assist the formation of a useful theory of response and may, more immediately, bring greater coherence and consistency to the almost daily act of commenting on student themes.

It is well to note at the outset that my concern here is with formative, not summative, evaluation. Determining a paper's grade and writing comments to explain or to justify that grade; deciding how well a paper measures up to one's expectations, fulfills the requirements of an assignment, meets certain criteria of good prose; in short, passing judgment, ranking: this is summative evaluation, which treats a text as a finished prod-

From *Rhetoric Review* 2 (January 1984): 136-56. Reprinted by permission of the publisher.

*References throughout are to items on the appended bibliography, which divides entries into three groups: theory and practice of written response, revision and its implications for response, and alternatives to written response.

uct and the student's writing ability as at least momentarily fixed. Formative evaluation, on the other hand, is intent on helping students improve their writing abilities; it approaches a paper "not in terms of what has been done, but of what needs to be done, what can be done . . . not to judge, but to identify problems and possibilities" (McDonald 1978). Formative evaluation treats a text as part of an ongoing process of skills acquisition and improvement, recognizing that what is being responded to is not a fixed but a developing entity.

A formative response can be of several kinds. It can, to follow Elaine O. Lees's taxonomy, (1) *correct,* supplying factual information but risking an undue, perhaps stifling emphasis on "the importance of editorial tidiness"; (2) *emote,* implying shared humanity but shifting the focus of attention from text to teacher, inviting the view that teacher responses are the irrelevant "crackpot reactions" of one reader; (3) *describe* textural features—how the paper is behaving—thus keeping attention focused on the text while supplying students with a set of critical terms, yet perhaps failing to help writers "produce a paper that may be described differently"; (4) *suggest* where changes might be made thereby addressing the writer's needs more directly than description alone permits, yet running the similar risk of providing comments too text-bound to prove generally useful; (5) *question,* forcing students to rethink material, thus encouraging further discovery; (6) *remind,* relating the text to class discussions so that comments and classwork reinforce each other; and (7) *assign,* creating a new writing task, "using what has been said already to discover how to say something new," thereby setting goals and emphasizing both writing and writing improvement as developmental processes. These seven possible response types suggest a hierarchy: the first three, says Lees, "put the burden of work on the teacher"; four through six "shift some of that burden to the student." But assigning, she implies, is best, for only assigning "provides a way to discover how much of the burden the student has taken": assigning creates a productive tension between what has been said and what may be said, implying that one can always write better than one has written, can say in the future what one has not yet said. (For more on the kinds and aims of response, see Beaven, Diederich 1974, Green, Irmscher, Kehl, Larson 1966, McDonald 1975, Sloan, Sternglass, Wells. For studies of teachers' habitual response practices, see Freedman; Searle and Dillon.)

Assigning, so conceived, attempts to improve student writing by providing feedback on skills seen as in transition toward greater competence. To take this notion of assigning a step further in the direction current research suggests (see bibliography, section two), we might add that responses to student writing prove most beneficial when each text is itself conceived as a work-in-progress amenable to revision. Offering formative evaluation of finished products is based on the assumption that students will learn from their mistakes, extrapolating advice useful in shaping better prose in future, unrelated writing situations. Certainly we must hope this is to some degree true; otherwise, we find ourselves as evaluators faced with the dilemma either of offering responses so tied to specific texts that students cannot discover in them general guidelines (the problem Lees would avoid) or of offering responses so generic they become what Nancy Sommers calls "rubber stamped"; being in no way "anchored in the particulars of the students' texts," such comments become "a series of vague directives" readily but uselessly interchangeable (Sommers 1982). Yet, as Sommers goes on to remark, despite the assumptions underlying response to finished products, what one says about a text as product differs from what one says about that text as process. For example, when an essay is treated as a finished product, comments will tend to judge, to describe, and to correct; but when an essay is treated as a draft to be revised, comments will tend to be suggestions, questions, reminders, and assignments: responses placing learning where it belongs—with the student.

Texts, then, should be treated as unfinished, each a stage in an ongoing process; students are encouraged to see revision as a desirable, necessary event that should occur. All "real" writers revise, and studies show that students who revise produce better work, by and large, than do nonrevisers, that a willingness to revise and the nature and extent of revision are factors distinguishing skilled from unskilled writers, poor writers from good (in section two of bibliography, see particularly Beach, Bridwell, Crabbe, Faigley and Witte, Murray 1978 [first entry], Sommers, Stallard). Among the advantages of responding to a text as in-process is that doing so helps bring students' writing behavior closer to that of professional, or skilled, writers (see Murray 1978 [first entry], Sommers 1980). Further, predicating responses upon subsequent revision occurring not only emphasizes composing as a process but also offers both formative evaluation founded concretely on the specifics of a text and the chance to act upon that response in its most immediately applicable context. The student is allowed to improve not only his general writing skills but also the text that most concerns him at the moment; if comments are made on the premise that revision will occur, then formative evaluation must precede summative with an intervening opportunity for students to act upon the responses. (For more on how to respond to student writing as unfinished products, see Flanigan and Menendez, Flower 1979, Kirby and Liner, Denise Lynch, McDonald 1978, Murray, Sommers 1982).

Approaching papers as in process also suggests a hierarchy of concerns and allows the instructor to provide a sequence of objectives the student can handle. She can first address large-scale problems with content, focus, organization, voice, logic, or purpose; then concentrate on transition, the adaptation of the text to a particular audience, style, and mechanical and usage violations that do not interfere with comprehension (overuse of dashes, random shifts in person, faulty capitalization, and such like). To put responses in sequence according to a hierarchy of concerns reduces the likelihood that a student will confuse revising with editing or proofreading, and allows her to work toward better prose via a series of manageable tasks, attainable goals (see especially Flower and Hayes 1979, McDonald 1978, Sommers 1982).

However, to encourage real revision, to challenge writers "to reconceptualize their discourses in ways that these writers may not have anticipated nor feel secure in attempting to engage in will lead to multi-leveled difficulties. . . . to expect, that risk-taking and improvement can occur simultaneously is unrealistic and inappropriate" (Onore). In short, a tolerance for error must be cultivated in both students and instructor, for revised texts need not necessarily be better, more successful texts when what has been at issue is a total reseeing of audience, content, voice, or organization. The simple avoidance of error should not be applauded, and errors themselves—whether a radical failure of message or the habitual misconstruction of a particular kind of sentence—must be treated as occasions for learning. (For brief introductions to error analysis, see Bartholomae, Kroll and Schafer, Rizzo.)

The nature and number of errors in any given student's work will, of course, vary with the assignment. A physics midterm, trouble at home, and other "extra-textual" problems in part account for a student's uneven production. But it is also important to remember that certain writing abilities may be task-specific: that there are students, for example, who can explain but cannot persuade. If comments cannot suffer transplantation to the student's mind where they may grow into larger truths, if comments cannot transcend the context in which they appear, they may prove to be of limited value. But comments must also take into account the full rhetorical context of the paper under the red pen, analyzing each part and feature as they relate to the essay's overall purpose, message, tone, mode, and targeted audience: "to respond to whole pieces as they are read . . . [which] might lead students to think that they should take composing seriously . . . as a transaction between human beings to whom writing matters, as the performing of an act that seeks to accomplish work in the world. . . ." So says Richard Larson (1974), who would likewise caution teachers against acting "as if the accumulation of varied comments on details constitutes an adequate response to [a] paper." This notion of the text-specificity of certain rhetorical features underlies Richard Lloyd-Jones's model for primary-trait scoring (see Cooper and Odell), which rejects the notion of "general writing ability" in favor of mode-specific competence. If one accepts Lloyd-Jones's assumptions, the rationale for offering text-specific comments within a full rhetorical context is further strengthened, although a distinction might be made between rhetorical and arhetorical features and it might then be argued that any writing sample provides a valid index of the writer's basic semantic and syntactic competence, of her grammatical proficiency and awareness of those rules of linguistic etiquette known as standard usage.

The evaluator's task, then, is to respond neither too vaguely nor too narrowly, to provide comments with "transfer value" (Larson 1966), keeping in mind the essay's full rhetorical situation and attempting to respond to it as an integrated work intent on accomplishing a certain aim—the student's intended aim—in the world. So to respond to a paper while simultaneously treating it as a draft to be revised means returning control of the text to the student. Grievances pile up, responses often become reprehensibly misdirected, and learning falters when students' texts are appropriated by teacher. If this happens, students may too readily conclude that success depends not upon fully realizing one's intentions, fully conveying one's meaning, fully expressing one's feelings or actualizing one's voice, but upon aping the teacher, upon successfully playing with what Gary Sloan calls the "Great Guessing Games," wherein "intellectual development and free expression are forgotten, and writing becomes an occasion for deceit." Co-opting students' authority over their own intended meanings "tends," according to Knoblauch and Brannon (1982), "to show students that the teacher's agenda is more important that their own, that what they wanted to say is less relevant than the teacher's impression of what they should have said," and this in turn tends to "remove the incentive to write and the motivation to improve skills." To return control of their texts to students, the authors continue, "teachers need to alter their traditional emphasis on a relationship between student texts and their own Ideal Text in favor of the relationship between what the writer meant to say and what the discourse actually manifests of that intention." (See too Sommers 1982.)

To Knoblauch and Brannon's injunction against commandeering students' texts may be added the implications of Richard Fulkerson's discussion of the four preeminent contemporary theories of composition, which taking his cue from M. H. Abrams, he labels the mimetic, the expressionistic, the formalist (objective), and the rhetorical (pragmatic). Although none of these theories is best, which theory one more-or-less consciously subscribes to determines one's notion of good prose, hence one's pedagogy. It is important, therefore, to be aware of the theory informing one's practice and to recognize that students operating under alternative theories may produce legitimate texts that from one blinkered perspective appear unacceptable. In practical terms, one implication of Fulkerson's work is that responses are misdirected when they assume that the teacher's intentions when making an assignment must necessarily be the student's intentions when fulfilling it: thus, if an assignment allows for either expressive or persuasive responses although the instructor intended to elicit only persuasive papers, she cannot fault the expressive student for a failure to persuade. The more general guideline inferable from Fulkerson is that students cannot be criticized for teacher-created problems.

Thus Fulkerson and Knoblauch and Brannon, like Sommers and Larson, caution against generic responses, against responding outside the assignment's context, against looking for things not asked for, against approaching texts with preconceptions regarding what and how they should be. They also remind instructors that styles, vocabularies, composing processes, problem-solving techniques, and so on differ: that students have different needs and need to hear different things at any given time; that one must start where the student is, dealing with the present state of his art; that comments, in short, must be student- as well as text-specific. (See too Hairston, Macrorie, Rogers.)

Inappropriate responses such as the above hardly exhaust the possibilities for wrong-headed commentary that confuses and alienates students, causing attitudes toward writing, class, and instructor to degenerate. Inappropriate as well are comments that "perpetuate archaic modes of expression, propagandize for a particular style, or coerce agreement with a sociopolitical attitude" (Sloan). Too much comment is counterproductive; too many helpful remarks serve finally to confuse, frustrate, and depress students unsure which aspects of their prose most need attention but quite sure they will never write well (see Harris and Shuman under *How to Handle the Paper Load,* Lamberg, Miller, and selections in Judine). To encourage the acquisition of self-editing skills, comments that do a student's work for her should be avoided, as, in the interests of experimentation, should comments that reward fidelity to a safe middle way. Also inappropriate are comments inhibiting or foreclosing effective revision, such as feedback on early drafts which focuses primarily on mechanical matters while ignoring larger problems, thus suggesting that a successful paper awaits the correction of spelling and punctuation errors. Such comments may cause students to confuse ends with means, seeing adherence to conventions not as a way of enhancing communication but as an end in itself, with substance, purpose, originality going out of focus as students concentrate on expressing vapid, trite ideas correctly. Moreover, such comments may cause revision to be seen as punishment, busy work, and/or a reductive, formulaic tidying up of already finalized messages and more concerned with incorporating teacher's changes than with productively reseeing one's content and aim. (See Sommers 1982 and section two of the bibliography, particularly Flanigan and Menendez, Murray, Onore.)

Perhaps most inappropriate of all are comments posing veiled attacks on the student, her opinions and interests, her worth as a writer. Students have a proprietary interest and emotional involvement in their writing because it is an intensely personal act in its revelation of character, intellect, belief, feeling. Therefore, responses manifesting scorn, hostility, condescension, flippancy, superficiality, or boredom are always out of line. Robert Gee conducted a study of student reactions to teacher comments, concluding that positive reinforcement is more conducive to healthy attitudes toward writing and so to writing improvement than is negative feedback or absence of response. (On the virtues of praise, see Diederich 1963, Irmscher, Johnston, Lynch and Klemans, Taylor and Hoedt.) Although dissenting opinion exists—that praise has no effect on improvement (see Clark; Jarabek and Dieterich), that a mix of praise and criticism works best (see Clarke), that written feedback of whatever kind may be of small value (see Arnold, Clark), that the effects of comments cannot be judged apart from the larger setting in which they occur and of which they are but one part (see Knoblauch and Brannon 1981)—certainly the evaluator's role as motivator is crucial, for it is in this role that the teacher speaks as the students' sincere (if somewhat artificial) friend, applauding their successes, empathizing with their difficulties, urging them to look forward to the effects certain remediations will have on their work, setting goals to strive toward, encouraging risk-taking, fostering the desire to write more and to write better.

There exists, in fact, a large body of opinions suggesting that formative responses assist the betterment of writing largely by creating an atmosphere conducive to learning. Sloan asserts that, at best, comments can encourage, evidencing a sympathetic understanding of the writer's intentions and helping her develop her own style. Carl Rogers has discussed the conditions under which learning best occurs, arguing persuasively that one cannot teach a person while simultaneously threatening either him or his values. And Mary Beaven, much indebted to Rogers, believes establishing "a climate of trust is essential" to encouraging learning's necessary risk-taking and advocates three kinds of positively conceived comments to help establish such a climate: requests for more information, paraphrases or reflections of students' thoughts, and the sharing of thoughts and experiences similar to those expressed in students' papers.

Avoiding negative, excessively judgmental response in favor of more positive annotation does not, however, prohibit criticism. A critical attitude such as that brought to a reading of John Updike or *Modern Fiction Studies* will indicate the intention to treat students' essays with equal seriousness and respect. Such an attitude will make clear to the student that what she wanted to say has been listened to, that authority over her text and a close, sympathetic attention have been granted. (For more on the importance of creating a motivating atmosphere, see Hairston, Hawkins 1980, Macrorie, entries on conferences and peer-evaluation in section three.)

Understanding how to reconcile one's role as motivator with one's role as critic means understanding all the roles played, preliminary to the final role of grader. Greg Cowan distinguishes among the experiencer, the examiner (essentially, the formative evaluator), and the evaluator (grader) of papers, with the student playing the corollary roles of sharer, apprentice, and candidate, respectively. Cowan notes the importance of keeping these pedagogical roles and relations vis-à-vis the student distinct, observing that role-confusion results in conflicting signals being sent or in the sending of signals inappropriate to the role assumed. He instructs evaluators to play all three roles and to offer apposite feedback for each, relative to the roles in which the student has been momentarily cast.

It may be fruitful to revise Cowan's division of the instructor's persona to distinguish the roles of editor, average reader, and more experienced writer in addition to those of summative evaluator and motivator/friend. As editor, the evaluator addresses all clear-cut errors and deficiencies, indicating the nature of the problem and what changes must occur. Editors do not, of course, habitually engage in selective criticism, yet the evaluator-as-editor must bear in mind the counter-productive consequence of over-response, of nitpicking "corrections" often founded primarily upon personal preference. No text is perfect: every style invites doctoring; almost any text could be made more readable; more descriptive or precise terms could be substituted for those found in many sentences; any argument could be strengthened; to any exposition, useful or interesting information could be added. Further, the evaluator-as-editor should remember that early in the semester, and with early drafts always, though grammar, usage, and mechanics problems should not be totally ignored, the focus of attention should be on large-scale problems so as not to smother the student's will or ability to improve beneath a breath-taking avalanche of commentary.

Against excessive response and an unreasonable preoccupation with relative minutia, the role of average reader serves as corrective. Joseph Williams argues convincingly that many "errors" if unlooked for would not be found, that they occur in published prose regularly and bother no one (see too Sloan). Thus, as average reader, the evaluator, though a captured audience, tries to respond as might a real-world reader, consequently not making overmuch of defensible fragments, slightly inexact word choices, contractions, split infinitives, and other slips of mind or pen that would not bother him if they were noticed elsewhere. In this role, responses will be offered concerning the paper's utility, humor, persuasiveness (whatever its intent might be); its readability and originality; its turns of phrase, plays on words, and so on. (See too Gibson, Lamberg, Maimon, Siegel; Hirsch's distinction between extrinsic and intrinsic textual features seems to jibe with the role distinctions here being made, extrinsic features being the concern of teacher-as-average-reader, intrinsic the concern of teacher-as-editor.)

Finally, mediating between these last two roles, the role of more experienced writer, coach, or guide affords the opportunity of offering advice, of suggesting options the student *might* have used and the effects on tone, effectiveness, content, and so forth these suggested options would have wrought. In short, as more experienced writer, the instructor offers techniques, tricks of the trade, that the student can add to her repertoire and elaborates upon why certain features of a text—figures used, words chosen, examples employed—worked as well as they did. (For more on response roles, students' attendant roles, and role interaction, see Cowan, Hairston, Hawkins 1980, Maimon, Rogers.)

A parting reminder: written teacher responses need not be students' only source of response; therefore, one needn't feel conscientiously impelled to offer every observation that comes to mind. Conferences, class discussion, small group work and written peer evaluation, tutors in writing labs, computers, and other strategies to get students to be self-editing all provide additional sources of response holding many advantages: they allow the student to receive more than one viewpoint on her work; they can reinforce classroom work and a teacher's written comments while reducing the time spent marking papers; peer-evaluation will involve students in the evaluation process and develop their skills as critical readers; discussion allows for more immediate reinforcement, for questioning, rebutting, and reconceiving reactions; and the use of self-evaluation techniques will hasten the time when students can self-edit and are able to read their own prose critically without a dependence on others. Though these supplemental response resources fall outside this essay's scope, section three of the bibliography offers an introductory reading list.

In fact, I would like to conclude this essay simply by referring the interested reader to the materials listed on the following pages, offered as a selective yet substantial guide to more thorough discussions of topics here raised.

Responding to Student Writing: A Selected Bibliography

I. Responding in Writing

Arnold, Louis V. "Writer's Cramp and Eyestrain—Are They Paying Off?" *English Journal,* 53 (January 1964), 10–15.
Reports a study concluding that students writing frequently and receiving intensive evaluation produce prose no better than that produced by students writing infrequently and receiving moderate feedback.

Bartholomae, David. "The Study of Error." *College Composition and Communication,* 31 (October 1980), 253–69.
Discusses the nature of error, techniques for analyzing error, and the teacher's appropriate response to error. Good introduction to error analysis.

Clark, William G. "An Evaluation of Two Techniques of Teaching Freshman Composition. Final Report." ERIC ED 053 142, 1968.
Reports a study concluding that extensive written response without class discussion of papers, class discussion without written response, discussion and response, and the absence of both all produced the same degree of writing improvement.

Clarke, Grace Allison. "Interpreting the Penciled Scrawl: A Problem in Teaching Theme Evaluation." ERIC ED 039 241, 1969.
Reports a study of the effects of teacher comments on student attitudes, finding positive reinforcement, accompanied but not outweighed by criticism, produces best attitudes, greatest writing confidence.

Cooper, Charles R. and Lee Odell. "Procedures for Evaluating Writing: Assumptions and Needed Research." *College English,* 42 (September 1980), 35–43.
Reviews the methodology, assumptions, strengths, and weaknesses of ETS General Impression (holistic) scoring, Paul Diederich's Analytic Scale, E. D. Hirsch's Relative Readability criteria, and Richard Lloyd-Jones's Primary Trait Scoring.

Cowan, Greg. "The Rhetorician's Personae." *College Composition and Communication,* 28 (October 1977), 259–62.
Discusses the three roles (experiencer, examiner, evaluator) played by the writing teacher.

Diederich, Paul B. "In Praise of Praise." *NEA Journal,* 52 (September 1963), 58–59.
Argues the need to find the good in student writing, that too many corrections are counterproductive (see Jarabek and Dieterich, below, for a report on studies concluding that praise has no effect on writing improvement).

———. *Measuring Growth in English.* Urbana, IL: National Council of Teachers of English, 1974. Pp. 85–99.
Offers ninety-six objectives the author feels should be the focus of paper comments, conferences, and class discussion.

Freedman, Sarah Warshauer. "Why Do Teachers Give the Grades They Do?" *College Composition and Communication,* 30 (May 1979), 161–64.
Discusses what teachers respond to when they respond to writing.

Fulkerson, Richard. "Four Philosophies of Composition." *College Composition and Communication,* 30 (December 1979), 343–48.
Explains how formalist, expressionist, mimetic, and rhetorical theories of composition each characterize good prose, noting consequent implications for responding to student texts.

Gee, Thomas C. "Students' Responses to Teacher Comments." *Research in the Teaching of English,* 6 (Fall 1972), 212–21.
The author reports his study of the effects on student attitudes toward and improvement in writing caused by teacher comments.

Gibson, Walker. "The Writing Teacher as a Dumb Reader." *College Composition and Communication,* 30 (May 1979), 192–95.
A reader-response approach to responding to "dramatize" for students "how *dumb* a reader is," how fraught with comprehension problems texts are. The author details six habitual problems undermining readability that are found in student texts.

Green, James L. "A Method for Writing Comments on Student Themes." *English Journal,* 57 (February 1968), 215–20).
Describes how Robert Gorrell's "commitment-response" approach to structure may guide the writing of end comments.

Griffin, C. W. "Theory of Responding to Student Writing: The State of the Art." *College Composition and Communication,* 33 (October 1982), 296–301.
Reviewing current literature, the author concludes that a theory of responding "will be concerned with three major components: our orientation, our verbal responses, and our students' reactions to our responses." This article contains much useful material here excluded.

Hairston, Maxine. "Carl Rogers' Alternative to Traditional Rhetoric." *College Composition and Communication,* 27 (December 1976), 373–77.
Rogers' opinions on learning lead the author to concur with his characterizations of proper teacher-student relations, productive learning environments, and the instructor's appropriate persona and responses.

Haswell, Richard H. "Minimal Marking." *College English,* 45 (October 1983), 600–604.
Advocates a method of marking surface errors in keeping with criteria for effective response set forth in Knoblauch and Brannon's "Teacher Commentary on Student Writing" (see below).

Hirsch, E. D. *The Philosophy of Composition.* Chicago, IL: University of Chicago Press, 1977. See particularly chapter 7, "The Valid Assessment of Writing Ability."
Discusses extrinsic (content, intent) vs. intrinsic (style, mechanical correctness, relative readability) textual features as focuses of response.

How to Handle the Paper Load. Classroom Practices in Teaching English 1979–1980, Gene Stanford, Chair, and the Committee on Classroom Practices. Urbana, IL: National Council of Teachers of English, 1979.

Section five, "Focused Feedback," offers advice on how best to comment on student writing without excessive expenditures of time:

Muriel Harris, "The Overgraded Paper: Another Case of More Is Less," advocates selective response, explaining why too much feedback is counterproductive.

R. Baird Shuman, "How to Grade Student Writing," urges marking "a limited number" of errors.

David A. England, "Objectives for Our Own Composing Processes—When We Respond to Students," outlines six objectives when marking a paper, offering sample responses.

Sheila Ann Lisman, "The Best of All Possible Worlds: Where X Replaces AWK," advocates replacing editing marks with X's.

See, too, Thomas Newkirk's "Read the Papers in Class," which offers lists of questions to ask while reading papers. Intended for use during student-teacher discussions, the questions—covering content, arrangement, style, and language—will work as well when the responses are written.

Irmscher, William F. *Teaching Expository Writing*. New York: Holt, Rinehart and Winston, 1979.

Chapter thirteen enters a plea for objective evaluation criteria and offers detailed advice on how to read, analyze, and respond to student themes, covering what to look for regarding content, form, diction, mechanics, and style. The author discusses kinds and tone of response, offers sample essays and passages with comments provided, and establishes criteria for determining grades.

Jarabek, Ross, and Daniel Dieterich. "Composition Evaluation: The State of the Art." *College Composition and Communication,* 26 (May 1975), 183–86.

Surveys recent opinion and research (including dissertations) on three questions: Should compositions be evaluated? What kinds of comments are best? and Are there better ways to respond? A useful bibliography is appended.

Johnston, Brian. "Non-Judgmental Responses to Students' Writing." *English Journal,* 71 (April 1982), 50–53.

Argues that "questioning to encourage self-assessment," "describing one's personal response," and "empathic response" motivate students more than do judgmental advice and evaluation.

Judine, M., ed. "A Guide for Evaluating Student Composition: Readings and Suggestions for the Teacher of English in the Junior and Senior High School." ERIC ED 033 948, 1965.

Twenty-four short essays offer advice on responding, discuss the psychological effects of comments on students, and remark upon the degree of thoroughness to be sought when responding. Sample essays with suggestions for evaluating them are included.

Judy, Stephen N., and Susan J. Judy. *An Introduction to the Teaching of Writing*. New York: John Wiley and Sons, 1981.

How to evaluate student themes is among the topics treated.

Kehl, D. G. "The Art of Writing Evaluative Comments on Student Themes." *English Journal,* 59 (October 1980), 972–80.
Offers advice on how to respond usefully, discussing voice, the causes of ineffective responses, the aims of summary comments, and the means of gaining a sense of an essay's overall intent. Examples provided. The author incorporates into his suggestions the work of Francis Christensen and Robert Gorrell.

Knoblauch, C. H., and Lil Brannon. "On Students' Rights to Their Own Texts: A Model of Teacher Response." *College Composition and Communication,* 33 (May 1982), 157–66.
The authors argue that, when responding to student writing, teachers must not appropriate their students' texts, explaining how to respond while allowing students to retain control over their messages, voices, and so on. Conferences stressed.

———. "Teacher Commentary on Student Writing: The State of the Art." *Freshman English News,* 10 (Fall 1981), 1–4.
Reviews research (mostly unpublished dissertations) into the effects of response on student writing, critiquing research assumptions and traditional response practices.

Kroll, Barry M., and John C. Schafer. "Error-Analysis and the Teaching of Composition." *College Composition and Communication,* 29 (October 1978), 242–48.
The authors explain why errors must be seen as necessary to the learning process, products of "intelligent cognitive strategies," detailing how to investigate error, disclosing error patterns (consistently misconceived strategies), and how to apply findings to help students improve their writing.

Larson, Richard L. "Training New Teachers of Composition in the Writing of Comments on Themes." *College Composition and Communication,* 17 (October 1966), 152–55.
Offers practical advice on the writing of marginal and end comments, and outlines a five-step procedure for handling papers. The importance of workshops for training new teachers is stressed.

———. "The Whole Is More than the Sum of Its Parts: Notes on Responding to Students' Papers." *Arizona English Bulletin,* 16 (February 1974), 175–81.
Explains how to focus on those features of a text that allow response to treat that text as a whole effort within a full rhetorical context.

Lees, Elaine O. "Evaluating Student Writing." *College Composition and Communication,* 30 (December 1979), 370–74.
Provides a taxonomy of possible kinds of response.

Lynch, Catherine, and Patricia Klemans . "Evaluating Our Evaluations." *College English,* 40 (October 1978), 166–80.
Reports basic writers' opinions on teacher commentary, explaining how they react to different kinds of annotations and how useful they find various sorts of feedback.

McDonald, W. U., Jr. "Grading Student Writing: A Plea for Change." *College Composition and Communication,* 24 (May 1975), 154–58.
Advocating a "commentary method" of summative evaluation, the author distinguishes between instructional evaluation and grading.

———. ''The Revising Process and the Marking of Student Papers.'' *College Composition and Communication,* 29 (May 1978), 167–70.
Explains how to respond to drafts: when to focus on what (and why).

Macrorie, Ken. *Uptaught.* New York: Hayden, 1980.
Post-Dartmouth Conference, Sixties-style advice on teaching English, containing much axiomatic advice on responding to student writing.

Maimon, Elaine. ''Talking to Strangers.'' *College Composition and Communication,* 30 (December 1979), 364–69.
Discusses the most effective roles played to help students learn to act like writers and when these roles give way to that of the ''critical stranger'' who assesses the finished product.

Miller, Robert K. ''The Selective Use of Criticism.'' *College Composition and Communication,* 33 (October 1982), 330–31.
Sketches a method and rationale for responding that allows students to choose the form of response they receive.

Rizzo, Betty. ''Systems Analysis for Correcting English Compositions.'' *College Composition and Communication,* 33 (October 1982), 320–22.
Presents a brief argument for error analysis as prerequisite to responding, particularly to grammar and usage errors.

Rogers, Carl R. *Freedom to Learn.* Columbus, OH: Charles E. Merrill, 1969.
Rogers discusses the conditions under which learning best occurs. One of the founders of the encounter group and hence an important source for the theory and practice behind small group work, Rogers articulates for composition teachers 1960's values. Viewing the teacher as facilitator, he details here the teacher's appropriate classroom and paper-marking persona.

Searle, Dennis, and David Dillon. ''The Message of Marking: Teacher Written Responses to Student Writing at Intermediate Grade Levels.'' *Research in the Teaching of English* 14 (October 1980), 233–42.
A study of the kinds of responses teachers make to papers reveals overwhelming attention to form paid through general evaluatory comments and corrections of mechanical errors.

Siegel, Muffy E. A. ''Responses to Student Writing from New Composition Faculty.'' *College Composition and Communication,* 33 (October 1982), 302–309.
Effective response strategies emerge incidentally from this study of how ''newly recruited'' non-English Department teachers drafted into a freshman composition program differ from ''very experienced teachers'' when responding to student themes. The study was done to determine what special training new teachers most require.

Sloan, Gary. ''The Perils of Paper Grading.'' *English Journal,* 66 (May 1977), 33–36.
Criticizes inappropriate responses: nitpicking attention to ''archaic conventions,'' compulsive revising of students' styles, and indignant partisan attacks on students' ideas. Appropriate responses suggested.

Sommers, Nancy. ''Responding to Student Writing.'' *College Composition and Communications,* 33 (May 1982), 148–56.
Discusses how to respond to papers as works in progress.

Sternglass, Marilyn. ''Applications of the Wilkinson Model of Writing Maturity to Col-

lege Writing.'' *College Composition and Communication,* 33 (May 1982), 167–75.
Describes a four-way model of stylistic, affective, cognitive, and moral development by which to analyze student writing more effectively and objectively.
Taylor, Winnifred F., and Kenneth C. Hoedt. ''The Effect of Praise upon the Quality and Quantity of Creative Writing.'' *Journal of Education Research,* 60 (October 1966), 80–83.
Studying the effects of praise and blame upon elementary school students ''supports the assumption that praise without correction . . . is superior to blame.''
Wells, Carlton, F. ''Ten Points for Composition Teachers.'' *English Journal,* 55 (November 1966), 1080–81.
Offers ten guidelines for reading and evaluating essays.
Williams, Joseph. ''The Phenomenology of Error.'' *College Composition and Communication,* 32 (May 1981), 152–68.
Argues that if teachers were not looking for usage and syntactic errors, they would not notice them.

II. Revision: Implications for Response

In discussing the nature of revision and its place in the composing process, the following articles, intended as an introductory sampling, offer or imply guidelines for responding to student writing as products in process.

Beach, Richard. ''Self-Evaluation Strategies of Extensive Revisers and NonRevisers.'' *College Composition and Communication,* 27 (May 1976), 160–64.
Informal study of extensive revisers' and nonrevisers' self-evaluation strategies.
Bridwell, Lillian S. ''Revising Strategies in Twelfth Grade Students' Transactional Writing.'' *Research in the Teaching of English,* 14 (October 1980), 197–222.
The author's study finds that better revisers are better writers.
Crabbe, Kathryn. ''The Composing Process of Mature Adults.'' ERIC ED 123 374, 1976.
Discusses the kinds of writing likeliest to be revised holistically.
Faigley, Lester, and Stephen Witte. ''Analyzing Revision.'' *College Composition and Communication,* 32 (December 1981), 400–407.
Explores how writers revise.
Flanigan, Michael C., and Diane Menendez. ''Perception and Change: Teaching Revision.'' *College English,* 42 (November 1980), 256–66.
Describes how to teach revision.
Flower, Linda. ''Writer-Based Prose: A Cognitive Basis for Problems in Writing.'' *College English,* 41 (September 1979), 19–37.
Describes the nature and heuristic uses of writer-based prose, suggesting the roles of response and revision in turning writer-based into reader-based prose.
Flower, Linda, and John R. Hayes. ''Problem Solving Strategies and the Writing Process.'' *College English,* 39 (December 1977), 449–61.
Studying good and poor writers, the authors conclude poor writers have a limited repertoire of techniques. Alternate problem solving techniques that can be offered to students are described.
———. ''Writing as Problem Solving.'' ERIC ED 172 202, 1979.
Implications for response are found in this model of the composing process that

sees writing as goal-directed (with goals possibly modified in process) and writing processes hierarchically organized and recursive.

Hawkins, Thom. "Intimacy and Audience: The Relation between Revision and the Social Dimension of Peer Tutoring." *College English,* 42 (September 1980), 64–68.
Discusses the importance of peer tutors and the personal, friendly, cooperative relations they establish with students to help them gain a sense of audience and use their oral skills to develop necessary revision abilities.

Kirby, Dan R., and Tom Liner. "Revision: Yes, They Do It. Yes, You Can Teach It." *English Journal,* 69 (March 1980), 41–45.
Discusses the principles underlying effective revision and offers guidelines for encouraging such revision.

McDonald, W. U., Jr. "The Revising Process and the Marking of Student Papers."
See above.

Murray, Donald M. "Internal Revision: A Process of Discovery." In *Research on Composing,* ed. Charles R. Cooper and Lee Odell. Urbana, IL: National Council of Teachers of English, 1978. Pp. 85–103.
Argues that revision is not editing or proofreading and should not be seen as punishment but as a necessary part of the discovery process. The author distinguishes internal revision (trying to discover one's message) from external (adapting one's text for an audience).

———. "The Maker's Eye: Revising Your Own Manuscripts." *The Writer,* 86 (October 1973), 14–16.
Offers a sequenced list of revision tasks pursued through critical rereading of one's text.

———. "Teach the Motivating Force of Revision." *English Journal,* 67 (October 1978), 56–60.
Discusses how to teach revision as a necessary, exciting part of writing as a process of discovery.

Onore, Cynthia S. "Revision, Learning, and the Myth of Improvement." CCCC Presentation, March 1982.
Notes that revision requires the reconstruction of the existing text, thus involving risk-taking. Therefore, to equate revision with improvement is wrong, for revisions need not be better.

Schwartz, Mimi. "Revision Profiles: Patterns and Implications." *College English,* 45 (October 1983), 549–58.
Guidelines for fostering "versatility and perspective in revision" are offered in the form of a typology of nine styles of revision good writers adopt.

Sommers, Nancy. "Revision Strategies of Student Writers and Experienced Adult Writers." *College Composition and Communication,* 31 (December 1980), 378–88.
Critiques linear models of the composing process that fail to consider the recursive shaping of thought and so make revision "both superfluous and redundant." Experienced writers hold a nonlinear sense of composing.

Stallard, Charles. "An Analysis of the Writing Behavior of Good Student Writers." *Research in the Teaching of English,* 8 (Fall 1974), 206–18.
Comparing good and poor writers, Stallard finds good writers revise more extensively than do poor writers.

Witte, Stephen P. "Topical Structure and Revision: An Exploratory Study." *College Composition and Communication,* 34 (October 1983), 313–41.
Explores "some of the textual causes and effects of revision which previous research has not examined."

III. Alternative Sources of Response

Beaven, Mary. "Individualized Goal Setting, Self-Evaluation, and Peer Evaluation." In *Evaluating Writing,* ed. Charles R. Cooper and Lee Odell. Urbana, IL: National Council of Teachers of English, 1977. Pp. 135–56.
The underlying assumptions or rationale, advantages, and disadvantages of these three methods for sharing with students the responsibility for describing and measuring their growth are detailed. Contains an extensive bibliography.

Beck, James P. "Asking Students to Annotate Their Own Papers." *College Composition and Communication,* 33 (October 1982), 322–26.
Describes a method for soliciting self-annotated work whereby students mark "which techniques they think they used, where, and . . . how well, or ill, they judge they used these techniques." Sample annotation instructions are given.

Bormann, Ernest, and Nancy Bormann. *Effective Small Group Communication.* Minneapolis, MN: Burgess, 1972.
Background reading on small group dynamics.

Bruffee, Kenneth A. "Two Related Issues in Peer Tutoring: Program Structure and Tutor Training." *College Composition and Communication,* 31 (February 1980), 76–80.
Covers two logistical issues regarding the use of peer tutoring.

Carnicelli, Thomas A. "The Writing Conference: A One-to-One Conversation." In *Eight Approaches to Teaching Composition,* ed. Timothy R. Donovan and B. W. McClelland, Urbana, IL: National Council of Teachers of English, 1974, Pp. 182–90.
Advice on conducting conferences.

Collins, James L. "Dialogue and Monologue and the Unskilled Writer." *English Journal,* 71 (April 1982), 84–86.
Explains the need for oral response to unskilled writers' work.

Dieterich, Daniel J. "Composition Evaluation: Options and Advice." *English Journal,* 61 (November 1972), 1264–71.
Discusses such alternatives to written teacher responses as use of lay readers, computer evaluation, taped response, peer criticism, and nonresponse. An annotated list of ten ERIC documents covering these and other related subjects (reliability; psychological effects of response on students) concludes the essay.

Elbow, Peter. *Writing Without Teachers.* New York: Oxford University Press, 1973. See particularly chapter four, "The Teacherless Writing Class."
Offers advice on how to respond to writing and how to receive comments in small groups. Among Elbow's goals: to make students self-editing.

Ellman, Neil. "Peer Evaluation and Peer Grading." *English Journal,* 64 (March 1975), 79–80.
Advocates keeping evaluation and grading separate, the latter to be discouraged.

Garrison, Roger H. "One-to-One Tutorial Instruction in Freshman Composition." *New Directions in Community Colleges,* 5 (1974), 55–84.

Details the theory and practice of the "Garrison method" of tutorial workshops.
Hardaway, Francine. "What Students Can Do to Take the Burden off You." *College English,* 36 (January 1975), 577–80.
Discusses advantages to students and to teachers of small group work, peer evaluation, and conferences.
Harris, Muriel. "Evaluation: The Process for Revision." *Basic Writing,* 1 (Spring/Summer 1978), 82–90.
Describes how to teach students to be self-editing by using peer and teacher responses in small groups, conferences, and paper annotations, explaining how to respond at prewriting, first draft, and revision stages of the composing process.
Hawkins, Thom. "Group Inquiry Techniques for Teaching Writing." ERIC ED 128 813, 1976.
Describes the theory, practice, and advantages of small group workshops, offering comments on teacher and student roles, task making, sources of response, grouping techniques, teaching strategies.
———. "Intimacy and Audience."
See above.
How to Handle the Paper Load. See above.
Sections two ("Teacher Involvement—Not Evaluation"), three ("Student Self-Editing"), and six ("Alternative Audiences") offer essays on how to respond to papers orally in class, how to use error analysis to enable students to recognize and correct their own errors, how to question students to get them to be self-evaluating and how to respond to their responses, how to set up and run small group workshops and other methods of eliciting peer responses—all with the aim of reducing the time spent marking papers.
Jacko, Carol H. "Small-Group Triad: An Instructional Mode for the Teaching of Writing." *College Composition and Communication,* 29 (October 1978), 290–92.
Describes a structured form of small group discussion.
Lamberg, Walter. "Self-provided and Peer-provided Feedback." *College Composition and Communication,* 31 (February 1980), 63–69.
Surveys studies (including dissertations) indicating the value of peer- and self-evaluation, offering ways of eliciting apposite feedback and a theory, holding "important implications for instruction," of how "feedback operates in a learning situation" to motivate students. Author explains why too much response is counterproductive. Endnotes references additional material on these alternate sources of response.
Lapidus-Saltz, Wendy. "The Effective Feedback Script: A Peer Response Procedure." *The Writing Instructor,* 1 (Fall 1981), 19–25.
The offered procedure follows a quick review of some studies and theories of peer response.
Lynch, Denise. "Easing the Process: A Strategy for Evaluating Compositions." *College Composition and Communication,* 33 (October 1982), 310–14.
Presents a method of peer- and self-evaluation to help "support students at critical points in the composing process: the pre-writing or thinking stage, the rough draft, the final draft, and the revision" via use of question lists and analytic scoring sheets. Audience analysis stressed.

Miller, Susan. ''How Writers Evaluate Their Own Writing.'' *College Composition and Communication,* 33 (May 1982), 176–83.
Discusses the importance of self-evaluation in the composing process and in the development of writing skills.

Murray, Donald M. ''Teaching the Other Self: The Writer's First Reader.'' *College Composition and Communication,* 33 (May 1982), 140–47.
Describes using conferences (and class discussions) to develop students' self-editing and evaluation abilities.

Peckham, Irvin. ''Peer Evaluation.'' *English Journal,* 67 (October 1978), 61–63.
The author describes his way of incorporating peer evaluation into his classes.

Rogers, Carl R.
See above.

Rose, Alan. ''Spoken versus Written Criticism of Student Writing: Some Advantages of the Conference Method.'' *College Composition and Communication,* 33 (October 1982), 326–31.
Details the advantages of conferences over written responses and offers excerpts from taped conferences to suggest how productive meetings may be conducted.

Ruin, Lois. ''Exploration of the Writing Experience: A Way to Improve Composing.'' *College Composition and Communication,* 34 (October 1983), 349–55.
Provides a program using diary-keeping and classroom instruction to encourage self-exploration to strengthen students' awareness of and attitudes toward composing.

Tirrell, Mary Kay. ''The Writing Conference and the Composing Process.'' *The Writing Instructor,* 1 (Fall 1981), 7–14.
Reviews the work of Roger Garrison, Charles Cooper, Donald Murray, Deanna M. Gutschow, Suzanne E. Jacobs and Adela B. Karliner on using conferences to respond to all stages of texts-in-process.

Tirrell, Mary Kay and Carolyn R. Brinkman. ''Useful Sources on Conferencing.'' *The Writing Instructor,* 1 (Fall 1981), 15–18.
Annotates twelve discussions on conferences as sources of response.

Jerry Farber

Learning How to Teach: A Progress Report

It was with sympathy and a certain nostalgia that I read Mary Rose O'Reilly's " 'Exterminate . . . the Brutes'—And Other Things That Go Wrong in Student-Centered Teaching," in which she looks retrospectively at my essay, "The Student as Nigger," in relation to her own history as a teacher. She took me back to a time—say, 1970—when many of us were trying to translate a radical critique of the education system into classroom practice. Free the slaves? Fine, no problem—on paper. But meanwhile the fall semester was starting, and just what was it exactly that we were supposed to be doing in there?

So my office mate, Jackie, walked into Intro to Lit and said, "OK, people, it's your class. What do you want to do with it?" But, of course, they didn't want to do anything with it. They didn't even want to be there. It was a GE class, for God's sake. Was she kidding?

Well, some teachers persisted—a few with methods so explosive that they were blown right out of the system. Others fell back to what was more or less the liberal position: business as usual, but conducted in a somewhat less arrogant, somewhat more humane manner. Padded handcuffs, you might say. Still others did a flip-flop: "Exterminate the brutes." Their short-lived radicalism went in the trash with the beads and the ankhs. Now, twenty years later, wouldn't you say it's the liberal position that has prevailed? Isn't that part of what grade inflation is about? On the one hand, grading itself isn't about to budge: it's not merely essential to the system; in a way, it is the system. But on the other hand, it just doesn't command the faith that it used to. Besides, who wants to cause students unnecessary grief? For many of us, cruelty, too, isn't the rush it used to be in antebellum days. So grades creep up. Not a courageous alternative, perhaps; certainly it doesn't address the real contradictions. Still, I suppose there's a certain wisdom to it.

But we're talking about 1970. And the problem, in a nutshell, was that we had no new body of technique to replace the old. We—those of us who persisted—had to invent it.

For a few years my own approach tended to be head-on and at full speed: "Your authenticity or your life!" Long, long silences. Sermons. Zen weirdness. Classes held in people's living rooms. Dancing. Darkness. It was a fruitful period actually. But there was this eternal focus on the process. I felt like those couples who do nothing but

From *College English* 52 (February 1990): 135-41.

discuss their own relationship; what I wanted to do was get on with it. What I wanted to do was teach comparative literature. And writing. And to do it as though the revolution had already been won.

A major obstacle was clearly the grading system. You can tell students anything you want about "taking responsibility" and "thinking for yourself." The grading system you employ—a middle finger extended before them—is always more eloquent still. It has nothing to do with literature, it has nothing to do with composition; that it appears to is only a measure of the way it has reshaped these subjects in its own image. For many of our students (and for some of us perhaps), grading constitutes a sort of Marxian base, a bedrock reality, compared to which the actual subject matter itself seems secondary and less real. The grading system tends to drive a class. What it does to students is in great part what "The Student as Nigger" is all about.

In an admirable earlier essay, "The Peaceable Classroom," O'Reilly suggests that "as teachers in the humanities, the most important thing we are able to do is encourage our students in their exploration of the inner life" (110). But when we grade a person C+ on her inner life, "encourage" takes on a more sinister meaning. I think of that admiral, executed, Voltaire tells us, *"pour encourager les autres."*

We complain about students' intellectual passivity, or even torpor. We mock their obsession with "what's going to be on the final." But that's like complaining about the laziness of slaves. Who do we think created these monsters? Is it really true that things have changed, that, as O'Reilly maintains, our students "have been dosed with progressive education since pre-school"? I'm not sure what she means by "progressive" or what may be happening in St. Paul, but in San Diego, where I have three children in the public school system and one in the university, there has been little structural change. I've seen good teaching occasionally, especially at the pre-school and elementary levels where there's a little more room to move. But over all, schools still foster that "slave mentality" which I described in 1967 and which O'Reilly says she continues to encounter. When are we teachers going to understand that much of what we deplore in students is actually functional, that they are responding with at least some intelligence, not to what we say, but to what we do?

This is not the place, however, to renew this attack (see Farber, *A Field Guide*, especially ch. 14: "Intro to Lit, MWF 9." For a revealing study of grading, including a review of research, see Milton, Pollio, and Eison). What I'm talking about is the search for alternatives.

By 1970 I had tried a number of unsatisfactory substitutes for traditional grading. I'd learned that you couldn't simply give grades away; in the high-pressure environment of a university, your course would cave in. Not only that, you would begin to attract the people you least wanted to see. Another possibility, having members of the class grade themselves, never did appeal to me, except as a sort of cruel screening test: OK, who here has brains enough to give themselves an automatic A?

One aspect of the grading system clearly seemed worth retaining; it made sense, at least in many instances, for a teacher to certify that a student had, in fact, completed a course: "Yes, this person has done beginning calculus"; "Yes, this person has completed two years of music theory." In the absence of GPAs, decisions about employment or admission to graduate school could be made more wisely on the basis of portfolios, letters of recommendation, lists of courses completed, and so on. Credit/no credit grading (*not* pass/fail), therefore, seemed like a reasonable compromise. But academic institutions have been reluctant to give students that much room, and credit/no credit grading, where it does appear, tends to be so hedged in with restrictions that it plays a negligible role in the overall process. Not only that, the competitive pressure of traditionally graded courses may draw a student's energy away from an isolated credit/no credit course.

My own solution was to develop, beginning in the early 70s, a sort of multileveled de facto credit/no credit system, similar to what is sometimes called contract grading. For each course, I would set up a group of core requirements corresponding to a C grade. In a literature course, this might include regular attendance and a reading journal—covering, say, nine or ten works—of a certain type and amplitude. Someone pursuing a B or an A would do additional projects. But nothing would be graded; projects were either acceptable or not. A reading journal, for example, would have to be original, of a certain length, reasonably specific in dealing with the work, and so on. But what they said in it was not graded, though I would respond as helpfully and supportively as I could in marginal notes. Grades below C tended to show up as Incompletes. As for tests, I simply stopped giving them.

What I was after was to de-emphasize grading as much as possible and get it out of our way, while avoiding the unpleasant consequences of a grade giveaway; to provide people with as much freedom as could be realized within this institutional setting; and to come up with the kinds of projects that would best help them learn, while reducing the anxiety that some people have about writing and reading literature. To get, in other words, the worm out of the rose.

This arrangement, however, is vulnerable to attack from two directions. From one side, someone could say, "Look, you're still pushing students around. You're still keeping them in bondage; they're just happier about it." To some extent, it's a fair criticism, I think. And yet, this system, which I've been using and refining for eighteen years, has made a spectacular difference, has been more liberating than my silences and sermons of twenty years back. People clearly thrive in this (relative) freedom. Whether in a Proust seminar or in a GE comp class, I'm continually seeing them discover their own motivation, their own authenticity, their own relation to the subject matter—to the point that, if I were required to go back to traditional grading, I couldn't and wouldn't do it; it would be worse than returning from the mountains to the smog.

But then there's the argument from the other side. "Exactly!" someone might say. "While you're playing Heidi in the mountains, we smog-bound Frankfurt folk are maintaining the productivity and the standards that keep civilization alive." This critic could point out that, though my approach to grading may well guarantee that anyone who gets through the course has met the core requirements, it doesn't adequately differentiate between A, B, and C students, since a very hard-working student might pick up an A without doing what might generally be regarded as A-level work. That's exactly right, and I could care less. As a matter of fact, I would be delighted to see that person working so hard. If 18 out of 25 students in a comp class opt for the A, it's going to make my grade distribution look immoral and set the red lights flashing in the SDSU data processing system; but, in fact, I'm going to be teaching more to more people than I ever could before. Does this lower standards—or raise them?

Unfortunately, this approach to grading, as my colleagues who have tried it have learned, tends to create more work for the teacher. That's one real drawback. Another, of course, is that, for a teacher, grading in this way is not exactly dressing for success. In most universities—until the climate changes—an untenured faculty member who wants to do this kind of grading may have to lay low for a few years. Because clearly it doesn't work to use this type of system and try at the same time to keep your class GPA respectable. I know people who have tried this, or who have tried to combine this system with traditional testing and grading. Their students tend to feel betrayed or confused. And with good reason.

Though it's by no means a trivial change, this alternative grading system hardly constitutes a full solution to the problems I raised in "The Student as Nigger." Schools aren't the way they are by accident. They fit the society and must be addressed in social, as well as in individual, terms. But still, what an individual teacher does matters.

Resolving the grading problem, if only partially, may allow us a clearer view of other teaching problems, one of which is raised by O'Reilly's use of the term "student centered." It was, she tells us, while developing a conference paper on "student-centered teaching" that she found herself confronting a class of freshmen who were "stonewalling" her—which led her to scrawl "Exterminate all the brutes" across her conference proposal and even to conclude, albeit somewhat whimsically, that the Heart of Darkness is "an appropriate metaphor for the classroom in more ways than one." And she tells us that "the natural work of young people is to subvert and challenge the authority of teacher and parent, no matter how 'enlightened.' . . . They are, in some strategic sense, the opposition" (" 'Exterminate . . . the Brutes' " 142, 145).

How discouraging it must be for a "student-centered" teacher to be forced to conclude that students are the opposition. "The horror!" But I would suggest that there is, in fact, a correspondence between these two terms: "student-centered" and "the opposition." One implies the other; both imply the *other*. I picture a weary, decent, even selfless person toiling among the—what? Disadvantaged? Natives? Savages? "Mr. Kurtz," O'Reilly writes, "like so many of us, begins as an idealist" (142).

But "students," I would argue, don't exist, unless we create them. How ironic that we who concern ourselves with language and who, in our sophistication, distance ourselves from so many categories, so many social constructs, should cling so tightly to this one. But of course we do; it entitles us—literally. We've been willing to deconstruct everything in sight except our own august podium. "Student" is not merely someone who happens to be, say, learning Italian. It is a member of a group onto which we shamelessly project our own corporate and individual mythologies and in relation to which we construct our own identity as teachers. Among us, the brutal overseer and the idealizing social worker are inhabiting the same metaphysic. Given the right circumstances one can convert to the other.

But if "students" don't exist, who is it that I'm teaching? Well, we would have to talk about the people in my classes this semester, wouldn't we? A dizzyingly diverse group of individuals. But I can tell you one thing: they're not the opposition. That is, I don't see them that way. Perhaps I'm lucky: during my first years of teaching I was immersed in the civil rights movement—and the people I was spending all my time with, partying with, sitting through meetings with, demonstrating with, going to jail with, were very much the same age as the people in my classes, in some cases *were* the people in my classes. For me and for other teachers in this position, it was simply not the environment for learning how to be a professor, not even a "student-centered" one.

What does become of "teacher" if there are no "students"—or if, to put it another way, we are all students, some of whom have made a career out of helping others learn? Well, as a relatively straightforward job description, "teacher" will hold. But the rest—the brahmin claptrap, the mystification, those stupid nostalgia-for-feudalism titles that we hide behind—none of that helps us teach; it gets in the way; it is to be outgrown. One advantage of an alternative grading system is that, as grading becomes less central, the traditional teaching persona begins to seem less and less appropriate. Not that students can't be accomplices in this whole con game. Especially graduate students, many of whom, I've found, love you to live out the whole stuffy charade. After all, they're buying into this hierarchy; they want the incense, they want the Latin mass.

This isn't an essay, really, about how to teach. The changes I'm advocating don't magically turn someone into an effective teacher. But they do create a new framework within which one can continually learn how to teach. Continually—because this learning isn't something you do once and for all, as though teaching were a solid piece of property that one could finally acquire and own. For three semesters now, some of us in my department have been meeting in a faculty teaching seminar—a series of informal and practical discussions, each focused on a specific teaching challenge or problem. Some in the group are fairly new to teaching; others have been at it a long time. Is there a point at which a teacher should lock it all up and say, "Now I know how"? I don't think so. If we approach teaching as a career-long process of constant renewal, we're going to have to work much harder at it, but, as students may learn when they take a fascinating and very demanding course, some things can be more work and yet less like work.

We dream of pinning a course down, putting it in the bank, so to speak. But that's not possible. Everything changes: the subject matter, the people in our classes, we ourselves. And even if everything stayed put, the course would still begin to die, a little more each term. A new set of insights that I roared into class with four semesters ago may by now be turning into little more than lecture notes, something joylessly to be covered, one step closer to the end of the term (though, fortunately, approaches that involve the whole class—in something other than fill-in-the-blanks fashion—tend to fade more slowly). Part of what is at issue here is presence, which is not, perhaps, the flavor of the month these days in literary theory, but which may be the most important, and elusive, dimension in teaching. The flagrant absence of presence is the tour guide syndrome. Minimum effort. Zombie city: a locale many of us will remember well from our undergraduate days.

In the conclusion of her essay, O'Reilly raises the "issues of love and power." "We have inherited our fathers' light saber," she writes, "and we have to learn how to use it. The worst thing we can do is pretend we don't have power." In this connection, she finds "disquieting" my contention that students should teach their teachers to thrive on love and admiration rather than fear and respect. "In practice," she says, "nothing messes up my classroom faster than love" (146).

But what "The Student as Nigger" was describing was power-tripping, a "disease of power." As teachers, we tend to grab power, to exercise it far beyond what is necessary; we get hung up on it, we misuse it. Too often we keep for ourselves the power that we should be helping others to develop (the way literature typically has been taught provides an excellent example of this). And all of this power-tripping, unfortunately, fits into larger patterns of social oppression.

No, we mustn't pretend that we don't have power. I can hardly see how power relations in themselves could be excluded from any notion of community, which is my own understanding of what a class or learning group, at its best, should approach. Nor are they incompatible with "love and admiration." Is love too strong a word here? Perhaps it is. But, though I have often been conspicuously angry at the foolishness of teachers, there have been teachers of mine whom I think I've loved: one whom I saw and heard once for a few hours, and that was all, and enough; one who used to work as a sort of housekeeper half the year and as a visiting professor the other half, moving from one university to another and leaving a trail of students behind her who would write her long letters and for whom she remained more present than the teachers whose classes they continued to attend; one who taught me English poetry and was so overwhelmingly my model of "the professor" that the enormous distance I have had to cover in moving away from him remains as a kind of tribute.

If our ambitions for teaching are modest, perhaps we can hope to exclude this messy and problematic mode and run things on a clear-cut business basis. Certainly, I'm all for demystifying teaching and for deflating the romanticized image that people can sometimes form of a teacher. But if teaching addresses and affects what matters most in human life, if it insists on overflowing the borders that are laid out for it, if it exists by virtue of the possibility of transformation, then, though we may well choose to play down love, to let it work in silence—I'm more than comfortable doing that—we shouldn't deceive ourselves about it. Reading Mary Rose O'Reilly's witty and compassionate essays, I realize how much I, as a student, would have appreciated her, and I think, "If love messes up your classroom, O'Reilly, then you're probably going to have to learn to live with the mess."

Works Cited

Farber, Jerry. *A Field Guide to the Aesthetic Experience.* North Hollywood: Foreworks, 1982.

———. "The Student as Nigger." *The Student as Nigger: Essays and Stories.* New York: Pocket, 1970, 90-100.

Milton, Ohmer, Howard R. Pollio, and James A. Eison. *Making Sense of College Grades: Why the Grading System Does Not Work and What Can Be Done About It.* San Francisco: Jossey-Bass, 1986.

O'Reilly, Mary Rose. " 'Exterminate . . . the Brutes'—And Other Things That Go Wrong in Student-Centered Teaching." *College English* 51 (1989): 142-146.

———. "The Peaceable Classroom." *College English* 46 (1984): 103-112.

AUDIENCES

Douglas B. Park

The Meanings of ''Audience''

As a subject for theory and for the teaching of writing, audience is obvious, crucial, and yet remarkably elusive. In Aristotelian terms ''audience'' points to the final cause for which form exists, to the purposefulness—or its lack—that makes a piece of prose shapely and full of possibility or aimless and empty. Within certain contexts, most notably those of classical rhetoric and argumentation, audience is relatively easy to talk about. Yet outside those contexts our grasp on it is much hazier than I suspect we would like to admit—especially given the importance the subject automatically assumes in any discussion of rhetoric. The very concept of audience applied to written discourse is itself far from straightforward. As Walter Ong has pointed out, writing can, in a strict sense, have only readers, not the collectivity of an audience (''The Writer's Audience Is Always a Fiction,'' *PMLA,* 90 [1975], 11). And, in fact, theoretical discussions of the nature of audience in written discourse are relatively rare. Evidently, though, since the term survives and flourishes, it carries important meanings that ''readers'' does not, although most talk and writing on the subject maintain the distinction between the ''audience'' and ''readers'' only tacitly and often ambiguously.

In practice, locating and discussing the audience for a given piece of prose can be frustrating. The familiar question, ''Who or what (a suggestively impersonal pronoun) is the audience for this piece?'' may prompt a ready answer, but equally often it suggests little, drawing especially blank looks from students. Usually the problem is one of finding or of pinning down satisfactory terms in which the question may be answered: ''People who do not approve of X''; ''readers of the *Atlantic Monthly*''; ''the layman''; ''the teacher.'' Each of these hypothetical answers assumes a different way of conceiving audience, a different principle of definition. How are these different ways of conceiving audience related? What makes one or another more or less useful for discussing a given piece of writing?

A more important question for teachers of writing: How does audience manifest itself to writers writing? The advice, ''Consider your audience,'' may appear to mean that the writer is to concentrate upon some particular person or persons, an implication that is patently far too simple. Only sometimes does considering audience mean directly considering particular people; more often it means something much hazier. Writers can attend to a number of different kinds of issues when they think about audience. And, further, for a writer writing a memo to the boss or a writer writing a column for the local newspaper ''considering audience'' may mean quite different things, not just differences of attending to different qualities of the audience but marked differences in the

From *College English* 44 (March 1982): 247-57.

matters attended to and the ways they are present to the writer. What are the different kinds of meanings "audience" can have for writers writing in different kinds of rhetorical situations?

Full answers to these questions are beyond the scope of this essay, but the questions and the absence of ready answers illustrate the elusiveness of audience in written discourse. The term "audience," old and powerful as it is in the rhetorical tradition, might almost be said to mean too much, to block thought by making us think we know what we are talking about when we often do not. In what follows, I want to open further this problem in meaning, to clarify some of the conceptual traps in the way "audience" is typically used, and to suggest some general reference points that may be useful in thinking about the theory and the teaching of audience.

One quick way to see how many things "audience" means is simply to note how we talk about what writers do. Writers, we most commonly say, adjust to audiences or accommodate them, but we also talk about writers aiming at, assessing, defining, internalizing, construing, representing, imagining, characterizing, inventing, and evoking audiences. Each of these verbs seems to capture some of the truth about audience in different situations or seen from different perspectives, yet the imaginative dynamics they suggest are fascinatingly different. Consider, for example, the different implications of two terms even as close together as "assessing" and "defining." One suggests that audience is a given to be carefully observed and analyzed; the other that audience is something unclear to be shaped or brought into clearer focus.

Even though the range of apparent conceptions of audience and of the writer's relationship to it is bafflingly rich, the range does have identifiable extremes. At one end words like "adjust" and "accommodate" convey the familiar notion of audience as something readily identifiable and external, requiring appropriate responses and strategies. Lloyd Bitzer's definition of the rhetorical situation is a useful reference point here, since it so unequivocally presents external circumstances as forming a defining context to which discourse must respond in fitting ways. ("The Rhetorical Situation," *Philosophy and Rhetoric,* 1 [1968], 1-15). The audience, in this view, is a defined presence outside the discourse with certain beliefs, attitudes, and relationships to the speaker or writer and to the situation that require the discourse to have certain characteristics in response. In Bitzer's terms the more structured the rhetorical situation, the more precise its characteristics, including those of the audience, the more it determines the specific features and content of the discourse.

At the other extreme, with verbs like "construe" and "invent," is the notion that Walter Ong explores at length in "The Writer's Audience is Always a Fiction." However real the readers are outside the text, the writer writing must represent an audience to consciousness in some fashion; and the results of that "fiction" appear in what the text appears to assume about the knowledge and attitudes of its readers and about their relationship to the writer and the subject matter. Particularly in cases where no clear structure, no readily identifiable audience exists—the situation free-lance writers and composition students face all the time—the writer must, in some sense, invent an audience. More accurately, the writer must create a context into which readers may enter and to varying degrees become the audience that is implied there. An article, let us say, on how to plant asparagus roots may evoke as its audience the dedicated home gardener filled with enthusiasm for hard work and fresh vegetables. Particular readers may fit this implied audience well; they may tolerate the definition to get the necessary information;

or they may read out of idle boredom as casual spectators amused at the eager enthusiasm of the implied audience. Readers may, in other words, be the ''audience'' to varying degrees, or not at all. In this sense the audience may be said to exist in the text—if it can actually be said to exist anywhere.

The meanings of ''audience,'' then, tend to diverge in two general directions: one toward actual people external to a text, the audience whom the writer must accommodate; the other toward the text itself and the audience implied there, a set of suggested or evoked attitudes, interests, reactions, conditions of knowledge which may or may not fit with the qualities of actual readers or listeners.

The first, most literal direction of meaning for ''audience'' more often than not occupies the center of attention because it is so concrete. After all, the basic image from which the concept of audience derives is that of a speaker addressing a group of people in some fairly well defined political, legal, or ceremonial situation. The group of people, the audience, listens intently because they have some specific involvement in the situation. They have a part to play. The speech shapes itself around the fact of their presence and their involvement. This basic image is powerful, easy to hold in mind, and therefore useful. But it also opens up a conceptual trap by making it easy to associate ''audience'' simply and literally with the people listening—all those folks out there in chairs. And in its most common, mundane sense that is all that ''audience'' means. Yet obviously one can listen to a speech or read a work of prose without being in any rhetorical sense a member of the audience. ''Audience'' as we use it in discussions of rhetoric means much more. Its essential rhetorical meaning is something like people-as-they-are-involved-in-a-rhetorical-situation. Further, that idea of people-as-they-are-involved has to do with final and formal cause—what a piece of discourse sets out to do and how it is shaped to accomplish that end. So ''audience'' really uses a very concrete image to evoke a much more abstract and dynamic concept. Whether we mean by ''audience'' primarily something in the text or something outside it, ''audience'' essentially refers not to people as such but to those apparent aspects of knowledge and motivation in readers and listeners that form the contexts for discourse and the ends of discourse.

Probably, it is precisely because ''audience'' has acquired so much abstract meaning that it is retained, however ambiguously, in discussions of written as well as of spoken discourse. ''Readers'' is too obviously literal. ''Audience'' by its literal inappropriateness is free to carry a much richer set of meanings. Note that we speak of how a discourse may affect its readers or of what a discourse assumes about its readers; but we speak of *the* audience of a discourse, by which we often mean an ideal conception, something akin to an informing principle in the work. For this reason we often speak of the audience impersonally as a thing: ''What is the audience?'' Even when we mean by it people outside the text, those people make up a collective entity, exist as an audience, only in terms of their relationship to the text and the relationship of the text to them.

Of course ''the reader'' has come, especially in literary criticism, to carry the abstract rhetorical meaning of ''audience.'' Note Wolfgang Iser's ''implied reader'' and the many recent discussions of the reader of the work. The overlap of different meanings in two terms that are often used interchangeably allows great confusion, especially in the classroom where an instructor might easily be using ''reader'' or ''audience'' in a largely rhetorical sense without being fully aware of it, while the students might be interpreting the terms in much more literal ways.

To specify the range of meaning a bit more at this point, the two general directions of meaning of ''audience''—outside the text and back into the text—divide into four more specific meanings:

1. Anyone who happens to listen to or to read a given discourse: ''The audience applauded.'' This meaning is inextricably rooted in common usage, but it is useless and misleading in serious rhetorical analysis.
2. External readers or listeners *as* they are involved in the rhetorical situation: ''The writer misjudged his audience.'' This meaning of ''audience'' comes into play in analyses of the historical situation in which a given discourse appeared or in studies of the actual effect of discourse upon an audience.
3. The set of conceptions or awareness in the writer's consciousness that shape the discourse as something to be read or heard. We try to get at this set of awarenesses in shorthand fashion when we ask, ''What audience do you have in mind?''
4. An ideal conception shadowed forth in the way the discourse itself defines and creates contexts for readers. We can come at this conception only through specific features of the text: ''What does this paragraph suggest about the audience?''

The last two meanings are obviously the most important for teachers or for anyone interested in forms of discourse. They also identify the aspects of audience that are the most elusive and that give rise to the questions posed earlier. What are the different considerations in writers' minds in different rhetorical situations when they come to terms with audience? What are the features of texts that we most appropriately use to define and discuss audience in different rhetorical situations?

Any systematic answers to these important questions will depend upon keeping in constant view the essential abstractness of the concept of audience. In the case of what Bitzer calls highly structured situations, this abstractness can be easy to miss because the more concrete referents of audience are so available: Ronald Reagan cuts subsidies for mass transit; a committee of mayors drafts a letter to argue for continued support. It is easy to say that Reagan is the audience for this hypothetical letter. But it is not Reagan as Reagan that the letter addresses but Reagan in his position as President and as representative of a set of attitudes on the subject of mass transit. Whatever other notions or knowledge of him as a person the writers may have in mind will have to be screened out as irrelevant. The conventions of the formal letter, levels of diction, tone, themselves help with the screening. As Walter Ong points out, such conventions establish roles for writers and for readers. Then within the limits of those roles, the mayors will proceed according to their estimation of Reagan's attitudes and political position. And more than that, they will try to create in the letter an image of their audience, the President, as they would like him to be—receptive, open-minded, concerned about the cities. So although it is easy, even inevitable, speaking in shorthand fashion, to identify Reagan as the audience for this letter, doing so tells us little. The audience as it exists in the writers' consciousness and as it shapes the text is a complex set of conventions, estimations, implied responses and attitudes.

In the case of unstructured situations where we would call the audience ''general,'' where no simple, concrete identifications of audience are possible, the whole concept becomes much more elusive. Consider a hypothetical article for the *Atlantic Monthly.* One might say that the audience is readers of the magazine, but this external identification says little except perhaps as it suggests certain demographic characteristics—social class, level of education, broad cultural attitudes—which form a vague outer boundary of possible contexts. One might go further and talk about the *Atlantic Monthly* Reader, an obvious fiction something akin to the Marlboro Man. This is a way of evoking the traditions of the magazine itself and, through that, the expectations readers have come to have of its articles. With some magazines, of course, such as the *New Yorker,* these traditions are strongly defined. But in the particular case of the *Atlantic,* the conventions of the particular article—literary essay, investigative reporting, political analysis—seem likely to be more determining.

Then, within the specific article, one must consider audience in terms of the particular contexts the writer will choose and shape. Here it becomes clear that "audience" is merely a rough way of pointing at that whole set of contexts. One can represent all that in shorthand fashion by saying that the audience is people who believe such and such, or who are interested in such and such, or who have a certain level of background knowledge. But a precise analysis of audience would have to examine, point by point, what is being assumed as understood, what is elaborated, what is assumed as the readers' range of attitudes or preconceptions about the subject at hand, and so on. And, in fact, expository prose written in unstructured situations often works by taking a series of bows in various directions, toward one set of possible attitudes in readers, then toward another, by assuming one set of understandings at one point and by filling in background information at another. Writers work, I suspect, on the basis of intuitions about the range of what most readers are likely to know. Individual readers match the assumptions of the discourse to varying degrees, overlooking or ignoring what does not fit or is not comprehensible. Seen from this perspective, the idea of the audience is not just an abstraction but essentially a metaphor. It uses the image of a group listening, immediately involved in a rhetorical situation, to evoke the totality of the assumptions the discourse makes about contexts. It evokes the form of the discourse in the guise of a set of ideal readers or listeners. But this image is itself only a fiction. Except to dramatize some major aspects of the rhetorical situation—"This piece is written for the layman"—there is little point in trying to actually describe *the* audience as an entity. Powerful as the idea of audience is, it may block thought to the extent that it presents as unified, single, locatable, something that, in fact, involves many different contexts dispersed through a text.

To learn how to systematically analyze audience in discourse, therefore, it seems best to avoid the metaphor, to replace the question, "Who is the audience?" with a set of more precise questions as to how the piece in question establishes or possesses the contexts that make it meaningful for readers. These contexts may perhaps be thought of as a set of overlapping boundaries which together delineate the territory we identify as the audience.

The outermost of these boundaries must involve a range of given conventions, from what must be some very fundamental assumptions underlying modern expository prose to more specific conventions governing, say, modern academic prose in the humanities or in the sciences, to even more specific matters such as editorial conventions for form and subject matter in particular journals, *PMLA, Sports Illustrated, National Geo-*

graphic. Some of these conventions are so totally accepted as to be invisible; some are more obvious; some explicitly artificial. All, as part of the occasion of "publication," help form the ground on which writers and readers meet with some shared understanding of means and ends to be served. Walter Ong describes them as defining in basic ways the nature of the relationship between writer, subject matter, and reader (pp. 16-19). For writers they provide some givens which can be assumed about readers' attitudes, even knowledge; for readers they define appropriate expectations. They are, therefore, inextricably involved in the definition of "audience" for a given piece of discourse.

For example, what Chaim Perelman identifies as the Universal Audience is clearly a set of conventionally accepted assumptions about the proper nature of argument: that it be controlled by reason, that all parties place a premium upon disinterestedness and tacitly agree that respect for truth is a prime measure of persuasiveness.[1] And behind these assumptions obviously lie others about the proper role of discourse in society—assumptions which place writers and readers under strong obligations. A reader who will not conform to these assumptions is not, in the most fundamental way, part of the audience to whom the work is directed.

Within the various boundaries of convention, then, occur the more specific audience contexts for any particular piece of discourse. These are the contexts that derive from the relationship of writer and readers to a particular subject matter. How these contexts take shape depends a great deal upon the contexts already established by the outer boundaries of conventions. In the above example, for instance, the conventions of the Universal Audience define some basic assumptions about the nature and social purpose of argument; and it is against the implicitly understood background of those assumptions that a particular argument will address the attitude of its audience on a particular subject. In some cases the occasion of publication itself can establish contexts that are quite specific to subject matter. Specialized magazines such as *Sunset Magazine,* for example, operate on the basis of a high level of interest in or knowledge concerning a particular subject. Accordingly, the articles appearing in such magazines can be seen to assume that certain elements of audience are givens, even though other elements of audience remain to be shaped and explicitly addressed.

To some extent, then, the task of analyzing audience is a matter of identifying the nature of the contexts that are already given by some aspect of the occasion of publication and of understanding the relationship between those that are given and those that must be more explicitly defined within the discourse itself. Another part of the task is understanding how particular contexts are created within the discourse. In general, the process is, as Walter Ong describes it, one of creating fictions. More specifically, in public prose, it is a matter of shaping into a rhetorical situation the potential bits of opinion, knowledge, motives for interest that lie about in the public domain in no particular form. The writer invents, so to speak, their significance and, in so doing, creates an audience. Often, especially in scholarly writing, an article will carefully define a public attitude or state of knowledge in the way that best creates an exigence for the argument to follow. And generally readers tolerate these fictions as long as they are not too obviously contrived or too much at odds with the real state of affairs as the readers perceive it.

---

1. Chaim Perelman and L. Olbrechts-Tyteca, *The New Rhetoric: A Treatise on Argumentation,* trans. by John Wilkinson and Purcell Weaver (Notre Dame, Ind.: University of Notre Dame Press, 1969), pp. 31-35.

We do not, I think, know enough to do either aspect of the task of analyzing audience as well as it might be done. The general importance of conventions in prose has been well described in general terms by Walter Ong and by James Britton in *The Development of Writing Abilities (11-18)* (Urbana, Ill.: National Council of Teachers of English, 1979), pp. 59-62; but, as far as I know, no one has tried to describe for ordinary expository prose what the exact conventions are and how they work. Nor do we know in any systematic way about the strategies by which specific contexts are created in expository prose—although every skilled reader knows intuitively and can do an ad hoc analysis of a particular piece of prose. Even though the variety of strategies is enormous and every piece of prose is, to some extent, unique, it ought to be possible, through critical analysis of a large sample of different kinds of prose, to describe a series of categories and models which would illustrate the range of possibilities and thereby serve as a frame of reference for the analysis of any given piece of discourse.

Identifying more systematically how many different issues lie within the metaphor of audience should also allow a clearer understanding of the subject from the writer's perspective. For writers writing, all things germane to audience can perhaps be described as a field of awareness that can manifest itself in different ways in different rhetorical situations. For example, writers are not, I am quite sure, fully aware of many of the conventions that govern a piece of writing. For a writer, operating comfortably within a given set of conventions is often a matter simply of being on sure ground, of being able to conjure up the right kind of voice, of knowing intuitively that that voice and way of proceeding have "listeners." Such an intuitive sense is surely describable as a sense of audience; we describe the apparent absence of it in a piece of prose as lack of a sense of audience. Yet that kind of intuitive grasp of convention is very little like a conscious focus on the imagined reactions of specific people. In fact, as William Irmscher suggested in a 1979 CCCC paper, a sure intuitive grasp of the appropriate conventions may allow a writer not to be overly concerned with and therefore inhibited by worries about audience. It may allow a more easy concentration upon the subject at hand. On the other hand, some aspects of convention such as specific format are very much a matter of conscious concern for writers, and such formats have a lot to do with audience in that they help identify and fulfill readers' expectations. But, again, writers in thinking about format are likely to be concentrating not upon audience in the conventional sense but upon following the format correctly.

I would guess that as a general rule it is only in highly structured situations or at particular times that writers consciously focus on audience as a discrete entity. Much more often writers, I suspect, think primarily in terms of shaping the material for appropriateness, clarity, accuracy. In so doing, they rely upon partly conscious, partly intuitive knowledge of common strategies for shaping contexts. Much of writers' actual deliberate "audience" analysis probably takes the form of attention on contexts for specific issues, as in the case of much scholarly writing, in which a great deal of analysis goes to determining and then operating within the context of what has been said before on a specific issue.

I would also guess that the crucial issues for a writer approaching a writing task can be categorized in terms of which of the contexts are given and which of them remain to be invented or given further shape. For example, some ways in which aspects of audience may be given, say in argumentative writing, put the writer in the position of responding to or appealing to audience; other ways in which they may be defined allow the writer to assume tacitly, such as with certain aspects of knowledge or motivation. On the other side, some ways in which aspects of knowledge may not be given put the writer in the position of explicitly defining motives for readers to read, or contexts of knowledge or attitude or even justifications for the act of writing itself.

Although the whole question of how writers perceive audience in the process of composing does not seem susceptible to highly detailed description, it ought to be possible, through introspection, interview, and case studies at least to describe the kinds of issues writers deal with consciously, as opposed to intuitively; to describe the kinds of questions writers ask themselves about audience and the kinds of solutions they propose in different rhetorical situations.[2]

Aside from its considerable intrinsic interest, the sort of mapping of the territory of audience described above should be of considerable practical usefulness. Most teaching of audience in composition courses is, I would judge from experience and from textbooks, relatively unsystematic, weak upon theory, heavily dependent upon ad hoc examples. We need to be able to approach the subject more systematically and precisely. But carried just to this point, the preceding analysis suggests some useful observations about the teaching of audience.

The major observation is simply that as teachers of writing we are too often unaware of the rich ambiguity of the term "audience" and depend too heavily upon the concrete image of audience as the readers external to the text. For example, teachers of writing often discuss the merits of using members of the class or the teacher as the audience for student papers. The former strategy, of course, means that students write knowing that their papers will be "published" in the classroom. This practice has obvious powerful effects on how students see the act of writing, but it can be said to provide an audience only in the commonly used sense of external listeners or readers. Students' reading of one another's writing does not provide that crucial ingredient, people rhetorically involved. The student writing for members of the class still has the problem of finding or inventing appropriate rhetorical contexts. In fact, useful as this strategy is, it may also create problems. Some discussions I have had with students in an advanced writing course suggest that their awareness of specific critical readers not sympathetic to the rhetorical situation they wish to create—readers who will not readily become their audience—can be inhibiting and complicates rather than simplifies the problem of dealing with audience.

Students writing for the teacher alone have a similar problem. Some students do make the teacher the exclusive audience in the essentially rhetorical meaning of "audience," but only when the prose becomes nothing more than a personal monologue on paper with all the context for the monologue assumed as given—"You asked me the question, and here's the answer"—or worse yet when the paper becomes a personal letter. Britton's study of audience in *The Development of Writing Abilities* has examples and descriptions of these sorts of situations (pp. 120-121). Obviously, by the time a student reaches secondary and post-secondary writing courses, teachers expect something more,

---

2. The possible fruitfulness of case studies for such research is suggested by the amount of material concerning audience in Linda Flower and John Hayes, "The Cognition of Discovery: Defining a Rhetorical Problem," *College Composition and Communication*, 31 (1980), 21-32.

even if that something remains implicit. They expect situations in which the student writes within some kind of rhetorical context and in which the teacher serves not as audience but as editor and judge of success.

Usually that something more is described, when it is noticed at all, as writing for other audiences, or wider audiences, or a variety of audiences. But it should be clear that these terms hide an enormous range of possibilities, of different kinds of rhetorical situations, of different problems for the writer. At this point, most teachers and textbooks rely upon a range of examples to convey to students the general idea of writing for an audience, and most often the examples tend to be—like mine of the hypothetical letter to Reagan—examples of highly structured situations where the contexts are relatively well defined and external to the text. But the writing tasks which can grow from examples tend to be limited to hypothetical cases: "Imagine that you are. . . ." Aside from the inherent limitations of such assignments, the fact is that most of the time we want students to learn to write for a "general" audience. That is to say, we want them to write in relatively unstructured situations where little is given in the way of context and much remains to be invented by the writer. And this is where most teaching about audience becomes the most ineffective.

The truth is that we demand from students—often without making it clear to them or to ourselves—a considerable rhetorical virtuosity in dealing with and inventing audience contexts. But beyond the rich connotations of "audience" we do not have a precise notion about the nature of that virtuosity. Case by case, assignment by assignment, paper by paper, we may be able to come up with useful suggestions. But what we need is the map of the territory of audience proposed above. We need to be able to place specific assignments or tasks of audience analysis within a larger frame of reference. We need to be able to break audience problems down into specific issues, to identify for students the ways and the strategies by which audience contexts exist in different kinds of prose. And we need to be able to give them better advice about what writers really do when they "consider audience" in different kinds of situations. How far all this is possible is hard to tell without more critical analysis of different kinds of prose and more research into the composing process.

The above analysis also raises a more fundamental issue for the teaching of audience in composition courses. It is not enough to talk only about specific strategies for dealing with or creating contexts for audience. Those contexts take shape against the background of the conventions appropriate for given kinds of writing. Much of the time it is not possible to separate a sense of audience from a sense of genre and convention. The student who tries to write with no clear sense of the kind of thing to be done, its social function, and the conventions appropriate to it flounders in a terrible vacuum. The results will be chaotic in many ways: limping "history of the world" introductions, scattered organization, and hollow diction. But the instructor who recognizes the problems as somehow symptomatic of a problem with audience may still not be able to help the student by talking about audience in any of the more obvious ways. The problem lies deeper than the metaphor "audience" implies.

Probably writers come to have an intimate sense of audience as convention by being readers of that kind of prose. I doubt that it can be taught directly or very quickly. But, in any case, the whole question of kinds of writing to be done is one that composition courses seldom face directly—partly because our profession has not studied seriously the conventions governing nonliterary prose, partly because composition courses do not descend strongly from the rhetorical tradition and have tended to be conceived as teaching general writing skills rather than particular kinds of writing. This conception is defensible, but I believe that it more often serves as a defense against the difficult questions that arise when one asks what kinds of writing are to be taught and what their different values and functions are.

At this point the subject of audience goes beyond the scope of this essay into questions of the history and sociology of our profession. But it seems important to note that audience is elusive in much teaching of writing not only because the concept itself is difficult. A fully serious art of rhetoric and a concomitant sophistication with audience—like that found in the classical rhetorics—must grow from a clear understanding of the kinds of discourse to be served and their purpose in society. Our composition courses generally do not operate with such an understanding.

Lisa Ede and Andrea Lunsford

Audience Addressed/Audience Invoked: The Role of Audience in Composition Theory and Pedagogy

One important controversy currently engaging scholars and teachers of writing involves the role of audience in composition theory and pedagogy. How can we best define the audience of a written discourse? What does it mean to address an audience? To what degree should teachers stress audience in their assignments and discussions? What *is* the best way to help students recognize the significance of this critical element in any rhetorical situation?

Teachers of writing may find recent efforts to answer these questions more confusing than illuminating. Should they agree with Ruth Mitchell and Mary Taylor, who so emphasize the significance of the audience that they argue for abandoning conventional composition courses and instituting a "cooperative effort by writing and subject instructors in adjunct courses. The cooperation and courses take two main forms. Either writing instructors can be attached to subject courses where writing is required, an organization which disperses the instructors throughout the departments participating; or the composition courses can teach students how to write the papers assigned in other concurrent courses, thus centralizing instruction for diversifying topics."[1] Or should teachers side with Russell Long, who asserts that those advocating greater attention to audience overemphasize the role of "observable physical or occupational characteristics" while ignoring the fact that most writers actually create their audiences. Long argues against the usefulness of such methods as developing hypothetical rhetorical situations as writing assignments, urging instead a more traditional emphasis on "the analysis of texts in the classroom with a very detailed examination given to the signals provided by the writer for his audience."[2]

To many teachers, the choice seems limited to a single option—to be for or against an emphasis on audience in composition courses. In the following essay, we wish to expand our understanding of the role audience plays in composition theory and peda-

From *College Composition and Communication* 35 (May 1984): 155-71.

1. Ruth Mitchell and Mary Taylor, "The Integrating Perspective: An Audience-Response Model for Writing," *CE*, 41 (November, 1979), 267. Subsequent references to this article will be cited in the text.
2. Russell C. Long, "Writer-Audience Relationships: Analysis or Invention," *CCC*, 31 (May, 1980), 223 and 225. Subsequent references to this article will be cited in the text.

gogy by demonstrating that the arguments advocated by each side of the current debate oversimplify the act of making meaning through written discourse. Each side, we will argue, has failed adequately to recognize 1) the fluid, dynamic character of rhetorical situations; and 2) the integrated, interdependent nature of reading and writing. After discussing the strengths and weaknesses of the two central perspectives on audience in composition—which we group under the rubrics of *audience addressed* and *audience invoked*[3]—we will propose an alternative formulation, one which we believe more accurately reflects the richness of "audience" as a concept.[4]

Audience Addressed

Those who envision audience as addressed emphasize the concrete reality of the writer's audience; they also share the assumption that knowledge of this audience's attitudes, beliefs, and expectations is not only possible (via observation and analysis) but essential. Questions concerning the degree to which this audience is "real" or imagined, and the ways it differs from the speaker's audience, are generally ignored or subordinated to a sense of the audience's powerfulness. In their discussion of "A Heuristic Model for Creating a Writer's Audience," for example, Fred Pfister and Joanne Petrik attempt to recognize the ontological complexity of the writer-audience relationship by noting that "students, like all writers, must fictionalize their audience."[5] Even so, by encouraging students to "construct in their imagination an audience that is as nearly a replica as is possible of *those many readers who actually exist in the world of reality,*" Pfister and Petrik implicitly privilege the concept of audience as addressed.[6]

Many of those who envision audience as addressed have been influenced by the strong tradition of audience analysis in speech communication and by current research in cognitive psychology on the composing process.[7] They often see themselves as reacting against the current-traditional paradigm of composition, with its a-rhetorical, product-oriented emphasis.[8] And they also frequently encourage what is called "real-world" writing.[9]

Our purpose here is not to draw up a list of those who share this view of audience but to suggest the general outline of what most readers will recognize as a central

---

3. For these terms we are indebted to Henry W. Johnstone, Jr., who refers to them in his analysis of Chaim Perelman's universal audience in *Validity and Rhetoric in Philosophical Argument: An Outlook in Transition* (University Park, PA: The Dialogue Press of Man & World, 1978), p. 105.

4. A number of terms might be used to characterize the two approaches to audience which dominate current theory and practice. Such pairs as identified/envisaged, "real"/fictional, or analyzed/created all point to the same general distinction as do our terms. We chose "addressed/invoked" because these terms most precisely represent our intended meaning. Our discussion will, we hope, clarify their significance; for the present, the following definitions must serve. The "addressed" audience refers to those actual or real-life people who read a discourse, while the "invoked" audience refers to the audience called up or imagined by the writer.

5. Fred R. Pfister and Joanne F. Petrik, "A Heuristic Model for Creating a Writer's Audience," *CCC,* 31 (May, 1980), 213.

6. Pfister and Petrik, 214; our emphasis.

7. See, for example, Lisa S. Ede. "On Audience and Composition," *CCC,* 30 (October, 1979), 291-295.

8. See, for example, David Tedlock, "The Case Approach to Composition," *CCC,* 32 (October, 1981), 253-261.

9. See, for example, Linda Flower's *Problem-Solving Strategies for Writers* (New York: Harcourt Brace Jovanovich, 1981) and John P. Field and Robert H. Weiss' *Cases for Composition* (Boston: Little Brown, 1979).

tendency in the teaching of writing today. We would, however, like to focus on one particularly ambitious attempt to formulate a theory and pedagogy for composition based on the concept of audience as addressed: Ruth Mitchell and May Taylor's "The Integrating Perspective: An Audience-Response Model for Writing." We choose Mitchell and Taylor's work because of its rhetorical richness and practical specificity. Despite these strengths, we wish to note several potentially significant limitations in their approach, limitations which obtain to varying degrees in much of the current work of those who envision audience as addressed.

In their article, Mitchell and Taylor analyze what they consider to be the two major existing composition models: one focusing on the writer and the other on the written product. Their evaluation of these two models seems essentially accurate. The "writer" model is limited because it defines writing as either self-expression or "fidelity to fact" (p. 255)—epistemologically naive assumptions which result in troubling pedagogical inconsistencies. And the "written product" model, which is characterized by an emphasis on "certain intrinsic features [such as a] lack of comma splices and fragments" (p. 258), is challenged by the continued inability of teachers of writing (not to mention those in other professions) to agree upon the precise intrinsic features which characterize "good" writing.

Most interesting, however, is what Mitchell and Taylor *omit* in their criticism of these models. Neither the writer model nor the written product model pays serious attention to invention, the term used to describe those "methods designed to aid in retrieving information, forming concepts, analyzing complex events, and solving certain kinds of problems."[10] Mitchell and Taylor's lapse in not noting this omission is understandable, however, for the same can be said of their own model. When these authors discuss the writing process, they stress that "our first priority for writing instruction at every level ought to be certain major tactics for structuring material because these structures are the most important in guiding the reader's comprehension and memory" (p. 271). They do not concern themselves with where "the material" comes from—its sophistication, complexity, accuracy, or rigor.

Mitchell and Taylor also fail to note another omission, one which might be best described in reference to their own model (Figure 1). This model has four components. Mitchell and Taylor use two of these, "writer" and "written product," as labels for the models they condemn. The third and fourth components, "audience" and "response," provide the title for their own "audience-response model for writing" (p. 249).

Mitchell and Taylor stress that the components in their model interact. Yet, despite their emphasis on interaction, it never seems to occur to them to note that the two other models may fail in large part because they overemphasize and isolate one of the four elements—wrenching it too greatly from its context and thus inevitably distorting the composing process. Mitchell and Taylor do not consider this possibility, we suggest, because their own model has the same weakness.

Mitchell and Taylor argue that a major limitation of the "writer" model is its empha-

---

10. Richard E. Young, "Paradigms and Problems: Needed Research in Rhetorical Invention," in *Research on Composing: Points of Departure,* ed. Charles R. Cooper and Lee Odell (Urbana, IL: National Council of Teachers of English, 1978), p. 32 (footnote #3).

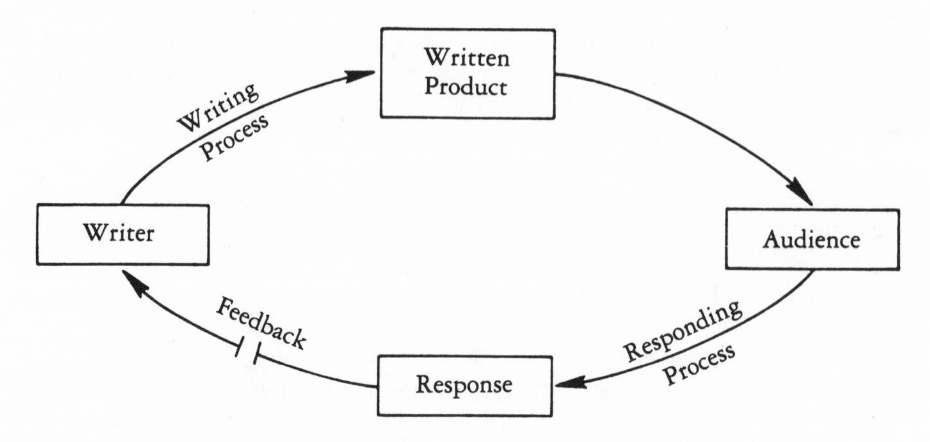

FIGURE 1. Mitchell and Taylor's "general model of writing" (p. 250)

sis on the self, the person writing, as the only potential judge of effective discourse. Ironically, however, their own emphasis on audience leads to a similar distortion. In their model, the audience has the sole power of evaluating writing, the success of which "will be judged by the audience's reaction: 'good' translates into 'effective,' 'bad' into 'ineffective.' " Mitchell and Taylor go on to note that "the audience not only judges writing; it also motivates it" (p. 250),[11] thus suggesting that the writer has less control than the audience over both evaluation and motivation.

Despite the fact that Mitchell and Taylor describe writing as "an interaction, a dynamic relationship" (p. 250), their model puts far more emphasis on the role of the audience than on that of the writer. One way to pinpoint the source of imbalance in Mitchell and Taylor's formulation is to note that they are right in emphasizing the creative role of readers who, they observe, "actively contribute to the meaning of what they read and will respond according to a complex set of expectations, preconceptions, and provocations" (p. 251), but wrong in failing to recognize the equally essential role writers play throughout the composing process not only as creators but also as *readers* of their own writing.

As Susan Wall observes in "In the Writer's Eye: Learning to Teach the Rereading/Revising Process," when writers read their own writing, as they do continuously while the compose, "there are really not one but two contexts for rereading: there is the writer-as-reader's sense of what the established text is actually saying, as of this reading; and there is the reader-as-writer's judgment of what the text might say or should say. . . ."[12] What is missing from Mitchell and Taylor's model, and from much work done from the perspective of audience as addressed, is a recognition of the crucial importance of this internal dialogue, through which writers analyze inventional problems and conceptualize patterns of discourse. Also missing is an adequate awareness that, no matter

11. Mitchell and Taylor do recognize that internal psychological needs ("unconscious challenges") may play a role in the writing process, but they cite such instances as an "extreme case (often that of the creative writer)" (p. 251). For a discussion of the importance of self-evaluation in the composing process see Susan Miller, "How Writers Evaluate Their Own Writing," *CCC*, 33 (May, 1982), 176-183.

12. Susan Wall, "In the Writer's Eye: Learning to Teach the Rereading/Revising Process," *English Education,* 14 (February, 1982), 12.

how much feedback writers may receive after they have written something (or in breaks while they write), as they compose writers must rely in large part upon their own vision of the reader, which they create, as readers do their vision of writers, according to their own experiences and expectations.

Another major problem with Mitchell and Taylor's analysis is their apparent lack of concern for the ethics of language use. At one point, the authors ask the following important question: "Have we painted ourselves into a corner, so that the audience-response model must defend sociologese and its related styles?" (p. 265). Note first the ambiguity of their answer, which seems to us to say no and yes at the same time, and the way they try to deflect its impact:

> No. We defend only the right of audiences to set their own standards and we repudiate the ambitions of English departments to monopolize that standard-setting. If bureaucrats and scientists are happy with the way they write, then no one should interfere.
>
> But evidence is accumulating that they are not happy. (p. 265)

Here Mitchell and Taylor surely underestimate the relationship between style and substance. As those concerned with Doublespeak can attest, for example, the problem with sociologese is not simply its (to our ears) awkward, convoluted, highly nominalized style, but the way writers have in certain instances used this style to make statements otherwise unacceptable to lay persons, to "gloss over" potentially controversial facts about programs and their consequences, and thus violate the ethics of language use. Hence, although we support Mitchell and Taylor when they insist that we must better understand and respect the linguistic traditions of other disciplines and professions, we object to their assumption that style is somehow value free.

As we noted earlier, an analysis of Mitchell and Taylor's discussion clarifies weaknesses, inherent in much of the theoretical and pedagogical research based on the concept of audience as addressed. One major weakness of this research lies in its narrow focus on helping students learn how to "continually modify their work with reference to their audience" (p. 251). Such a focus, which in its extreme form becomes pandering to the crowd, tends to undervalue the responsibility a writer has to a subject and to what Wayne Booth in *Modern Dogma and the Rhetoric of Assent* calls "the art of discovering good reasons."[13] The resulting imbalance has clear ethical consequences, for rhetoric has traditionally been concerned not only with the effectiveness of a discourse, but with truthfulness as well. Much of our difficulty with the language of advertising, for example, arises out of the ad writer's powerful concept of audience as addressed divorced from a corollary ethical concept. The toothpaste ad that promises improved personality, for instance, knows too well how to address the audience. But such ads ignore ethical questions completely.

Another weakness in research done by those who envision audience as addressed suggests an oversimplified view of language. As Paul Kameen observes in "Rewording the Rhetoric of Composition," "discourse is not grounded in forms or experience or

---

13. Wayne Booth, *Modern Dogma and the Rhetoric of Assent* (Chicago: The University of Chicago Press, 1974), p. xiv.

audience; it engages all of these elements simultaneously.''[14] Ann Berthoff has persistently criticized our obsession with one or another of the elements of discourse, insisting that meaning arises out of their synthesis. Writing is more, then, than ''a means of acting upon a receiver'' (Mitchell and Taylor, p. 250); it is a means of making meaning for writer *and* reader.[15] Without such a unifying, balanced understanding of language use, it is easy to overemphasize one aspect of discourse, such as audience. It is also easy to forget, as Anthony Petrosky cautions us, that ''reading, responding, and composing are aspects of understanding, and theories that attempt to account for them outside of their interaction with each other run the serious risk of building reductive models of human understanding.''[16]

Audience Invoked

Those who envision audience as invoked stress that the audience of a written discourse is a construction of the writer, a ''created fiction'' (Long, p. 225). They do not, of course, deny the physical reality of readers, but they argue that writers simply cannot know this reality in the way that speakers can. The central task of the writer, then, is not to analyze an audience and adapt discourse to meet its needs. Rather, the writer uses the semantic and syntactic resources of language to provide cues for the reader—cues which help to define the role or roles the writer wishes the reader to adopt in responding to the text. Little scholarship in composition takes this perspective; only Russell Long's article and Walter Ong's ''The Writer's Audience Is Always a Fiction'' focus centrally on this issue.[17] If recent conferences are any indication, however, a growing number of teachers and scholars are becoming concerned with what they see as the possible distortions and oversimplifications of the approach typified by Mitchell and Taylor's model.[18]

Russell Long's response to current efforts to teach students analysis of audience and adaptation of text to audience is typical: ''I have become increasingly disturbed not only about the superficiality of the advice itself, but about the philosophy which seems to lie beneath it'' (p. 211). Rather than detailing Long's argument, we wish to turn to Walter Ong's well-known study. Published in *PMLA* in 1975, ''The Writer's Audience Is Always a Fiction'' has had a significant impact on composition studies, despite the fact that its major emphasis is on fictional narrative rather than expository writing. An analysis of Ong's argument suggests that teachers of writing may err if they uncritically accept Ong's statement that ''what has been said about fictional narrative applies ceteris paribus to all writing'' (p. 17).

Ong's thesis includes two central assertions: ''What do we mean by saying the audience is a fiction? Two things at least. First, that the writer must construct in his imagina-

---

14. Paul Kameen, ''Rewording the Rhetoric of Composition,'' *Pre/Text,* 1 (Spring-Fall, 1980), 82.

15. Mitchell and Taylor's arguments in favor of adjunct classes seem to indicate that they see writing instruction, wherever it occurs, as a skills course, one instructing students in the proper use of a tool.

16. Anthony R. Petrosky. ''From Story to Essay: Reading and Writing,'' *CCC,* 33 (February, 1982), 20.

17. Walter J. Ong, S. J., ''The Writer's Audience Is Always a Fiction,'' *PMLA,* 90 (January, 1975), 9-21. Subsequent references to this article will be cited in the text.

18. See, for example, William Irmscher, ''Sense of Audience: An Intuitive Concept,'' unpublished paper delivered at the CCCC in 1981; Douglas B. Park, ''The Meanings of Audience: Pedagogical Implications,'' unpublished paper delivered at the CCCC in 1981; and Luke M. Reinsma, ''Writing to an Audience: Scheme or Strategy?'' unpublished paper delivered at the CCCC in 1982.

tion, clearly or vaguely, an audience cast in some sort of role. . . . Second, we mean that the audience must correspondingly fictionalize itself'' (p. 12). Ong emphasizes the creative power of the adept writer, who can both project and alter audiences, as well as the complexity of the reader's role. Readers, Ong observes, must learn or ''know how to play the game of being a member of an audience that 'really' does not exist'' (p. 12).

On the most abstract and general level, Ong is accurate. For a writer, the audience is not *there* in the sense that the speaker's audience, whether a single person or a large group, is present. But Ong's representative situations—the orator addressing a mass audience versus a writer alone in a room—oversimplify the potential range and diversity of both oral and written communication situations.

Ong's model of the paradigmatic act of speech communication derives from traditional rhetoric. In distinguishing the terms audience and reader, he notes that ''the orator has before him an audience which is a true audience, a collectivity. . . . Readers do not form a collectivity, acting here and now on one another and on the speaker as members of an audience do'' (p. 11). As this quotation indicates, Ong also stresses the potential for interaction among members of an audience, and between an audience and a speaker.

But how many audiences are actually collectives, with ample opportunity for interaction? In *Persuasion: Understanding, Practice, and Analysis,* Herbert Simons establishes a continuum of audiences based on opportunities for interaction.[19] Simons contrasts commercial mass media publics, which ''have little or no contact with each other and certainly have no reciprocal awareness of each other as members of the same audience'' with ''face-to-face work groups that meet and interact continuously over an extended period of time.'' He goes on to note that: ''Between these two extremes are such groups as the following: (1) the *pedestrian audience,* persons who happen to pass a soap box orator . . . ; (2) the *passive, occasional audience,* persons who come to hear a noted lecturer in a large auditorium . . . ; (3) the *active, occasional audience,* persons who meet only on specific occasions but actively interact when they do meet'' (pp. 97-98).

Simons' discussion, in effect, questions the rigidity of Ong's distinctions between a speaker's and a writer's audience. Indeed, when one surveys a broad range of situations inviting oral communication, Ong's paradigmatic situation, in which the speaker's audience constitutes a ''collectivity, acting here and now on one another and on the speaker'' (p. 11), seems somewhat atypical. It is certainly possible, at any rate, to think of a number of instances where speakers confront a problem very similar to that of writers: lacking intimate knowledge of their audience, which comprises not a collectivity but a disparate, and possibly even divided, group of individuals, speakers, like writers, must construct in their imaginations ''an audience cast in some sort of role.''[20] When President Carter announced to Americans during a speech broadcast on television, for instance, that his program against inflation was ''the moral equivalent of warfare,'' he was doing more than merely characterizing his economic policies. He was providing an important cue to his audience concerning the role he wished them to adopt as listeners—that of a people braced for a painful but necessary and justifiable battle. Were we to

---

19. Herbert W. Simons, *Persuasion: Understanding, Practice, and Analysis* (Reading, MA: Addison-Wesley, 1976).

20. Ong, p. 12. Ong recognizes that oral communication also involves role-playing, but he stresses that it ''has within it a momentum that works for the removal of masks'' (p. 20). This may be true in certain instances, such as dialogue, but does not, we believe, obtain broadly.

examine his speech in detail, we would find other more subtle, but equally important, semantic and syntactic signals to the audience.

We do not wish here to collapse all distinctions between oral and written communication, but rather to emphasize that speaking and writing are, after all, both rhetorical acts. There are important differences between speech and writing. And the broad distinction between speech and writing that Ong makes is both commonsensical and particularly relevant to his subject, fictional narrative. As our illustration demonstrates, however, when one turns to precise, concrete situations, the relationship between speech and writing can become far more complex than even Ong represents.

Just as Ong's distinction between speech and writing is accurate on a highly general level but breaks down (or at least becomes less clear-cut) when examined closely, so too does his dictum about writers and their audiences. Every writer must indeed create a role for the reader, but the constraints on the writer and the potential sources of and possibilities for the reader's role are both more complex and diverse than Ong suggests. Ong stresses the importance of literary tradition in the creation of audience: "If the writer succeeds in writing, it is generally because he can fictionalize in his imagination an audience he has learned to know not from daily life but from earlier writers who were fictionalizing in their imagination audiences they had learned to know in still earlier writers, and so on back to the dawn of written narrative" (p. 11). And he cites a particularly (for us) germane example, a student "asked to write on the subject to which schoolteachers, jaded by summer, return compulsively every autumn: 'How I Spent My Summer Vacation' " (p. 11). In order to negotiate such an assignment successfully, the student must turn his real audience, the teacher, into someone else. He or she must, for instance, "make like Samuel Clemens and write for whomever Samuel Clemens was writing for" (p. 11).

Ong's example is, for his purposes, well-chosen. For such an assignment does indeed require the successful student to "fictionalize" his or her audience. But why is the student's decision to turn to a literary model in this instance particularly appropriate? Could one reason be that the student knows (consciously or unconsciously) that his English teacher, who is still the literal audience of his essay, appreciates literature and hence would be entertained (and here the student may intuit the assignment's actual aim as well) by such a strategy? In Ong's example the audience—the "jaded" schoolteacher—is not only willing to accept another role but, perhaps, actually yearns for it. How else to escape the tedium of reading 25, 50, 75 student papers on the same topic? As Walter Minot notes, however, not all readers are so malleable:

> In reading a work of fiction or poetry, a reader is far more willing to suspend his beliefs and values than in a rhetorical work dealing with some current social, moral, or economic issue. The effectiveness of the created audience in a rhetorical situation is likely to depend on such constraints as the actual identity of the reader, the subject of the discourse, the identity and purpose of the writer, and many other factors in the real world.[21]

An example might help make Minot's point concrete.

Imagine another composition student faced, like Ong's, with an assignment. This

21. Walter S. Minot, "Response to Russell C. Long," *CCC*, 32 (October, 1981), 337.

student, who has been given considerably more latitude in her choice of a topic, has decided to write on an issue of concern to her at the moment, the possibility that a home for mentally-retarded adults will be built in her neighborhood. She is alarmed by the strongly negative, highly emotional reaction of most of her neighbors and wishes in her essay to persuade them that such a residence might not be the disaster they anticipate.

This student faces a different task from that described by Ong. If she is to succeed, she must think seriously about her actual readers, the neighbors to whom she wishes to send her letter. She knows the obvious demographic factors—age, race, class—so well that she probably hardly needs to consider them consciously. But other issues are more complex. How much do her neighbors know about mental retardation, intellectually or experientially? What is their image of a retarded adult? What fears does this project raise in them? What civic and religious values do they most respect? Based on this analysis—and the process may be much less sequential than we describe here—she must, of course, define a role for her audience, one congruent with her persona, arguments, the facts as she knows them, etc. She must, as Minot argues, *both* analyze and invent an audience.[22] In this instance, after detailed analysis of her audience and her arguments, the student decided to begin her essay by emphasizing what she felt to be the genuinely admirable qualities of her neighbors, particularly their kindness, understanding, and concern for others. In so doing, she invited her audience to see themselves as *she* saw them: as thoughtful, intelligent people who, if they were adequately informed, would certainly not act in a harsh manner to those less fortunate than they. In accepting this role, her readers did not have to "play the game of being a member of an audience that 'really' does not exist" (Ong, "The Writer's Audience," p. 12). But they did have to recognize in themselves the strengths the student described and to accept her implicit linking of these strengths to what she hoped would be their response to the proposed "home."

When this student enters her history class to write an examination she faces a different set of constraints. Unlike the historian who does indeed have a broad range of options in establishing the reader's role, our student has much less freedom. This is because her reader's role has already been established and formalized in a series of related academic conventions. If she is a successful student, she has so effectively internalized these conventions that she can subordinate a concern for her complex and multiple audiences to focus on the material on which she is being tested and on the single audience, the teacher, who will respond to her performance on the test.[23]

We could multiply examples. In each instance the student writing—to friend, em-

---

22. We are aware that the student actually has two audiences, her neighbors and her teacher, and that this situation poses an extra constraint for the writer. Not all students can manage such a complex series of audience constraints, but it is important to note that writers in a variety of situations often write for more than a single audience.

23. In their paper on "Student and Professional Syntax in Four Disciplines" (unpublished paper delivered at the CCCC in 1981), Ian Pringle and Aviva Freedman provide a good example of what can happen when a student creates an aberrant role for an academic reader. They cite an excerpt from a third year history assignment, the tone of which "is essentially the tone of the opening of a television travelogue commentary" and which thus asks the reader, a history professor, to assume the role of the viewer of such a show. The result is as might be expected: "Although the content of the paper does not seem significantly more abysmal than other papers in the same set, this one was awarded a disproportionately low grade" (p. 2).

ployer, neighbor, teacher, fellow readers of her daily newspaper—would need, as one of the many conscious and unconscious decisions required in composing, to envision and define a role for the reader. But *how* she defines that role—whether she relies mainly upon academic or technical writing conventions, literary models, intimate knowledge of friends or neighbors, analysis of a particular group, or some combination thereof—will vary tremendously. At times the reader may establish a role for the reader which indeed does not "coincide[s] with his role in the rest of actual life" (Ong, p. 12). At other times, however, one of the writer's primary tasks may be that of analyzing the "real life" audience and adapting the discourse to it. One of the factors that makes writing so difficult, as we know, is that we have no recipes: each rhetorical situation is unique and thus requires the writer, catalyzed and guided by a strong sense of purpose, to reanalyze and reinvent solutions.

Despite their helpful corrective approach, then, theories which assert that the audience of a written discourse is a construction of the writer present their own dangers.[24] One of these is the tendency to overemphasize the distinction between speech and writing while undervaluing the insights of discourse theorists, such as James Moffett and James Britton, who remind us of the importance of such additional factors as distance between speaker or writer and audience and levels of abstraction in the subject. In *Teaching the Universe of Discourse,* Moffett establishes the following spectrum of discourse: recording ("the drama of what is happening"), reporting ("the narrative of what happened"), generalizing ("the exposition of what happens") and theorizing ("the argumentation of what will, may happen").[25] In an extended example, Moffett demonstrates the important points of connection between communication acts at any one level of the spectrum, whether oral or written:

> Suppose next that I tell the cafeteria experience to a friend some time later in conversation. . . . Of course, instead of recounting the cafeteria scene to my friend in person I could write it in a letter to an audience more removed in time and space. Informal writing is usually still rather spontaneous, directed at an audience known to the writer, and reflects the transient mood and circumstances in which the writing occurs. Feedback and audience influence, however, are delayed and weakened. . . . *Compare in turn now the changes that must occur all down the line when I write about this cafeteria experience in a discourse destined for publication and distribution to a mass, anonymous audience of present and perhaps, unborn people.* I cannot allude to things and ideas that only my friends know about. I must use a vocabulary, style, logic, and rhetoric that anybody in that mass audience can understand and respond to. I must name and organize what happened during those moments in the cafeteria that day in such a way that this mythical average reader can relate what I say to some primary moments of experience of his own. (pp. 37-38; our emphasis)

---

24. One danger which should be noted is a tendency to foster a questionable image of classical rhetoric. The agonistic speaker-audience relationship which Long cites as an essential characteristic of classical rhetoric is actually a central point of debate among those involved in historical and theoretical research in rhetoric. For further discussion, see: Lisa Ede and Andrea Lunsford, "On Distinctions Between Classical and Modern Rhetoric," in *Classical Rhetoric and Modern Discourse: Essays in Honor of Edward P. J. Corbett,* ed. Robert Connors, Lisa Ede, and Andrea Lunsford (Carbondale, IL: Southern Illinois University Press, 1984).

25. James Moffett, *Teaching the Universe of Discourse* (Boston: Houghton Mifflin, 1968), p. 47. Subsequent references will be mentioned in the text.

Though Moffett does not say so, many of these same constraints would obtain if he decided to describe his experience in a speech to a mass audience—the viewers of a television show, for example, or the members of a graduating class. As Moffett's example illustrates, the distinction between speech and writing is important; it is, however, only one of several constraints influencing any particular discourse.

Another weakness of research based on the concept of audience as invoked is that it distorts the processes of writing and reading by overemphasizing the power of the writer and undervaluing that of the reader. Unlike Mitchell and Taylor, Ong recognizes the creative role the writer plays as reader of his or her own writing, the way the writer uses language to provide cues for the reader and tests the effectiveness of these cues during his or her own rereading of the text. But Ong fails adequately to recognize the constraints placed on the writer, in certain situations, by the audience. He fails, in other words, to acknowledge that readers' own experiences, expectations, and beliefs do play a central role in their reading of a text, and that the writer who does not consider the needs and interests of his audience risks losing that audience. To argue that the audience is a "created fiction" (Long, p. 225), to stress that the reader's role "seldom coincides with his role in the rest of actual life" (Ong, p. 12), is just as much an oversimplification, then, as to insist, as Mitchell and Taylor do, that "the audience not only judges writing, it also motivates it" (p. 250). The former view overemphasizes the writer's independence and power; the latter, that of the reader.

Rhetoric and Its Situations[26]

If the perspectives we have described as audience addressed and audience invoked represent incomplete conceptions of the role of audience in written discourse, do we have an alternative? How can we most accurately conceive of this essential rhetorical element? In what follows we will sketch a tentative model and present several defining or constraining statements about this apparently slippery concept, "audience." The result will, we hope, move us closer to a full understanding of the role audience plays in written discourse.

Figure 2 represents our attempts to indicate the complex series of obligations, resources, needs, and constraints embodied in the writer's concept of audience. (We emphasize that our goal here is *not* to depict the writing process as a whole—a much more complex task—but to focus on the writer's relation to audience.) As our model indicates, we do not see the two perspectives on audience described earlier as necessarily dichotomous or contradictory. Except for past and anomalous audiences, special cases which we describe paragraphs hence, all of the audience roles we specify—self, friend, colleague, critic, mass audience, and future audience—may be invoked or addressed.[27]

---

26. We have taken the title of this section from Scott Consigny's article of the same title, *Philosophy and Rhetoric,* 7 (Summer, 1974), 175-186. Consigny's effort to mediate between two opposing views of rhetoric provided a stimulating model for our own efforts.

27. Although we believe that the range of audience roles cited in our model covers the general spectrum of options, we do not claim to have specified all possibilities. This is particularly the case since, in certain instances, these roles may merge and blend—shifting subtly in character. We might also note that other terms for the same roles might be used. In a business setting, for instance, colleague might be better termed co-worker; critic, supervisor.

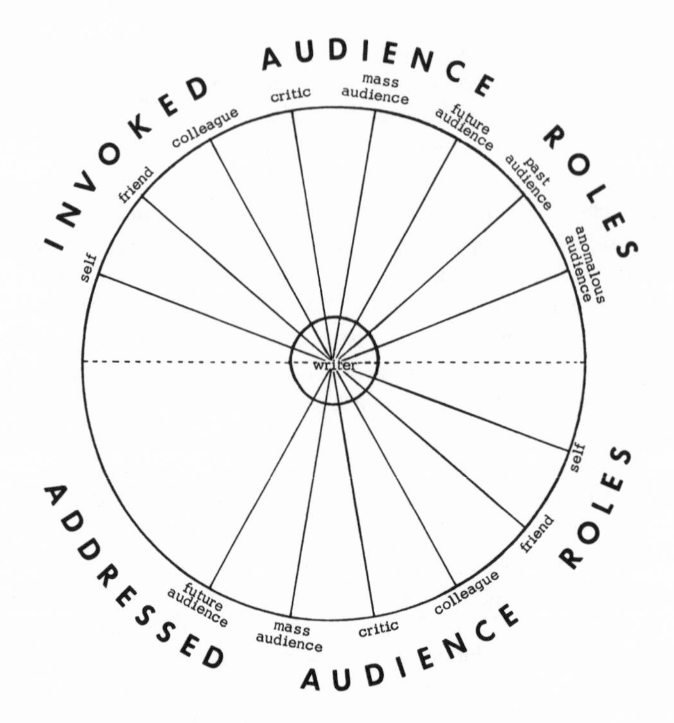

FIGURE 2. The concept of audience

It is the writer who, as writer and reader of his or her own text, one guided by a sense of purpose and by the particularities of a specific rhetorical situation, establishes the range of potential roles an audience may play. (Readers may, of course, accept or reject the role or roles the writer wishes them to adopt in responding to a text.)

Writers who wish to be read must often adapt their discourse to meet the needs and expectations of an addressed audience. They may rely on past experience in addressing audiences to guide their writing, or they may engage a representative of that audience in the writing process. The latter occurs, for instance, when we ask a colleague to read an article intended for scholarly publication. Writers may also be required to respond to the intervention of others—a teacher's comments on an essay, a supervisor's suggestions for improving a report, or the insistent, catalyzing questions of an editor. Such intervention may in certain cases represent a powerful stimulus to the writer, but it is the writer who interprets the suggestions—or even commands—of others, choosing what to accept or reject. Even the conscious decision to accede to the expectations of a particular addressed audience may not always be carried out; unconscious psychological resistance, incomplete understanding, or inadequately developed ability may prevent the writer from following through with the decision—a reality confirmed by composition teachers with each new set of essays.

The addressed audience, the actual or intended readers of a discourse, exists outside of the text. Writers may analyze these readers' needs, anticipate their biases, even defer to their wishes. But it is only through the text, through language, that writers embody or give life to their conception of the reader. In so doing, they do not so much create a role for the reader—a phrase which implies that the writer somehow creates a mold to which the reader adapts—as invoke it. Rather than relying on incantations, however, writers conjure their vision—a vision which they hope readers will actively come to share as they read the text—by using all the resources of language available to them to establish a broad, and ideally coherent, range of cues for the reader. Technical writing conventions, for instance, quickly formalize any of several writer-reader relationships, such as colleague to colleague or expert to lay reader. But even comparatively local semantic decisions may play an equally essential role. In "The Writer's Audience Is Always a Fiction," Ong demonstrates how Hemingway's use of definite articles in *A Farewell to Arms* subtly cues readers that their role is to be that of a "companion in arms . . . a confidant" (p. 13).

Any of the roles of the addressed audience cited in our model may be invoked via the text. Writers may also invoke a past audience, as did, for instance, Ong's student writing to those Mark Twain would have been writing for. And writers can also invoke anomalous audiences, such as a fictional character—Hercule Poirot perhaps. Our model, then, confirms Douglas Park's observation that the meanings of audience, though multiple and complex, "tend to diverge in two general directions: one toward actual people external to a text, the audience whom the writer must accommodate; the other toward the text itself and the audience implied there: a set of suggested or evoked attitudes, interests, reactions, conditions of knowledge which may or may not fit with the qualities of actual readers or listeners."[28] The most complete understanding of audience thus involves a synthesis of the perspectives we have termed audience addressed, with its focus on the reader, and audience invoked, with its focus on the writer.

One illustration of this constantly shifting complex of meanings for "audience" lies in our own experiences writing this essay. One of us became interested in the concept of audience during an NEH Seminar, and her first audience was a small, close-knit seminar group to whom she addressed her work. The other came to contemplate a multiplicity of audiences while working on a textbook; the first audience in this case was herself, as she debated the ideas she was struggling to present to a group of invoked students. Following a lengthy series of conversations, our interests began to merge: we shared notes and discussed articles written by others on audience, and eventually one of us began a draft. Our long distance telephone bills and the miles we travelled up and down I-5 from Oregon to British Columbia attest most concretely to the power of a co-author's expectations and criticisms and also illustrate that one person can take on the role of several different audiences: friend, colleague, and critic.

As we began to write and re-write the essay, now for a particular scholarly journal, the change in purpose and medium (no longer a seminar paper or a textbook) led us to new audiences. For us, the major "invoked audience" during this period was Richard Larson, editor of this journal, whose questions and criticisms we imagined and tried to anticipate. (Once this essay was accepted by *CCC,* Richard Larson became for us an addressed audience: he responded in writing with questions, criticisms, and suggestions, some of which we had, of course, failed to anticipate.) We also thought of the readers of *CCC* and those who attend the annual CCCC, most often picturing you as members of our own departments, a diverse group of individuals with widely varying degrees of

28. Douglas B. Park, "The Meanings of 'Audience,' " *CE,* 44 (March, 1982), 249.

interest in and knowledge of composition. Because of the generic constraints of academic writing, which limit the range of roles we may define for our readers, the audience represented by the readers of *CCC* seemed most vivid to us in two situations: (1) when we were concerned about the degree to which we needed to explain concepts or terms; and (2) when we considered central organizational decisions, such as the most effective way to introduce a discussion. Another, and for us extremely potent, audience was the authors—Mitchell and Taylor, Long, Ong, Park, and others—with whom we have seen ourselves in silent dialogue. As we read and reread their analyses and developed our responses to them, we felt a responsibility to try to understand their formulations as fully as possible, to play fair with their ideas, to make our own efforts continue to meet their high standards.

Our experience provides just one example, and even it is far from complete. (Once we finished a rough draft, one particular colleague became a potent but demanding addressed audience, listening to revision upon revision and challenging us with harder and harder questions. And after this essay is published, we may revise our understanding of audiences we thought we knew or recognize the existence of an entirely new audience. The latter would happen, for instance, if teachers of speech communication for some reason found our discussion useful.) But even this single case demonstrates that the term *audience* refers not just to the intended, actual, or eventual readers of a discourse, but to *all* those whose image, ideas, or actions influence a writer during the process of composition. One way to conceive of "audience," then, is as an overdetermined or unusually rich concept, one which may perhaps be best specified through the analysis of precise, concrete situations.

We hope that this partial example of our own experience will illustrate how the elements represented in Figure 2 will shift and merge, depending on the particular rhetorical situation, the writer's aim, and the genre chosen. Such an understanding is critical: because of the complex reality to which the term audience refers and because of its fluid, shifting role in the composing process, any discussion of audience which isolates it from the rest of the rhetorical situation or which radically overemphasizes or underemphasizes its function in relation to other rhetorical constraints is likely to oversimplify. Note the unilateral direction of Mitchell and Taylor's model (p. 246), which is unable to represent the diverse and complex role(s) audience(s) can play in the actual writing process—in the creation of meaning. In contrast, consider the model used by Edward P. J. Corbett in his *Little Rhetoric and Handbook.*[29] This representation, which allows for interaction among all the elements of rhetoric, may at first appear less elegant and predictive than Mitchell and Taylor's. But it is finally more useful since it accurately represents the diverse range of potential interrelationships in any written discourse.

We hope that our model also suggests the integrated, interdependent nature of reading and writing. Two assertions emerge from this relationship. One involves the writer as reader of his or her own work. As Donald Murray notes in "Teaching the Other Self: The Writer's First Reader," this role is critical, for "the reading writer—the map-maker and map-reader—reads the word, the line, the sentence, the paragraph, the page, the entire text. This constant back-and-forth reading monitors the multiple complex relation-

29. Edward P. J. Corbett, *The Little Rhetoric & Handbook,* 2nd edition (Glenview, IL: Scott, Foresman, 1982), p. 5.

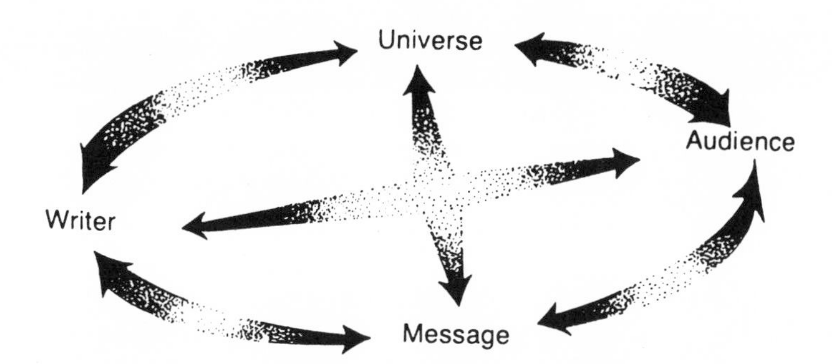

FIGURE 3. Corbett's model of "The Rhetorical Interrelationships" (p. 5)

ships between all the elements in writing."[30] To ignore or devalue such a central function is to risk distorting the writing process as a whole. But unless the writer is composing a diary or journal entry, intended only for the writer's own eyes, the writing process is not complete unless another person, someone other than the writer, reads the text also. The second assertion thus emphasizes the creative, dynamic duality of the process of reading and writing, whereby writers create readers and readers create writers. In the meeting of these two lies meaning, lies communication.

A fully elaborated view of audience, then, must balance the creativity of the writer with the different, but equally important, creativity of the reader. It must account for a wide and shifting range of roles for both addressed and invoked audiences. And, finally, it must relate the matrix created by the intricate relationship of writer and audience to all elements in the rhetorical situation. Such an enriched conception of audience can help us better understand the complex act we call composing.

30. Donald M. Murray, "Teaching the Other Self: The Writer's First Reader," *CCC*, 33 (May, 1982), 142.

Peter Elbow

Closing My Eyes as I Speak: An Argument for Ignoring Audience

Very often people don't listen to you when you speak to them. It's only when you talk to yourself that they prick up their ears.

John Ashberry

When I am talking to a person or a group and struggling to find words or thoughts, I often find myself involuntarily closing my eyes as I speak. I realize now that this behavior is an instinctive attempt to blot out awareness of audience when I need all my concentration for just trying to figure out or express what I want to say. Because the audience is so imperiously *present* in a speaking situation, my instinct reacts with this active attempt to avoid audience awareness. This behavior—in a sense impolite or antisocial—is not so uncommon. Even when we write, alone in a room to an absent audience, there are occasions when we are struggling to figure something out and need to push aside awareness of those absent readers. As Donald Murray puts it, "My sense of audience is so strong that I have to suppress my conscious awareness of audience to hear what the text demands" (Berkenkotter and Murray 171). In recognition of how pervasive the role of audience is in writing, I write to celebrate the benefits of ignoring audience.[1]

It will be clear that my argument for writing without audience awareness is not meant to undermine the many good reasons for writing *with* audience awareness some of the time. (For example, that we are liable to neglect audience because we write in solitude; that young people often need more practice in taking into account points of view different from their own; and that students often have an impoverished sense of writing as communication because they have only written in a school setting to teachers.) Indeed

From *College English* 49 (January 1987): 50-69.

1. There are many different entities called audience: (a) The actual readers to whom the text will be given; (b) the writer's conception of those readers—which may be mistaken (see Ong; Park; Ede and Lunsford); (c) the audience that the text implies—which may be different still (see Booth); (d) the discourse community or even genre addressed or implied by the text (see Walzer); (e) ghost or phantom "readers in the head" that the writer may unconsciously address or try to please (see Elbow, *Writing with Power* 186ff. Classically, this is a powerful former teacher. Often such an audience is so ghostly as not to show up as actually "implied" by the text). For the essay I am writing here, these differences don't much matter: I'm celebrating the ability to put aside the needs or demands of *any* or all of these audiences. I recognize, however, that we sometimes cannot fight our way free of unconscious or tacit audiences (as in b or e above) unless we bring them to greater conscious awareness.

I would claim some part in these arguments for audience awareness—which now seem to be getting out of hand.

I start with a limited claim: even though ignoring audience will usually lead to weak writing at first—to what Linda Flower calls "writer-based prose"—this weak writing can help us in the end to better writing than we would have written if we'd kept readers in mind from the start. Then I will make a more ambitious claim: writer-based prose is sometimes better than reader-based prose. Finally I will explore some of the theory underlying these issues of audience.

A Limited Claim

It's not that writers should never think about their audience. It's a question of when. An audience is a field of force. The closer we come—the more we think about these readers—the stronger the pull they exert on the contents of our minds. The practical question, then, is always whether a particular audience functions as a helpful field of force or one that confuses or inhibits us.

Some audiences, for example, are *inviting* or *enabling*. When we think about them as we write, we think of more and better things to say—and what we think somehow arrives more coherently structured than usual. It's like talking to the perfect listener: we feel smart and come up with ideas we didn't know we had. Such audiences are helpful to keep in mind right from the start.

Other audiences, however, are powerfully *inhibiting*—so much so, in certain cases, that awareness of them as we write blocks writing altogether. There are certain people who always make us feel dumb when we try to speak to them: we can't find words or thoughts. As soon as we get out of their presence, all the things we wanted to say pop back into our minds. Here is a student telling what happens when she tries to follow the traditional advice about audience:

> You know——[author of a text] tells us to pay attention to the audience that will be reading our papers, and I gave that a try. I ended up without putting a word on paper until I decided the hell with——; I'm going to write to who I damn well want to; otherwise I can hardly write at all.

Admittedly, there are some occasions when we benefit from keeping a threatening audience in mind from the start. We've been putting off writing that letter to that person who intimidates us. When we finally sit down and write *to* them—walk right up to them, as it were, and look them in the eye—we may manage to stand up to the threat and grasp the nettle and thereby find just what we need to write.

Most commonly, however, the effect of audience awareness is somewhere between the two extremes: the awareness disturbs or disrupts our writing and thinking without completely blocking it. For example, when we have to write to someone we find intimidating (and of course students often perceive teachers as intimidating), we often start thinking wholly defensively. As we write down each thought or sentence, our mind fills with thoughts of how the intended reader will criticize or object to it. So we try to qualify or soften what we've just written—or write out some answer to a possible objection. Our writing becomes tangled. Sometimes we get so tied in knots that we cannot even figure out what we *think*. We may not realize how often audience awareness has this effect on our students when we don't see the writing processes behind their papers: we just see texts that are either tangled or empty.

Another example. When we have to write to readers with whom we have an awkward relationship, we often start beating around the bush and feeling shy or scared, or start to write in a stilted, overly careful style or voice. (Think about the cute, too-clever style of many memos we get in our departmental mailboxes—the awkward self-consciousness academics experience when writing to other academics.) When students are asked to write to readers they have not met or cannot imagine, such as "the general reader" or "the educated public," they often find nothing to say except cliches they know *they* don't even quite believe.

When we realize that an audience is somehow confusing or inhibiting us, the solution is fairly obvious. We can ignore that audience altogether during the *early* stages of writing and direct our words only to ourselves or to no one in particular—or even to the "wrong" audience, that is, to an *inviting* audience of trusted friends or allies. This strategy often dissipates the confusion; the clenched, defensive discourse starts to run clear. Putting audience out of mind is of course a traditional practice: serious writers have long used private journals for early explorations of feeling, thinking, or language. But many writing teachers seem to think that students can get along without the private writing serious writers find so crucial—or even that students will *benefit* from keeping their audience in mind for the whole time. Things often don't work out that way.

After we have figured out our thinking in copious exploratory or draft writing—perhaps finding the right voice or stance as well—*then* we can follow the traditional rhetorical advice: think about readers and revise carefully to adjust our words and thoughts to our intended audience. For a particular audience it may even turn out that we need to *disguise* our point of view. But it's hard to disguise something while engaged in trying to figure it out. As writers, then, we need to learn when to think about audience and when to put readers out of mind.

Many people are too quick to see Flower's "writer-based prose" as an analysis of what's wrong with this type of writing and miss the substantial degree to which she was celebrating a natural, and indeed developmentally enabling, response to cognitive overload. What she doesn't say, however, despite her emphasis on planning and conscious control in the writing process, is that we can *teach* students to notice when audience awareness is getting in their way—and when this happens, consciously to put aside the needs of readers for a while. She seems to assume that when an overload occurs, the writer-based gear will, as it were, automatically kick into action to relieve it. In truth, of course, writers often persist in using a malfunctioning *reader*-based gear despite the overload—thereby mangling their language or thinking. Though Flower likes to rap the knuckles of people who suggest a "correct" or "natural" order for steps in the writing process, she implies such an order here: when attention to audience causes an overload, start out by ignoring them while you attend to your thinking; after you work out your thinking, turn your attention to audience.

Thus if we ignore audience while writing on a topic about which we are not expert or about which our thinking is still evolving, we are likely to produce exploratory writing that is unclear to anyone else—perhaps even inconsistent or a complete mess. Yet by doing this exploratory "swamp work" in conditions of safety, we can often coax our thinking through a process of new discovery and development. In this way we can end up with something better than we could have produced if we'd tried to write to our audience all along. In short, ignoring audience can lead to worse drafts but better revisions. (Because we are professionals and adults, we often write in the role of expert: we may know what we think without new exploratory writing; we may even be able to speak confidently to critical readers. But students seldom experience this confident professional stance in their writing. And think how much richer *our* writing would be if we defined ourselves as *in*expert and allowed ourselves private writing for new explorations of those views we are allegedly sure of.)

Notice then that two pieties of composition theory are often in conflict:

1. Think about audience as you write (this stemming from the classical rhetorical tradition).
2. Use writing for *making new meaning,* not just transmitting old meanings already worked out (this stemming from the newer epistemic tradition I associate with Ann Berthoff's classic explorations).

It's often difficult to work out new meaning while thinking about readers.

A More Ambitious Claim

I go further now and argue that ignoring audience can lead to better writing—immediately. In effect, writer-based prose can be *better* than reader-based prose. This might seem a more controversial claim, but is there a teacher who has not had the experience of struggling and struggling to no avail to help a student untangle his writing, only to discover that the student's casual journal writing or freewriting is untangled and strong? Sometimes freewriting is stronger than the essays we get only because it is expressive, narrative, or descriptive writing and the student was not constrained by a topic. But teachers who collect drafts with completed assignments often see passages of freewriting that are strikingly stronger *even* when they are expository and constrained by the assigned topic. In some of these passages we can sense that the strength derives from the student's unawareness of readers.

It's not just unskilled, tangled writers, though, who sometimes write better by forgetting about readers. Many competent and even professional writers produce mediocre pieces *because* they are thinking too much about how their readers will receive their words. They are acting too much like a salesman trained to look the customer in the eye and to think at all times about the characteristics of the "target audience." There is something too staged or planned or self-aware about such writing. We see this quality in much second-rate newspaper or magazine or business writing: "good-student writing" in the awful sense of the term. Writing produced this way reminds us of the ineffective actor whose consciousness of self distracts us: he makes us too aware of his own awareness of us. When we read such prose, we wish the writer would stop thinking about us—would stop trying to "adjust" or "fit" what he is saying to our frame of reference. "Damn it, put all your attention on what you are saying," we want to say, "and forget about us and how we are reacting."

When we examine really good student or professional writing, we can often see that its goodness comes from the writer's having gotten sufficiently wrapped up in her meaning and her language as to forget all about audience needs: the writer manages to "break through." The Earl of Shaftesbury talked about writers needing to escape their audience in order to find their own ideas (Cooper 1:109; see also Griffin). It is characteristic of much truly good writing to be, as it were, on fire with its meaning. Consciousness of readers is burned away; involvement in subject determines all. Such writing is analogous to the performance of the actor who has managed to stop attracting attention to her awareness of the audience watching her.

The arresting power in some writing by small children comes from their obliviousness to audience. As readers, we are somehow sucked into a more-than-usual connection with the meaning itself because of the child's gift for more-than-usual concentration on what she is saying. In short, we can feel some pieces of children's writing as being very writer-based. Yet it's precisely that quality which makes it powerful for us as readers. After all, why should we settle for a writer's entering our point of view, if we can have the more powerful experience of being sucked out of our point of view and into her world? This is just the experience that children are peculiarly capable of giving because they are so expert at total absorption in their world as they are writing. It's not just a matter of whether the writer "decenters," but of whether the writer has a sufficiently strong focus of attention to make the *reader* decenter. This quality of concentration is what D. H. Lawrence so admires in Melville:

> [Melville] was a real American in that he always felt his audience in front of him. But when he ceases to be American, when he forgets all audience, and gives us his sheer apprehension of the world, then he is wonderful, his book *[Moby Dick]* commands a stillness in the soul, an awe. (158)

What most readers value in really excellent writing is not prose that is right for readers but prose that is right for thinking, right for language, or right for the subject being written about. If, in addition, it is clear and well suited to readers, we appreciate that. Indeed we feel insulted if the writer did not somehow try to make the writing *available* to us before delivering it. But if it succeeds at being really true to language and thinking and "things," we are willing to put up with much difficulty as readers:

> [G]ood writing is not always or necessarily an adaptation to communal norms (in the Fish/Bruffee sense) but may be an attempt to construct (and instruct) a reader capable of reading the text in question. The literary history of the "difficult" work—from Mallarme to Pound, Zukofsky, Olson, etc.—seems to say that much of what we value in writing we've had to learn to value by learning how to read it. (Trimbur)

The effect of audience awareness on *voice* is particularly striking—if paradoxical. Even though we often develop our voice by finally "speaking up" to an audience or "speaking out" to others, and even though much dead student writing comes from students' not really treating their writing as a communication with real readers, nevertheless, the opposite effect is also common: we often do not really develop a strong, authentic voice in our writing till we find important occasions for *ignoring* audience—saying, in effect, "To hell with whether they like it or not. I've got to say this the way *I* want to say it." Admittedly, the voice that emerges when we ignore audience is sometimes odd or idiosyncratic in some way, but usually it is stronger. Indeed, teachers sometimes complain that student writing is "writer-based" when the problem is simply the idiosyncracy—and sometimes in fact the *power*—of the voice. They would value this odd but resonant voice if they found it in a published writer (see "Real Voice," Elbow, *Writing with Power*). Usually we cannot *trust* a voice unless it is unaware of us and our needs and speaks out in its own terms (see the Ashberry epigraph). To celebrate writer-based prose is to risk the charge of *romanticism:* just warbling one's woodnotes wild. But my position also contains the austere *classic* view that we must nevertheless *revise* with conscious awareness of audience in order to figure out which pieces of writer-based prose are good as they are—and how to discard or revise the rest.

To point out that writer-based prose can be *better* for readers than reader-based prose is to reveal problems in these two terms. Does *writer-based* mean:

1. That the text doesn't work for readers because it is too much oriented to the writer's point of view?
2. Or that the writer was not thinking about readers as she wrote, although the text *may* work for readers?

Does *reader-based* mean:

3. That the text works for readers—meets their needs?
4. Or that the writer was attending to readers as she wrote although her text may *not* work for readers?

In order to do justice to the reality and complexity of what actually happens in both writers and readers, I was going to suggest four terms for the four conditions listed above, but I gradually realized that things are even too complex for that. We really need to ask about what's going on in three dimensions—in the *writer,* in the *reader,* and in the *text*—and realize that the answers can occur in virtually any combination:

—Was the *writer* thinking about readers or oblivious to them?

—Is the *text* oriented toward the writer's frame of reference or point of view, or oriented toward that of readers? (A writer may be thinking about readers and still write a text that is largely oriented toward her own frame of reference.)

—Are the readers' needs being met? (The text may meet the needs of readers whether the writer was thinking about them or not, and whether the text is oriented toward them or not.)

Two Models of Cognitive Development

Some of the current emphasis on audience awareness probably derives from a model of cognitive development that needs to be questioned. According to this model, if you keep your readers in mind as your write, you are operating at a higher level of psychological development than if you ignore readers. Directing words to readers is "more mature" than directing them to no one in particular or to yourself. Flower relates writer-based prose to the inability to "decenter" which is characteristic of Piaget's early stages of development, and she relates reader-based prose to later more mature stages of development.

On the one hand, of course this view must be right. Children do decenter as they develop. As they mature they get better at suiting their discourse to the needs of listeners, particularly to listeners very different from themselves. Especially, they get better at doing so *consciously*—thinking *awarely* about how things appear to people with different viewpoints. Thus much unskilled writing is unclear or awkward *because* the writer was doing what it is so easy to do—unthinkingly taking her own frame of reference for granted and not attending to the needs of readers who might have a different frame of reference. And of course this failure is more common in younger, immature, "egocentric" students (and also more common in writing than in speaking since we have no audience present when we write).

But on the other hand, we need the contrary model that affirms what is also obvious once we reflect on it, namely that the ability to *turn off* audience awareness—especially when it confuses thinking or blocks discourse—is also a "higher" skill. I am talking about an ability to use language in "the desert island mode," an ability that tends to require learning, growth, and psychological development. Children, and even adults who have not learned the art of quiet, thoughtful, inner reflection, are often unable to get much cognitive action going in their heads unless there are other people present to have action *with*. They are dependent on live audience and the social dimension to get their discourse rolling or to get their thinking off the ground.

For in contrast to a roughly Piagetian model of cognitive development that says we start out as private, egocentric little monads and grow up to be public and social, it is important to invoke the opposite model that derives variously from Vygotsky, Bakhtin, and Meade. According to this model, we *start out* social and plugged into others and only gradually, through learning and development, come to "unplug" to any significant degree so as to function in a more private, individual and differentiated fashion: "Development in thinking is not from the individual to the socialized, but from the social to the individual" (Vygotsky 20). The important general principle in this model is that we tend to *develop* our important cognitive capacities by means of social interaction with others, and having done so we gradually learn to perform them alone. We fold the "simple" back-and-forth of dialogue into the "complexity" (literally, "foldedness") of individual, private reflection.

Where the Piagetian (individual psychology) model calls our attention to the obvious need to learn to enter into viewpoints other than our own, the Vygotskian (social psychology) model calls our attention to the equally important need to learn to produce good thinking and discourse *while alone*. A rich and enfolded mental life is something that people achieve only gradually through growth, learning, and practice. We tend to associate this achievement with the fruits of higher education.

Thus we see plenty of students who lack this skill, who have nothing to say when asked to freewrite or to write in a journal. They can dutifully "reply" to a question or a topic, but they cannot seem to *initiate* or *sustain* a train of thought on their own. Because so many adolescent students have this difficulty, many teachers chime in: "Adolescents have nothing to write about. They are too young. They haven't had significant experience." In truth, adolescents don't lack experience or material, no matter how "sheltered" their lives. What they lack is practice and help. Desert island discourse is a learned cognitive process. It's a mistake to think of private writing (journal writing and freewriting) as merely "easy"—merely a relief from trying to write right. It's also hard. Some exercises and strategies that help are Ira Progoff's "Intensive Journal" process, Sondra Perl's "Composing Guidelines," or Elbow's "Loop Writing" and "Open Ended Writing" processes (*Writing with Power* 50-77).

The Piagetian and Vygotskian developmental models (language-begins-as-private vs. language-begins-as-social) give us two different lenses through which to look at a common weakness in student writing, a certain kind of "thin" writing where the thought is insufficiently developed or where the language doesn't really explain what the writing implies or gestures toward. Using the Piagetian model, as Flower does, one can specify the problem as a weakness in audience orientation. Perhaps the writer has immaturely taken too much for granted and unthinkingly assumed that her limited explanations carry as much meaning for readers as they do for herself. The cure or treatment is for the writer to think more about readers.

Through the Vygotskian lens, however, the problem and the "immaturity" look altogether different. Yes, the writing isn't particularly clear or satisfying for readers, but this alternative diagnosis suggests a failure of the private desert island dimension: the writer's explanation is too thin because she didn't work out her train of thought fully enough *for herself*. The suggested cure or treatment is *not* to think more about readers but to think more for herself, to practice exploratory writing in order to learn to engage in that reflective discourse so central to mastery of the writing process. How can she engage readers more till she has engaged herself more?

The current emphasis on audience awareness may be particularly strong now for being fueled by *both* psychological models. From one side, the Piagetians say, in effect, "The egocentric little critters, we've got to *socialize* 'em! Ergo, make them think about audience when they write!" From the other side, the Vygotskians say, in effect, "No wonder they're having trouble writing. They've been bamboozled by the Piagetian heresy. They think they're solitary individuals with private selves when really they're just congeries of voices that derive from their discourse community. Ergo, let's intensify the social context—use peer groups and publication: make them think about audience when they write! (And while we're at it, let's hook them up with a better class of discourse community.)" To advocate ignoring audience is to risk getting caught in the crossfire from two opposed camps.

Two Models of Discourse: Discourse as Communication and Discourse as Poesis or Play

We cannot talk about writing without at least implying a psychological or developmental model. But we'd better make sure it's a complex, paradoxical, or spiral model. Better yet, we should be deft enough to use two contrary models or lenses. (Bruner pictures the developmental process as a complex movement in an upward reiterative spiral—not a simple movement in one direction.)

According to one model, it is characteristic of the youngest children to direct their discourse to an audience. They learn discourse *because* they have an audience; without an audience they remain mute, like "the wild child." Language is social from the start.

But we need the other model to show us what is also true, namely that it is characteristic of the youngest children to use language in a *non-social* way. They use language not only because people talk to them but also because they have such a strong propensity to play and to build—often in a *non*-social or non-audience-oriented fashion. Thus although one paradigm for discourse is social communication, another is private exploration or solitary play. Babies and toddlers tend to babble in an exploratory and reflective way—to themselves and not to an audience—often even with no one else near. This archetypally private use of discourse is strikingly illustrated when we see a pair of toddlers in "parallel play" alongside each other—each busily talking but not at all trying to communicate with the other.

Therefore, when we choose paradigms for discourse, we should think not only about children using language to communicate, but also about children building sandcastles or drawing pictures. Though children characteristically show their castles or pictures to others, they just as characteristically trample or crumple them before anyone else can see them. Of course sculptures and pictures are different from words. Yet discourse implies more media than words; and even if you restrict discourse to words, one of our most mature uses of language is for building verbal pictures and structures for their own sake—not just for communicating with others.

Consider this same kind of behavior at the other end of the life cycle: Brahms staggering from his deathbed to his study to rip up a dozen or more completed but unpublished and unheard string quartets that dissatisfied him. How was he relating to audience here—worrying too much about audience or not giving a damn? It's not easy to say. Consider Glenn Gould deciding to renounce performances before an audience. He used his private studio to produce recorded performances for an audience, but to produce ones that satisfied *himself* he clearly needed to suppress audience awareness. Consider the more extreme example of Kerouac typing page after page—burning each as soon as he completed it. The language behavior of humans is slippery. Surely we are well advised to avoid positions that say it is "always X" or "essentially Y."

James Britton makes a powerful argument that the "making" or poesis function of language grows out of the expressive function. Expressive language is often for the sake of communication with an audience, but just as often it is only for the sake of the speaker—working something out for herself (66-67, 74ff). Note also that "writing to learn," which writing-across-the-curriculum programs are discovering to be so important, tends to be writing for the self or even for no one at all rather than for an outside reader. You throw away the writing, often unread, and keep the mental changes it has engendered.

I hope this emphasis on the complexity of the developmental process—the limits of our models and of our understanding of it—will serve as a rebuke to the tendency to label students as being at a lower stage of cognitive development just because they don't yet write well. (Occasionally they *do* write well—in a way—but not in the way that the labeler finds appropriate.) Obviously the psychologistic labeling impulse started out charitably. Shaughnessy was fighting those who called basic writers *stupid* by saying they weren't dumb, just at an earlier developmental stage. Flower was arguing that writer-based prose is a natural response to a cognitive overload and indeed developmentally enabling. But this kind of talk can be dangerous since it labels students as literally "retarded" and makes teachers and administrators start to think of them as such. Instead of calling poor writers *either* dumb or slow (two forms of blaming the victim), why not simply call them poor writers? If years of schooling haven't yet made them good writers, perhaps they haven't gotten the kind of teaching and support they need. Poor students are often deprived of the very thing they need most to write well (which is given to good students): lots of extended and adventuresome writing for self and for audience. Poor students are often asked to write *only* answers to fill-in exercises.

As children get older, the developmental story remains complex or spiral. Though the first model makes us notice that babies start out with a natural gift for using language in a social and communicative fashion, the second model makes us notice that children and adolescents must continually learn to relate their discourse better to an audience—must struggle to decenter better. And though the second model makes us notice that babies also start out with a natural gift for using language in a *private,* exploratory and playful way, the first model makes us notice that children and adolescents must continually learn to master this solitary, desert island, poesis mode better. Thus we mustn't think of language only as communication—nor allow communication to claim dominance either as the earliest or as the most "mature" form of discourse. It's true that language is inherently communicative (and without communication we don't develop language), yet language is just as inherently the stringing together of exploratory discourse for the self—or for the creation of objects (play, poesis, making) for their own sake.

In considering this important poesis function of language, we need not discount (as Berkenkotter does) the striking testimony of so many witnesses who think and care most about language: professional poets, writers, and philosophers. Many of them maintain that their most serious work is *making,* not *communicating,* and that their commitment is to language, reality, logic, experience, not to readers. Only in their willingness to cut loose from the demands or needs of readers, they insist, can they do their best work. Here is William Stafford on this matter:

> I don't want to overstate this . . . but . . . my impulse is to say I don't think of an audience at all. When I'm writing, the satisfactions in the process of writing are my satisfactions in dealing with the language, in being surprised by phrasings that occur to me, in finding that this miraculous kind of convergent focus begins to happen. That's my satisfaction, and to think about an audience would be a distraction. I try to keep from thinking about an audience. (Cicotello 176)

And Chomsky:

> I can be using language in the strictest sense with no intention of communicating. . . . As a graduate student, I spent two years writing a lengthy manuscript, assuming throughout that it would never be published or read by anyone. I meant everything I wrote, intending nothing as to what anyone would [understand], in fact taking it for granted that there would be no audience. . . . [C]ommunication is only one function of language, and by no means an essential one. (Quoted in Feldman 5-6).

It's interesting to see how poets come together with philosophers on this point—and even with mathematicians. All are emphasizing the "poetic" function of language in its literal sense—"poesis" as "making." They describe their writing process as more like "getting something right" or even "solving a problem" for its own sake than as communicating with readers or addressing an audience. The task is not to satisfy readers but to satisfy the rules of the system: ''[T]he writer is not thinking of a reader at all; he makes it 'clear' as a contract with *language*'' (Goodman 164).

Shall we conclude, then, that solving an equation or working out a piece of symbolic logic is at the opposite end of the spectrum from communicating with readers or addressing an audience? No. To draw that conclusion would be to fall again into a one-sided position. Sometimes people write mathematics *for* an audience, sometimes not. The central point in this essay is that we cannot answer audience questions in an *a priori* fashion based on the ''nature'' of discourse or of language or of cognition—only in terms of the different *uses* or *purposes* to which humans put discourse, language, or cognition on different occasions. If most people have a restricted repertoire of uses for writing—if most people use writing-only to send messages to readers, that's no argument for constricting the *definition* of writing. It's an argument for helping people expand their repertoire of uses.

The value of learning to ignore audience while writing, then, is the value of learning to cultivate the private dimension: the value of writing in order to make meaning to oneself, not just to others. This involves learning to free oneself (to some extent, anyway) from the enormous power exerted by society and others, to unhook oneself from external prompts and social stimuli. We've grown accustomed to theorists and writing teachers puritanically stressing the *problem* of writing: the tendency to neglect the needs of readers because we usually write in solitude. But let's also celebrate this same feature of writing as one of its glories: writing *invites* disengagement too, the inward turn of mind, and the dialogue with self. Though writing is deeply social and though we usually help things by enhancing its social dimension, writing is also the mode of discourse best suited to helping us develop the reflective and private dimension of our mental lives.

''But Wait a Minute, ALL Discourse Is Social''

Some readers who see *all* discourse as social will object to my opposition between public and private writing (the ''trap of oppositional thinking'') and insist that *there is no such thing as private discourse*. What looks like private, solitary mental work, they would say, is really social. Even on the desert island I am in a crowd.

> [B]y ignoring audience in the conventional sense, we return to it in another sense. What I get from Vygotsky and Bakhtin is the notion that audience is not really out there at all but is in fact ''always already'' (to use that poststructuralist mannerism . . .) inside, interiorized in the conflicting languages of others—parents, former teachers, peers, prospective readers, whomever—that writers have to negotiate to write, and that we do negotiate when we write whether we're aware of it or not. The audience we've got to satisfy in order to feel good about our writing is as much in the past as in the present or future. But we experience it (it's so internalized) as *ourselves*. (Trimbur)

(Ken Bruffee likes to quote from Frost: '' 'Men work together, . . . //Whether they work together or apart' '' [''The Tuft of Flowers''].) Or—putting it slightly differently—when I engage in what seems like private non-audience-directed writing, I am really engaged in communication with the ''audience of self.'' For the self is multiple, not single, and discourse to self is communication from one entity to another. As Feldman argues, "The self functions as audience in much the same way that others do" (290).

Suppose I accept this theory that all discourse is really social—including what I've been calling "private writing" or writing I don't intend to show to any reader. Suppose I agree that all language is essentially communication directed toward an audience—whether some past internalized voice or (what may be the same thing) some aspect of the self. What would this theory say to my interest in "private writing"?

The theory would seem to destroy my main argument. It would tell me that there's no such thing as "private writing"; it's impossible *not* to address audience; there are no vacations from audience. But the theory might try to console me by saying not to worry, because we don't *need* vacations from audience. Addressing audience is as easy, natural, and unaware as breathing—and we've been at it since the cradle. Even young, unskilled writers are already expert at addressing audiences.

But if we look closely we can see that in fact this theory doesn't touch my central practical argument. For even if all discourse is naturally addressed to *some* audience, it's not naturally addressed to the *right* audience—the living readers we are actually trying to reach. Indeed the pervasiveness of past audiences in our heads is one more reason for the difficulty of reaching present audiences with our texts. Thus even if I concede the theoretical point, there still remains an enormous practical and phenomenological difference between writing "public" words for others to read and writing "private" words for no one to read.

Even if "private writing" is "deep down" social, the fact remains that, as we engage in it, we don't have to worry about whether it works on readers or even makes sense. We can refrain from doing all the things that audience-awareness advocates advise us to do ("keeping our audience in mind as we write" and trying to "decenter"). Therefore this social-discourse theory doesn't undermine the benefits of "private writing" and thus provides no support at all for the traditional rhetorical advice that we should "always try to think about (intended) audience as we write."

In fact this social-discourse theory reinforces two subsidiary arguments I have been making. First, even if there is no getting away from *some* audience, we can get relief from an inhibiting audience by writing to a more inviting one. Second, audience problems don't come only from *actual* audiences but also from phantom "audiences in the head" (Elbow, *Writing with Power* 186ff). Once we learn how to be more aware of the effects of both external and internal readers and how to direct our words elsewhere, we can get out of the shadow even of a troublesome phantom reader.

And even if all our discourse is *directed to* or *shaped by* past audiences or voices, it doesn't follow that our discourse is *well directed to* or *successfully shaped for* those audiences or voices. Small children *direct* much talk to others, but that doesn't mean they always *suit* their talk to others. They often fail. When adults discover that a piece of their writing has been "heavily shaped" by some audience, this is bad news as much as good: often the writing is crippled by defensive moves that try to fend off criticism from this reader.

As teachers, particularly, we need to distinguish and emphasize "private writing" in order to teach it, to teach that crucial cognitive capacity to engage in extended and productive thinking that doesn't depend on audience prompts or social stimuli. It's sad to see so many students who can reply to live voices but cannot engage in productive dialogue with voices in their heads. Such students often lose interest in an issue that had intrigued them—just because they don't find other people who are interested in talking about it and haven't learned to talk reflectively to *themselves* about it.

For these reasons, then, I believe my main argument holds force even if I accept the theory that all discourse is social. But, perhaps more tentatively, I resist this theory. I don't know all the data from developmental linguistics, but I cannot help suspecting that babies engage in *some* private poesis—or "play-language"—some private babbling in addition to social babbling. Of course Vygotsky must be right when he points to so much social language in children, but can we really trust him when he denies *all* private or nonsocial language (which Piaget and Chomsky see)? I am always suspicious when someone argues for the total nonexistence of a certain kind of behavior or event. Such an argument is almost invariably an act of definitional aggrandizement, not empirical searching. To say that *all* language is social is to flop over into the opposite one-sidedness that we need Vygotsky's model to save us from.

And even if all language is *originally* social, Vygotsky himself emphasizes how "inner speech" becomes more individuated and private as the child matures. "[E]gocentric speech is relatively accessible in three-year-olds but quite inscrutable in seven-year-olds: the older the child, the more thoroughly has his thought become inner speech" (Emerson 254; see also Vygotsky 134). "The inner speech of the adult represents his 'thinking for himself' rather than social adaptation. . . . Out of context, it would be incomprehensible to others because it omits to mention what is obvious to the 'speaker' " (Vygotsky 18).

I also resist the theory that all private writing is really communication with the *"audience of self."* ("When we represent the objects of our thought in language, we intend to make use of these representations at a later time. . . . [T]he speaker-self must have audience directed intentions toward a listener-self" [Feldman 289].) Of course private language often *is* a communication with the audience of self:

—When we make a shopping list. (It's obvious when we can't decipher that third item that we're confronting *failed* communication with the self.)

—When we make a rough draft for ourselves but not for others' eyes. Here we are seeking to clarify our thinking with the leverage that comes from standing outside and reading our own utterance as audience—experiencing our discourse as receiver instead of as sender.

—When we experience ourselves as slightly split. Sometimes we experience ourselves as witness to ourselves and hear our own words from the outside—sometimes with great detachment, as on some occasions of pressure or stress.

But there are other times when private language is *not* communication with audience of self:

—Freewriting to no one: for the *sake* of self but not *to* the self. The goal is not to communicate but to follow a train of thinking or feeling to see where it leads. In doing this kind of freewriting (and many people have not learned it), you don't particularly plan to come back and read what you've written. You just write along and the written product falls away to be ignored, while only the "real product"—any new perceptions, thoughts, or feelings produced in the mind by the freewriting—is saved and looked at again. (It's not that you don't experience your words *at all* but you experience them only as speaker, sender, or emitter—not as receiver or audience. To say that's the same as being audience is denying the very distinction between 'speaker' and 'audience.')

As this kind of freewriting actually works, it often *leads* to writing we look at. That is, we freewrite along to no one, following discourse in hopes of getting somewhere, and then at a certain point we often sense that we have *gotten* somewhere: we can tell (but not because we stop and read) that what we are now writing seems new or intriguing or important. At this point we may stop writing; or we may keep on writing, but in a new audience-relationship, realizing that we *will* come back to this passage and read it as audience. Or we may take a new sheet (symbolizing the new audience-relationship) and try to write out for ourselves what's interesting.

—Writing as exorcism is a more extreme example of private writing *not* for the audience of self. Some people have learned to write in order to get rid of thoughts or feelings. By freewriting what's obsessively going round and round in our head we can finally let it go and move on.

I am suggesting that some people (and especially poets and freewriters) engage in a kind of discourse that Feldman, defending what she calls a "communication-intention" view, has never learned and thus has a hard time imagining and understanding. Instead of always using language in an audience-directed fashion for the sake of communication, these writers unleash language for its own sake and let *it* function a bit on its own, without much *intention* and without much need for *communication,* to see where it leads—and thereby end up with some intentions and potential communications they didn't have before.

It's hard to turn off the audience-of-self in writing—and thus hard to imagine writing to no one (just as it's hard to turn off the audience of *outside* readers when writing an audience-directed piece). Consider "invisible writing" as an intriguing technique that helps you become less of an audience-of-self for your writing. Invisible writing prevents you from seeing what you have written: you write on a computer with the screen turned down, or you write with a spent ballpoint pen on paper with carbon paper and another sheet underneath. Invisible writing tends to get people not only to write faster than they normally do, but often better (see Blau). I mean to be tentative about this slippery issue of whether we can really stop being audience to our own discourse, but I cannot help drawing the following conclusion: just as in freewriting, suppressing the *other* as audience tends to enhance quantity and sometimes even quality of writing; so in invisible writing, suppressing the *self* as audience tends to enhance quantity and sometimes even quality.

Contraries in Teaching

So what does all this mean for teaching? It means that we are stuck with two contrary tasks. On the one hand, we need to help our students enhance the social dimension of writing: to learn to be *more* aware of audience, to decenter better and learn to fit their discourse better to the needs of readers. Yet it is every bit as important to help them learn the private dimension of writing: to learn to be *less* aware of audience, to put audience needs aside, to use discourse in the desert island mode. And if we are trying to advance contraries, we must be prepared for paradoxes.

For instance if we emphasize the social dimension in our teaching (for example, by getting students to write to each other, to read and comment on each others' writing in pairs and groups, and by staging public discussions and even debates on the topics they are to write about), we will obviously help the social, public, communicative dimension of writing—help students experience writing not just as jumping through hoops for a grade but rather as taking part in the life of a community of discourse. But "social discourse" can also help private writing by getting students sufficiently involved or invested in an issue so that they finally want to carry on producing discourse alone and in private—and for themselves.

Correlatively, if we emphasize the private dimension in our teaching (for example, by using lots of private exploratory writing, freewriting, and journal writing and by helping students realize that of course they may need practice with this "easy" mode of discourse before they can use it fruitfully), we will obviously help students learn to write better reflectively for themselves without the need for others to interact with. Yet this private discourse can also help public, social writing—help students finally feel full enough of their *own* thoughts to have some genuine desire to *tell* them to others. Students often feel they "don't have anything to say" until they finally succeed in engaging themselves in private desert island writing for themselves alone.

Another paradox: whether we want to teach greater audience awareness or the ability to ignore audience, we must help students learn not only to "try harder" but also to "just relax." That is, sometimes students fail to produce reader-based prose because they don't *try* hard enough to think about audience needs. But sometimes the problem is cured if they just relax and write *to* people—as though in a letter or in talking to a trusted adult. By unclenching, they effortlessly call on social discourse skills of immense sophistication. Sometimes, indeed, the problem is cured if the student simply writes in a more social *setting*—in a classroom where it is habitual to share lots of writing. Similarly, sometimes students can't produce sustained private discourse because they don't try hard enough to keep the pen moving and forget about readers. They must persist and doggedly push aside those feelings of, "My head is empty, I have run out of anything to say." But sometimes what they need to learn through all that persistence is how to relax and let go—to unclench.

As teachers, we need to think about what it means to *be an audience* rather than just be a teacher, critic, assessor, or editor. If our only response is to tell students what's strong, what's weak, and how to improve it (diagnosis, assessment, and advice), we actually *undermine* their sense of writing as a social act. We reinforce their sense that writing means doing school exercises, producing for authorities what they already know—*not* actually trying to say things to readers. To help students experience us as *audience* rather than as assessment machines, it helps to respond by "replying" (as in a letter) rather than always "giving feedback."

Paradoxically enough, one of the best ways teachers can help students learn to turn off audience awareness and write in the desert island mode—to turn off the babble of outside voices in the head and listen better to quiet inner voices—is to be a special kind of private audience to them, to be a reader who nurtures by trusting and believing in the writer. Britton has drawn attention to the importance of teacher as "trusted adult" for school children (67-68). No one can be good at private, reflective writing without some *confidence and trust in self.* A nurturing reader can give a writer a kind of permission to forget about other readers or to be one's own reader. I have benefitted from this special kind of audience and have seen it prove useful to others. When I had a teacher who believed in me, who was interested in me and interested in what I had to say, I wrote well. When I had a teacher who thought I was naive, dumb, silly, and in need of being "straightened out," I wrote badly and sometimes couldn't write at all. Here is an interestingly paradoxical instance of the social-to-private principle from Vygotsky and Meade: we learn to listen better and more trustingly to *ourselves* through interaction with trusting *others.*

Look for a moment at lyric poets as paradigm writers (instead of seeing them as aberrant), and see how they heighten *both* the public and private dimensions of writing. Bakhtin says that lyric poetry implies "the absolute certainty of the listener's sympathy" (113). I think it's more helpful to say that lyric poets learn to create more than usual privacy in which to write *for themselves*—and then they turn around and let *others overhear.* Notice how poets tend to argue for the importance of no-audience writing, yet they are especially gifted at being public about what they produce in private. Poets are revealers—sometimes even grandstanders or showoffs. Poets illustrate the need for opposite or paradoxical or double audience skills: on the one hand, the ability to be private and solitary and tune out others—to write only for oneself and not give a damn about readers, yet on the other hand, the ability to be more than usually interested in audience and even to be a ham.

If writers really need these two audience skills, notice how bad most conventional schooling is on both counts. Schools offer virtually no privacy for writing: everything students write is collected and read by a teacher, a situation so ingrained students will tend to complain if you don't collect and read every word they write. Yet on the other hand, schools characteristically offer little or no social dimension for writing. It is *only* the teacher who reads, and students seldom feel that in giving their writing to a teacher they are actually communicating something they really want to say to a real person. Notice how often they are happy to turn in to teachers something perfunctory and fake that they would be embarrassed to show to classmates. Often they feel shocked and insulted if we want to distribute to classmates the assigned writing they hand in to us. (I think of Richard Wright's realization that the naked white prostitutes didn't bother to cover themselves when he brought them coffee as a black bellboy because they didn't really think of him as a man or even a person.) Thus the conventional school setting for writing tends to be the least private and the least public—when what students need, like all of us, is practice in writing that is the most private and also the most public.

Practical Guidelines About Audience

The theoretical relationships between discourse and audience are complex and paradoxical, but the practical morals are simple:

1. Seek ways to heighten both the *public* and *private* dimensions of writing. (For activities, see the previous section.)

2. When working on important audience-directed writing, we must try to emphasize audience awareness *sometimes*. A useful rule of thumb is to start by putting the readers in mind and carry on as long as things go well. If difficulties arise, try putting readers out of mind and write either to no audience, to self, or to an inviting audience. Finally, always *revise* with readers in mind. (Here's another occasion when orthodox advice about writing is wrong—but turns out right if applied to revising.)

3. Seek ways to heighten awareness of one's writing process (through process writing and discussion) to get better at taking control and deciding when to keep readers in mind and when to ignore them. Learn to discriminate factors like these:

(a) The writing task. Is this piece of writing *really* for an audience? More often than we realize, it is not. It is a draft that only we will see, though the final version will be for an audience; or exploratory writing for figuring something out; or some kind of personal private writing meant only for ourselves.

(b) Actual readers. When we put them in mind, are we helped or hindered?

(c) One's own temperament. Am I the sort of person who tends to think of what to say and how to say it when I keep readers in mind? Or someone (as I am) who needs long stretches of forgetting all about readers?

(d) Has some powerful "audience-in-the-head" tricked me into talking to it when I'm really trying to talk to someone else—distorting new business into old business? (I may be an inviting teacher-audience to my students, but they may not be able to pick up a pen without falling under the spell of a former, intimidating teacher.)

(e) Is *double audience* getting in my way? When I write a memo or report, I probably have to suit it not only to my "target audience" but also to some colleagues or supervisor. When I write something for publication, it must be right for readers, but it won't be published unless it is also right for the editors—and if it's a book it won't be much read unless it's right for reviewers. Children's stories won't be bought unless they are right for editors and reviewers *and* parents. We often tell students to write to a particular "real-life" audience—or to peers in the class—but of course they are also writing for us as graders. (This problem is more common as more teachers get interested in audience and suggest "second" audiences.)

(f) Is *teacher-audience* getting in the way of my students' writing? As teachers we must often read in an odd fashion: in stacks of 25 or 50 pieces all on the same topic; on topics we know better than the writer; not for pleasure or learning but to grade or find problems (see Elbow, *Writing with Power* 216-236).

To list all these audience pitfalls is to show again the need for thinking about audience needs—yet also the need for vacations from readers to think in peace.

Works Cited

Bakhtin, Mikhail. "Discourse in Life and Discourse in Poetry." Appendix. *Freudianism: A Marxist Critique*. By V. N. Volosinov. Trans. I. R. Titunik. Ed. Neal H. Bruss. New York: Academic, 1976. (Holquist's attribution of this work to Bakhtin is generally accepted.)

Berkenkotter, Carol, and Donald Murray. "Decisions and Revisions: The Planning Strategies of a Publishing Writer and the Response of Being a Rat—or Being Protocoled." *College Composition and Communication* 34 (1983): 156-172.

Blau, Sheridan. "Invisible Writing." *College Composition and Communication* 34 (1983): 297-312.

Booth, Wayne. *The Rhetoric of Fiction*. Chicago: U of Chicago P, 1961.

Britton, James. *The Development of Writing Abilities*, 11-18. Urbana: NCTE, 1977.

Bruffee, Kenneth A. "Liberal Education and the Social Justification of Belief." *Liberal Education* 68 (1982): 95-114.

Bruner, Jerome. *Beyond the Information Given: Studies in the Psychology of Knowing*. Ed. Jeremy Anglin. New York: Norton, 1973.

———.*On Knowing: Essays for the Left Hand*. Expanded ed. Cambridge: Harvard UP, 1979.

Chomsky, Noam. *Reflections on Language*. New York: Random, 1975.

Cicotello, David M. "The Art of Writing: An Interview with William Stafford." *College Composition and Communication* 34 (1983): 173-177.

Clarke, Jennifer, and Peter Elbow. "Desert Island Discourse: On the Benefits of Ignoring Audience." *The Journal Book*. Ed. Toby Fulwiler. Montclair, NJ: Boynton, 1987.

Cooper, Anthony Ashley, 3rd Earl of Shaftesbury. *Characteristics of Men, Manners, Opinions, Times, Etc*. Ed. John M. Robertson. 2 vols. Gloucester, MA: Smith, 1963.

Ede, Lisa, and Andrea Lunsford. "Audience Addressed/Audience Invoked: The Role of Audience in Composition Theory and Pedagogy." *College Composition and Communication* 35 (1984): 140-154.

Elbow, Peter. *Writing with Power*. New York: Oxford UP, 1981.

———.*Writing Without Teachers*. New York: Oxford UP, 1973.

Emerson, Caryl. "The Outer Word and Inner Speech: Bakhtin, Vygotsky, and the Internalization of Language." *Critical Inquiry* 10 (1983): 245-264.

Feldman, Carol Fleisher. "Two Functions of Language." *Harvard Education Review* 47 (1977): 282-293.

Flower, Linda. "Writer-Based Prose: A Cognitive Basis for Problems in Writing." *College English* 41 (1979): 19-37.

Goodman, Paul. *Speaking and Language: Defense of Poetry*. New York: Random, 1972.

Griffin, Susan. "The Internal Voices of Invention: Shaftesbury's Soliloquy." Unpublished. 1986.

Lawrence, D. H. *Studies in Classic American Literature*. Garden City: Doubleday, 1951.

Ong, Walter. "The Writer's Audience Is Always a Fiction." *PMLA* 90 (1975): 9-21.

Park, Douglas B. "The Meanings of 'Audience.' " *College English* 44 (1982): 247-257.

Perl, Sondra. ''Guidelines for Composing.'' Appendix A. *Through Teachers' Eyes: Portraits of Writing Teachers at Work.* By Sondra Perl and Nancy Wilson. Portsmouth, NH: Heinemann, 1986.

Progoff, Ira. *At a Journal Workshop.* New York: Dialogue, 1975.

Shaughnessy, Mina. *Errors and Expectations: A Guide for the Teacher of Basic Writing.* New York: Oxford UP, 1977.

Trimbur, John. Letter to the author. September 1985.

———.''Beyond Cognition: Voices in Inner Speech.'' *Rhetoric Review* 5 (1987): 211-221.

Vygotsky, L. S. *Thought and Language.* Trans. and ed. E. Hanfmann and G. Vakar. 1934. Cambridge: MIT P, 1962.

Walzer, Arthur E. ''Articles from the 'California Divorce Project': A Case Study of the Concept of Audience.'' *College Composition and Communication* 36 (1985): 150-159.

Wright, Richard. *Black Boy.* New York: Harper, 1945.

I benefited from much help from audiences in writing various drafts of this piece. I am grateful to Jennifer Clarke, with whom I wrote a collaborative piece containing a case study on this subject. I am also grateful for extensive feedback from Pat Belanoff, Paul Connolly, Sheryl Fontaine, John Trimbur, and members of the Martha's Vineyard Summer Writing Seminar.

STYLES

Robert J. Connors

Static Abstractions and Composition

The continuing revitalization of rhetoric that has made it such an exciting subject in our day has given us many novel pedagogical techniques, and as new teaching methods slowly displace the hoary tenets of current-traditional rhetoric, we have a tendency to cast the older ways aside without examining them carefully. This is, I believe, a mistake. Current-traditional rhetoric has much to tell us about how we have come to be who we are as teachers of writing, and it well repays historical and analytical study. If we are to avoid the pitfalls that caught our disciplinary forebears in ineffective and even harmful teaching methods and philosophies, we must examine what they did and try to understand why they did it. If we can thus determine what sorts of theoretical roads lead to dead ends, we can begin to learn from the mistakes of our progenitors. With this approach in mind I would like to examine one of the important historical components of the current-traditional rhetoric we have inherited from the nineteenth century: the element of ''static abstractions.''

Static Abstractions: Nature and Early History

First of all, a definition is necessary. What are these ''static abstractions'' and what have they to do with the teaching of composition?

''Static abstractions'' is a term coined by Albert R. Kitzhaber in his seminal Ph.D. dissertation, *Rhetoric in American Colleges, 1850-1900,* to refer to the famous triad of ''Unity, Coherence, and Emphasis.''[1] I have broadened the term to include *any pseudo-heuristic listing of derived nominals—abstract adjective-based nouns—whose purpose is to define good structure in prose writing.* Common examples of such terms are Unity, Variety, Precision, Energy, Clearness, Order, Economy, etc. Such terms are *static* because they exist in an unchanging theoretical context, are presented as paradigmatic and absolute, and are concerned with labeling finished texts. They are *abstractions* because they exist only as general terms, without qualifiers or specificity. Static abstractions, of which ''Unity, Coherence, and Emphasis'' are merely the best-known examples, have been an important element in composition teaching for the past one hundred years. They first appeared during the mid-nineteenth century, attained almost absolute acceptance by the end of the century, and have remained an important element in composition teaching

From *Freshman English News* 12 (Spring 1983): 1-4, 9-12. Reprinted by permission of the publisher.

1. Albert R. Kitzhaber, *Rhetoric in American Colleges,* 1850-1900 (Doctoral dissertation, Univ. of Washington, 1953), p. 221.

down to our own time. They have been the revered "master terms" of hundreds of composition courses; they have been the touchstones against which thousands of students have been instructed to compare their writing. And they evidence, I believe, an essentially flawed conception of how to teach writing, a conception that has broken down almost completely in our day. Let us examine the use of these terms in teaching.

Although the use of lists of static descriptive nouns might seem to be a traditional part of discourse education (indeed, one 1963 text defends Unity, Coherence, and Emphasis as parts of classical rhetorical theory!), important use of static abstractions in the teaching of rhetoric is less than 100 years old. Like so many elements of current-traditional rhetoric, static abstractions can be traced back no farther than the post-Civil War period in America, a time when the novel discipline of "composition" was striving to define itself. In order to understand how the use of static abstractions (hereafter designated as SAs) evolved and grew, we must look at the forces that shaped rhetorical education in the nineteenth century, both before and after the "composition revolution" of the period 1880-1900.

From 1790 through the 1850's, discourse education in America took place at colleges that were small, private, and usually religiously-affiliated. Rhetorical training at these schools consisted primarily of training in oratory (often pulpit oratory) and in style, taste, literary appreciation, and the writing of fine sentences. Classical rhetorical theory was important to such courses, but far more important was the work of the great English rhetoricians of the eighteenth century, George Campbell and Hugh Blair. Campbell's *Philosophy of Rhetoric* of 1776 and Blair's *Lectures on Rhetoric and Belles-lettres* of 1783 were reprinted scores of times in America during the period 1800-1860, and the rhetorical teaching of the time reflected their work—especially the work of Blair—faithfully.

Neither Campbell nor Blair can be said to make real use of SAs as I have defined them. Their concerns were primarily with style and persuasion, and though much of their discussion was informed by Aristotle's "four virtues" of style—clarity, propriety, dignity, and correctness—Campbell and Blair did not structure their treatments of style around these terms, and thus never turned the terms into prescriptive SAs. Campbell indeed spoke often of Purity, Perspicuity, and Vivacity, but never in any mechanical way. Blair mentioned "the properties most essential to a perfect sentence" as "1. Clearness and Precision. 2. Unity. 3. Strength. 4. Harmony."[2] This was, however, as far as Blair's *Lectures* went in listing adjective-based descriptions; neither Blair nor Campbell was interested in lists for their own sake.

But rhetorical education was changing, especially in the United States. With the passage of the Morrill Act in 1862, Congress created the land-grant Agricultural and Mechanical Colleges that were to grow into the major state universities. The era of mass higher education was at hand, and with the increase in the number of college students after 1870 came a need for new sorts of courses. The older type of rhetorical education, with its stress on oratory, stylistic beauties, fine sentences, and analysis of belletristic literature, had served well enough for small private schools turning out doctors, lawyers, teachers, and ministers. Students at the new A&M colleges, however, needed something else. Agriculturists had little use for stylistic analysis; businessmen were uninterested in

2. Hugh Blair, *Lectures on Rhetoric and Belles-Lettres* (Philadelphia: James Kay & Bro., 1844), p. 114.

pulpit oratory; engineers did not much appreciate sentence-by-sentence analysis of the Beauties of Johnson. The once-popular rhetoric of Blair and Campbell began to seem extraneous to the curricula of many students.

These new A&M students, however, had other needs that were to call forth a new sort of rhetorical education. Many of them were from public schools and had seldom been asked to write compositions of any extended length. Many were studying technical or professional fields and needed to be taught all-purpose forms of written communication to meet their career plans. Some students, to the chagrin and shock of older professors, could not write with perfect grammatical correctness—or indeed with any fluency at all. To meet the needs of these students in both the new and the more established colleges (which also felt the effects of the ongoing Industrial Revolution), a new sort of rhetoric course was developed. This course stressed written discourse and ignored oral; it concentrated on mechanical correctness and conventional usage over stylistic beauty and sentence analysis; it demanded short "themes" written to order from its students rather than relying upon examinations; and it was concerned with structure and predictability in writing rather than with the development of an individual stylistic voice or persuasive ability. This new course was, of course, the prototypical freshman English course.

The use of SAs in these early composition courses grew, as Kitzhaber points out, from the earlier Blairian tradition of stylistic rhetoric. Although Blair's use of SA-like terms in his *Lectures* was not structurally important, American rhetoricians influenced by Blair can be seen developing a "list-consciousness" of the sort that would lead naturally to SAs as early as the 1820's. Much of this fascination with abstract descriptive terms was probably a result of the popularity of Lindley Murray's *English Grammar,* first published in 1795 and reprinted countless times in abridged form during the period 1805-1860. Murray made the application of the terms "Purity, Propriety, Precision, Clearness, Strength, Unity" much more mechanical than had any of the rhetorical theorists, and his delight in listings soon found followers in America. Samuel Newman in his *A Practical System of Rhetoric* of 1827 lists the "qualities of a good style" as "Correctness, Perspicuity, Vivacity, Euphony, and Naturalness." In his *Advanced Course of Composition and Rhetoric* of 1854, George Quackenbos follows Lindley Murray, listing the "essential properties of style" as "Purity, Propriety, Precision, Perspicuity (Clearness), Strength, Harmony, and Unity." Indeed, the alliterative euphony of the first four terms made them very popular during the midcentury period; Purity, Propriety, and Precision, sometimes supplemented by Perspicuity, are often seen linked in the texts of the time.

We must, however, carefully discriminate these lists of stylistic qualities from listings of static abstractions, which are concerned more with the structure of writing than with its style. Stylistic qualities are concerned with word-choice only, while static abstractions are meant to be applied to more complex structures like paragraphs and whole discourses. They are "assembly-words." At the sentence level there may be some overlap between stylistic qualities and static abstractions; for instance, such static abstractions as "coherence," "economy," and "order" might be applied to the sentence, but not very usefully. Application at the paragraph and whole-discourse level is the test of SAs; such stylistic qualities as "purity" and "precision" are meaningless above the sentence level. Stylistic qualities were usually expressed as adjectives by the end of the nineteenth century, while static abstractions are by nature always nouns. Although lists of stylistic qualities such as those of Newman and Quackenbos never really died out of composition texts, after 1880 or so they were vastly less important in providing overall organization for textbooks. SAs, it can be argued, evolved from the practice of piling up descriptive stylistic adjectives, but they persisted long after stylistic rhetoric had died out.

There is no doubt about who introduced the first true SAs into American composition. Henry N. Day wrote 14 books on rhetoric and composition between 1844 and his death in 1890, and it is to his books *Rhetorical Praxis* (1861), *Art of English Composition* (1861) and *Art of Discourse* (1867) that we must look for the first popular use of SAs. Day's rhetoric provides a bridge between the earlier and later forms of the American composition course; it does contain a heavy dose of stylistic material, but in its love for lists, rules, and Laws it presages the schematic structural rhetoric of the eighties and nineties. In addition to the first listing of the "methods of exposition," which were to become so important in the current-traditional rhetoric of the twentieth century, Day's texts present his protypical SAs, which he called "Laws of Discourse." Day posited four Laws that operated throughout all explanatory writing: "the Law of Unity, the Law of Selection, the Law of Method, and the Law of Completeness." Students were expected to observe each "Law" in the writing they did. Without going deeply into the details of each Law, we can note Day's presentation of them. They are clearly not stylistic descriptions like Purity, Propriety, and Precision. They are primarily prescriptive, not descriptive; they are apodictic, not inductive. What we have in Day's Laws is a genuinely new sensibility, an approach to teaching composition that marks a radical departure from descriptive belletristic-stylistic rhetoric: Day is the first important rhetorician attempting to make composition "rational" and "scientific" through the introduction of strict definitions and comprehensive rules (or Laws).

He was, of course, not the last. This was a time when the names of Darwin, Huxley, and Spencer were in the air, and the possibility of a truly scientific rhetoric based upon discoverable laws of discourse seemed real in the burgeoning scientific atmosphere of the day. Early psychological explanations of the "laws" governing mental operations sounded impressive; surely, theorized midcentury rhetorical writers, these laws of mind were connected to discoverable laws of discourse. A number of early treatises posited the existence of these laws and some actually listed versions of them; undoubtedly the most complete, serious, and arcane of these was David J. Hill's *Science of Rhetoric* in 1877, which assumed that composition was as much a science as botany or chemistry.

The Growth of Reductionism

In conflict with the growing scientism of some theorists of rhetoric during the post-Civil War period was the attitude of many teachers who were becoming painfully aware of the ineffectiveness of any theoretical approach to teaching students to write. It had been, in fact, the essentially theoretical nature of Blairian stylistic rhetoric that had made it useless in the new "theme-writing" courses. Erastus O. Haven expressed the attitudes of many of the new practitioners in the Preface to his 1869 *Rhetoric: A Text-Book:*

> Abstruse arguments about style and oratory, about the conflicting theories of taste and beauty, about conviction and persuasion and the laws of mind, and the philosophy of

> language, are all good and valuable in their place; but a student may read and repeat them with but little more effect on his own habits of speaking or power to write well, than he would receive from an equal amount of study in mathematics, medicine, or law, or any other subject.[3]

Instructors in the classroom wanted a teaching rhetoric, one that would make their intercourse with their charges easier and more effective. They needed teachable and seemingly practical course-content. The Laws of Day and Processes of Hill were a step in the right direction, but too much of their concern remained stylistic and theoretical. Teachers needed a teachable method, and in response to their need, one was developed.

Alexander Bain, who is the single largest contributor to current-traditional rhetoric as it was established in this century, was only an indirect contributor to the establishment of SAs in composition. His *English Composition and Rhetoric* of 1866, though it contained some stylistic terms, did not enunciate any SA-like terms. It did, however, establish a further expansion of the "list-sensibility" that would provide such a fertile soil for others' static abstractions. Especially influential, of course, were Bain's listing of the four "forms of discourse" and his list of the six rules governing construction of paragraphs.[4] These rules, transmogrified, were to be the basis of the most popular SAs ever developed—the fabled trinity of Unity, Coherence, and Emphasis.

The leap from Bain's six paragraph rules to Unity, Coherence, and Emphasis would not take place for twenty years, and during that period the demand for simplicity in theory came to fore. After Day's Laws there were, however, no SA-like terms that gained general popularity for a time. Adams Sherman Hill presented the terms "Clearness, Force and Elegance" in his popular *Principles of Rhetoric* in 1878, but these terms, though they achieved a good deal of currency, were essentially stylistic; although they were presented prescriptively, Hill's terms were concerned with word-choice rather than with structure. The great leap from the six "rules" of Bain to full-fledged SAs did not take place until 1886, when John Genung of Amherst College published his extremely influential *Practical Elements of Rhetoric*.

Genung's text proposed three "fundamental qualities" of the paragraph, qualities that were obviously drawn from Bain's rules but which supplied a simpler heuristic form. Bain's rules were as follows: 1. Explicit Reference 2. Parallel Construction 3. Topic Sentence 4. Consecutive Arrangement 5. Overall Unity 6. Subordination. Genung's simplification of these rules boiled them down to three static abstractions: a paragraph, said Genung, must be distinguished by *Unity,* by *Continuity,* and by *Proportion.* These three terms structured and illuminated the entire paragraph chapter in *Practical Elements.* Likewise the sentence, in Genung's view, had two general qualities: *Unity* and *Emphasis.* Large sections of the text were devoted to these qualities, and this approach proved very popular; Genung's book became the best-selling composition text of the late eighties.

Throughout this period the struggle to define "composition" that had been going on since the early sixties intensified. The remnants of the Blairian stylistic tradition were hardening and atrophying; what had once been a wide-ranging discussion of stylistic

3. Erastus O. Haven, *Rhetoric: A Text-Book* (New York: Harper & Bros., 1869), p. vii.
4. Alexander Bain, *English Composition and Rhetoric* (New York: D. Appleton, 1867), pp. 142-152.

effects had by the late 1880's become a series of prescriptive demands for the "Four P's"—Purity, Propriety, Precision, and Perspicuity.[5] It began to seem that textbooks would be successful to the degree that they could reduce rhetorical theory to lists of prescriptive concepts, and thus SAs and the "Four P's" proliferated. A teaching rhetoric was under construction, and by 1890 the stage was set for a truly radical simplification of the theory of composition.

In 1891 it arrived. Barrett Wendell, an "English A" instructor at Harvard, published in that year *English Composition,* a textbook based on eight lectures he had given at the Lowell Institute in 1890. *English Composition* was immediately popular despite the fact that it broke the usual textbook mold, containing no exercises or supporting examples from literature. It was popular because it brilliantly reduced all the complex principles of the chaotic postwar rhetorical tradition to a few easily remembered and universally applicable terms. Wendell himself states his case best as he explains the problems he found in the composition texts of his time:

> These books consist chiefly of directions as to how one who would write should set about composing. Many of these suggestions are extremely sensible, many very suggestive. But in every case these directions are appallingly numerous. It took me some years to discern that all which have so far come to my notice could be grouped under one of three very simple heads, each of which might be phrased as a simple proposition. Various as they are, all these directions concern either what may be included in a given composition (a sentence, a paragraph, or a whole); or what I may call the outline, or perhaps better, the mass of the composition—in other words, where the chief parts may most conveniently be placed; or finally, the internal arrangement of the composition in detail. In brief, I may phrase these three principles of composition as follows: (1) Every composition should group itself about one central idea; (2) The chief parts of every composition should be so placed as readily to catch the eye; (3) Finally, the relation of each part of a composition to its neighbors should be unmistakable. The first of these principles may conveniently be named the principle of Unity; the second, the principle of Mass; the third, the principle of Coherence.[6]

Here, in a fine example of Wendell's assured, convincing style, we have the prototype of the master-SAs of modern current-traditional rhetoric.

Wendell went on in his gracefully-written and extremely influential book to apply his principles of Unity, Mass, and Coherence to the three "levels" of discourse into which textbooks increasingly segmented writing after 1880: the sentence, the paragraph, and the whole theme. Greatly weakened by Wendell's time was the decaying tradition of stylistic rhetoric that had made even Hill and Genung open their books with a "diction" section; Wendell plunged immediately into a discussion of the discourse levels of written composition.

It is not hard in retrospect to see what Wendell accomplished; his ideas were hardly

---

5. See, for instance, James De Mille, *The Elements of Rhetoric* (New York: Harper & Bros., 1878), Virginia Waddy, *Elements of Rhetoric and Composition* (New York: American Book Co., 1889), and William Williams, *Composition and Rhetoric by Practice* (Boston: D. C. Heath, 1890).

6. Barrett Wendell, *English Composition* (New York: Scribners, 1891), pp. 28-29.

new, and the stunning popularity of his text (the only composition text of the nineties which is still available today in two different editions) must have surprised even him. He had achieved this amazing success by taking Genung's three-term application of Bain's six paragraph rules, changing two of the terms slightly—"Continuity" became "Coherence" and "Proportion" became "Mass"—and by proposing to apply the resulting heuristic to all levels of writing. In an educational milieu seeking ever simpler, ever more teachable concepts, it was a reduction that could not have come at a more propitious time. *English Composition* established the use of SAs as a central tenet of all the texts that followed its lead—and 90% of all texts did.

Derivation and Stagnation

The years between 1891 and 1910 saw some struggles of nomenclature, but in general Wendell's concepts swept the field, encountering only one important challenge. Wendell's term "Mass" was early seen as problematical; it was rather too vague even for a static abstraction, and in 1893 George Rice Carpenter suggested in his otherwise unexceptional *Exercises in Rhetoric and English Composition* that the trinity be changed to Unity, Coherence, and Emphasis (UCE). This formulation proved acceptable, and from that point on UCE were the main SAs in pedagogy.

There were several half-hearted attempts during the nineties to add to the canon of Wendell's terms or to change them further. Fred N. Scott and Joseph Denney's *Paragraph-Writing* of 1893 proposed the "general laws of the paragraph" as Unity, Selection, Proportion, Sequence, and Variety, but despite their book's importance in popularizing the Bainian organic paragraph, their "general laws" fell flat. In their *Composition-Rhetoric* of 1897, Scott and Denney gave in and used UCE as part of their exercises. Alphonso Newcomer introduced Sequence, Clearness, and Effectiveness on top of UCE, and his book sank without a trace. Wendell's terms as modified by Carpenter were showing themselves hard to compete against.

Not all textbooks went along with this gathering flood of SAs, however; older and more established text authors found Wendell's pill particularly bitter to swallow. John Genung resolutely kept his structural discussion concrete in his texts of 1893 and 1900, neither of which was as popular as *The Practical Elements* had been. A. S. Hill stuck with his stylistically-based "qualities of expression" (Clearness, Force, and Elegance, which Wendell had, in fact, utilized in the stylistic section of *English Composition*) in all his later books. But these were the voices of the past; the newer derivative textbooks of the late nineties and the early twentieth century latched onto Unity, Coherence, and Emphasis as they had onto the modes of discourse, the organic paragraph, sentence grammar, and the other ritual treatments of an increasingly standardized teaching rhetoric.

Not all texts of this period, I must hasten to say, used UCE on all three levels. The terms were nearly always used to organize large sections of textbooks, and were most common at the paragraph level (from whence, of course, they had been drawn) or at the level of the whole discourse; UCE, being true static abstractions, were never very effective at the sentence level, and other terms were often tried in discussions of the sentence. (Failure to note the difference between SAs and stylistic terms has been a continuing confusion in current-traditional rhetoric.) Alternative SA formulations never really died out even in the heyday of UCE: there was Albert Granberry Reed's formula of Unity, Coherence, Effectiveness, Variety and Progression; Hammond Lamont proposed Unity, Order, Proportion, Clearness, and Interest; Edwin C. Woolley, in his *Handbook of Composition: A Compendium of Rules* of 1907 (which may well be the first dogmatic handbook), proffered Unity, Organization, and Coherence. C. S. Baldwin, perhaps the most influential text-author of the first fifth of the new century, had championed UCE in his 1902 *College Manual of Rhetoric,* but in 1909 he turned from the true faith and demoted UCE to secondary status behind *his* new master SAs, Clearness and Interest, in his *Composition: Oral and Written.* Throughout this period, however, Wendell's terms remained the clear favorites for most textbook writers.

Beginning around 1910, UCE had achieved such a clear victory that few new SAs appeared to challenge them. An avalanche of textbooks using the terms appeared: Canby's *English Composition in Theory and Practice,* Maxcy's *Rhetorical Principles of Narration,* Hanson's *Two-Years' Course,* Slater's *Freshman Rhetoric*—the list goes on and on. Rival classifications never really died out completely, and there were always those authors who could not bear conformity and so changed "Emphasis" to "Proportion" or made other changes equally substantive, but after 1910 the concept of SAs meant for many years the Holy Trinity of Unity, Coherence, and Emphasis.

If the teens established UCE as the Master-SAs of the century, the decade of the 1920's was the great flowering of SA-consciousness.[7] It saw the most sterile and mechanical use of the terms—for instance, Charles Harvey Raymond's *Essentials of English Composition* of 1923, which reduced writing to 90 rigid rules, all based on UCE—as well as their spread to nearly all textbooks. But the twenties also saw some changes that would eventually weaken the grip of UCE and then of SAs in general. Several novel sets of SAs appeared, indicating that even in their heyday UCE were not completely satisfactory to teachers. These new sets of terms are not remarkable because they displaced UCE; rather they are notable for their fatuity, the sense they give that the SA paradigm itself was coming to the end of its tether. Widtsoe and Lewis' *Effective Writing* of 1923, for instance, offered Clearness, Interest, and Effectiveness as its major terms. The Widtsoe and Lewis SAs are all obvious examples of what I call "tautological SAs"—terms that are so general, circular, and self-referential as to be almost completely uninformative. And Rankin, Thorpe, and Solve's *College Composition* in 1929 created a novel system of SAs that will probably never be bested for sheer mechanism. Their "Qualities of Good Writing" were Interest, Clearness, Organic Unity, Originality, and Sincerity and Restraint. All these terms were exhaustively detailed and were followed up by rigorous exercises—including "Exercises in Sincerity and Restraint"—an activity which illustrates how far divorced from reality SA-based teaching was be-

---

7. The degree to which SAs had become part of the conceptual baggage of the composition course can be seen in student annotations of textbooks from this period. Even in texts that use UCE very heavily, flyleaves and margins often show that students were expected to write them again and again. The list was obviously *drummed into* freshman heads. Even the process of *seeking* SA-type terms was encouraged. For instance, in a copy of Thomas Rankin's 1917 text *The Method and Practice of Exposition* in the Middleton Library at LSU, the student annotations tell a strange story. Rankin, a student of the iconoclastic Fred Newton Scott at Michigan, totally avoided SAs in this, his first textbook. And yet, on a page in the discursive section of the book in which Rankin describes the nature of exposition, the student owner of the text has numbered and underlined *every* SA-type noun: "1. Truth, 2. clear, 3. interesting, 4. convincing, 5. controlled, 6. ethical." For that student (and for that teacher), SAs were *the* way in which good writing was conceptualized.

coming. Again, Rankin, Thorpe, and Solve's terms were largely tautological. SAs had become, in Richard Lanham's words, "a tedious repetitive, unoriginal body of dogma,"[8] and one is left with the impression that text-authors dissatisfied with UCE had few roads to travel that did not lead to inanity.

Gradual Breakdown of Static Abstractions

We cannot point to a date at which it can confidently be said that static abstractions as important elements of composition teaching began to die. Unlike the modes of discourse, which were suffocated by the growth of expository writing, SAs never drowned in their own waste products; they are in fact still extant today in some texts, though their power has been vastly reduced. We can trace a few currents in charting the gradual fade-out of SAs from textbooks, but they continue to hold sway today in the minds of many teachers, and the use of SAs in writing classes will probably not die out completely in this century.

We do know that the 1930's, which was a revolutionary time for education at all levels in America, started in motion forces that would undercut SAs by reducing the need for them as content in writing courses. The influence of John Dewey's educational theories began to be important in the early thirties and became more so as the decade advanced. Texts began to appear that were psychologically more sophisticated, taking a more process-oriented approach to writing, paying more attention to students and their everyday experiences, attempting to link writing with vocational training and the "real world." Many of these Deweyite books did incorporate UCE or other SAs, but the terms were no longer central to the purpose of the texts as they had been in the teens and twenties.

As the thirties advanced, more and more texts were "thesis texts," which put forward as the controlling theme of the whole book a single central idea about how writing should be approached, and these theses seldom had anything to do with SAs, further reducing their importance in the eyes of the newer practitioners.[9] A whole sub-genre of thesis texts was the "American writing" text—Whitford and Foster's *American Standards of Writing,* Davidson's *American Composition and Rhetoric,* Watt's *An American Rhetoric,* most of which were consciously anti-traditional and many of which avoided UCE in specific or SAs in general.

These trends, important though they were, should not be seen as destroying the power of SAs; UCE in particular kept rolling along in derivative composition texts and by the thirties were usually referred to as the "fundamental laws" of writing. (Like many elements of late nineteenth-century rhetoric, UCE became so widely accepted that their origins were lost by the 1930's and they were generally presumed to be part of the primeval wisdom of the race.) The degree to which the terms had become dogma can be seen amusingly in William C. Hoffman and Roy Davis' *Write and Speak Better,* a Dale Carnegie-influenced advice-to-a-businessman treatise from 1937. In a folksy, no-

---

8. Richard A. Lanham, *Style: An Anti-Textbook* (New Haven: Yale University Press, 1974), p. 19.

9. For more on the concept of thesis texts, see Robert J. Connors, "The Rise and Fall of the Modes of Discourse," *CCC* 33 (Dec. 1981), 451. It is ironic that Wendell, who originated the UCE trinity, was also the author of the first of the thesis texts that would eventually help to weaken his SA theory.

intellectuals-here style, the authors put forward the usual nostrums for business success through good communication, but then they stop, shuffle for a moment, and admit that . . . uh . . . "a message will be well-put only when it conforms to the old-fashioned requirements of unity, coherence, emphasis, and harmony."

> Now, it may be that you associate these terms with rather depressing experiences . . . you remember unity, coherence, emphasis and harmony from the "dear old school days" when you "took" doses of English composition and speech training . . . it is not sufficient to keep your eye on the golf ball (unity) and "follow through" on your drive (coherence). You must "smack the pill" hard enough to make it go, and the force of the smack (Emphasis) should be governed by the distance you wish the ball to cover.[10]

As we can see, reliance upon SAs had become pervasive in all sorts of instructional situations.

In the late thirties UCE began to break down in some texts into less general terms. It was obvious to many teachers by this time that UCE did not work well on every level of discourse, and some authors began to experiment with more specific SAs derived from UCE. John Kierzek's first *Macmillan Handbook* of 1939 allowed for Unity, Order, and Proportion on the theme level, Unity, Order, Arrangement, and Transitions on the paragraph level, and Clearness, Order, and Effectiveness on the sentence level. Donald Davidson's *American Composition and Rhetoric* in the same year broke UCE down completely into concrete discussions of how proportion can be achieved in writing, how to use transitions, etc. Other texts responded to the weakening of UCE by going the opposite way: instead of avoiding SA terms they piled them up with desperate abandon like sandbags on an eroding levee. Harry Shaw's *Complete Course* of 1940 was built around the terms Unity, Completeness, Clarity, Effectiveness, Order, Proportion, Length, and Transition. Obviously the radical simplification that had been the great appeal of Wendell's three terms was on the way out.

The 1940's brought more problems for UCE, as the General Education "communications" movement and the General Semantics movement became powerfully influential on composition courses. Most "communications" texts did not use SAs at all, since they were concerned with all four of the "communications skills"—listening, reading, speaking, and writing. Writing was, after all, only one-fourth of the purview of such courses, and "communications" texts had no dearth of material to cover without UCE. In addition, most "communications" texts reflected the beliefs of the Deweyite General Education movement: they were audience- and process-oriented rather than being concerned exclusively with the written product. General Semantics texts were too much interested in words and their meanings to pay attention to prescriptive abstractions; abstraction, in fact, was one of the elements distrusted in communication. Thus, the two great educational fads of the forties and early fifties, though they themselves did not last, helped to weaken many aspects of the older current-traditional rhetoric that preceded them.

After the 1940's, composition texts changed. Thesis texts came more and more to be

---

10. William C. Hoffman and Roy Davis, *Write and Speak Better* (New York: Whittlesey House, 1937), pp. 21-23.

the standard texts, and as a result textbook organization, which had once been ritually predictable, became experimental and arbitrary. Authors seemed to have given up on the ''scientific'' nature of composition; books after the late forties tended to be discursive, filled with many short subsections, process- and audience-oriented in novel ways. It was a period of change, and though the use of SAs remained widespread, their positions in the books that used them were no longer so central. Even traditional books like Brooks and Warren's *Modern Rhetoric* of 1950 (which was modern only by the standards of the twenties) and Richard Weaver's *Composition* of 1957 devoted only a few pages to UCE. Experimental texts used SAs even less.

The ''Back to the Basics'' movement of the mid-fifties, followed by the launching of Sputnik and the resulting ''education race'' with Russia, put an end to the period of experimentation of 1945-55. But although the obvious experimentation ended and current-traditional rhetoric again put its stress on ''traditional,'' the damage to SAs had been done. After 1960, only a few textbooks would use SAs to organize whole chapters, and none would use them to organize whole books. Thesis texts typically devoted five or six pages to UCE and then went on; the terms were often found, but their power as a heuristic list was diminishing. Throughout the fifties and sixties, and even the early seventies, SAs hung on in traditional textbooks. UCE was still the most popular formulation, but such old friends as Continuity, Movement, Variety, Effectiveness, and Economy were found as well. Through the great theoretical upheaval of the sixties, SAs continued to be seen in texts, and it is only in the mid-seventies, when a crop of textbooks appeared that were informed by the ''new rhetoric'' of the sixties, that we can see the deathgrip of SAs relax. Beginning around 1975, almost as if by agreement, new-textbook authors stopped using UCE and SAs in general. Even texts that were largely traditional did away with the old Wendell trinity and did not replace it with other SA terms. The possible reason for this sudden rejection of SAs we will examine in a moment.

Today SAs are found only in texts that have atrophied, like the older handbooks and McCrimmon's *Writing with a Purpose*. Newer texts have newer nostrums to offer—better systems with which to organize essays, more informed models of the writing process, more information on invention processes. Static abstractions as teaching tools seem to have come close to the end of their rope, and they hang on only contingently, soon to be sent to join sentence diagramming and the subset outline in whatever afterlife there may be for discredited current-traditional myths.

Epilogue: The Bloom and Blast of a Myth

Before I go on to discuss the problems that SAs present as a teaching method, let me make one point clear. I am not proposing that good writing does not have the qualities of Unity, Coherence, and Emphasis. Of course it does. I hope this essay does. In addition, I hope it has the qualities of Clearness, Purity, Force, Beauty, Completeness, Effectiveness, Energy, Correctness, Smoothness, Dignity, Variety, Economy, Order, and Ease. Not to mention, of course, Sincerity and Restraint. This sheer piling up of SAs begins to make my case against them; while no one would deny that all these fine terms are indeed desirable qualities in writing, they are devalued by their sheer numbers.

These terms may indeed describe good writing, but they are useless, I will argue, in helping students create it.

The use of SAs in teaching writing, like so many aspects of current-traditional rhetoric, became popular primarily because it appealed to teachers—not because it aided students. Casting about for something concrete to teach in a rhetoric course that was obviously cut off from traditional oratorical and belletristic concerns, the composition teachers of the late nineteenth century created a mythology: the modes of discourse, the methods of exposition, the organic paragraph, and static abstractions. These were the components, assembled over the course of approximately 25 years, of the "discourse theory" of current-traditional rhetoric. I have here omitted many other important elements, particularly the obsession with correct mechanics and usage and with pure diction, since these concerns are at or below the sentence level.) I call these theories mythology not because they are fantastic or absolutely untrue. They are myth because they were passed on unquestioningly by true believers, because their theoretical origins were lost and they atrophied into dogma, and because they grew to have a powerful symbolic value for those who believed in them. Without any clear idea of how to teach a contentless course, early composition teachers were forced to create content from their subjective perceptions of what was important in writing; thus necessity became the mother of the invention of "composition theory." Like other aspects of early current-traditional rhetoric, lists of SAs sounded complete, sounded helpfully descriptive of good writing, even sounded scientific to some. They offered teachers a simple, easily taught, and seemingly enforceable heuristic, and for these reasons SAs thrived for many years.

We do not, however, have to look very far to find problems with SAs as teaching tools. It was recognized as early as 1910 that the use of such abstract terms was not helpful to students. In Francis Berkeley's *A College Course in Writing from Models,* she attacked the use of SAs unequivocally:

> It is my firm belief that no student ever yet learned to write by means of studying rules and abstract principles from a textbook on rhetoric. After a fairly long experience with the endeavours of Freshmen and Sophomores, I feel absolutely sure that, to these long-suffering youngsters, "unity, mass, coherence," and all their works remain "miching mallecho" to the end of the chapter. Still another objection to this sort of teaching, besides its abstractness and consequent unintelligibility to the undergraduate mind, is that such teaching is mainly destructive.[11]

By "destructive," of course, Berkeley means that the UCE formula was used mainly to cast doubts on already written student work, to force students to constantly view their own work through the fun-house mirror of SA terms.

A more specific criticism of SAs was levelled by Henry Burrowes Lathrop in an article in *Wisconsin Studies in Language and Literature* in 1918 and extended in the preface to his 1920 text *Freshman Composition.* Though Lathrop believed that Wendell's *English Composition* was "the beginning of a rational treatment of structure in

---

11. Frances Campbell Berkeley, *A College Course in Writing from Models* (New York: Henry Holt, 1910.), p. ix.

English composition,'' he also complained that Wendell was ''inaccurate in various ways'':

> The first is the erection of three apparently coordinate principles of structure. There is one principle—unity; and to consider unity, coherence, and emphasis as things in any way independent tends to make students try, as one of mine put it, to ''apply'' them successively to their work. The result is a definite rigidity, a tendency to mechanism in style.[12]

Lathrop's complaint goes directly to the heart of the problem of using SAs in composition classes: students must try to ''apply'' them sequentially to their writing, to make linear a process we know to be recursive. (It is for this reason that the use of SAs is one of the defining elements of the ''product-orientation'' of current-traditional rhetoric.) This process of attempting to apply SAs while writing has never been very helpful to students.

One of the main reasons for the failure of SAs to work well, as these early authors recognized, was the very abstractness of the terms. Such terms might create a neat descriptive list, but their generality made them useless as prescription. Such general terms live for highly educated people in a thick nutrient stew of examples, synonyms, antonyms, experiences, experiments, consideration. But unless SA-type terms are informed by this sort of background knowledge they remain useless buzzwords. I. A. Richards was not directly addressing the question of SAs in his *Philosophy of Rhetoric* in 1936, but his thoughts on this question are helpful. ''The stability of the meaning of a word,'' said Richards, ''comes from the constancy of the contexts that give it its meaning.''

> You will see, I hope, that these criteria—*precision, vividness, expressiveness, clarity, beauty* are representative instances of them—are misleading and unprofitable to study unless we use them with a due recognition of this interdependence among the words we use . . . and an alert distrust of our habit of taking words and their meanings for examination in isolation. The isolation is never complete, of course; a completely isolated word would be meaningless. The detachment we attempt is by means of a supposed standard setting, an imaginary schematic context which is assumed to be representative.[13]

When teachers look at such terms as ''Coherence'' or ''Precision'' in isolation, they may seem useful and even genuinely descriptive, but only because of the informed ''interdependence'' that exists in our own minds. Students very seldom possess that informed context for such terms, and as a result, such abstract descriptors are indeed often ''miching mallecho'' to them.

The closest analogy we have for the pedagogical use of SAs on the level of discourse structure is the use of what Louis Milic has called ''metaphysical descriptions'' on the level of style. For most of history, says Milic, style study has languished under the same sort of general descriptive terms we have been discussing. As we saw during the breakdown of stylistic rhetoric during the nineteenth century, such terms as ''clear,''

12. Henry Burrowes Lathrop, *Freshman Composition* (New York: Century Co., 1920), p. x.
13. I. A. Richards, *The Philosophy of Rhetoric* (New York: Oxford Univ. Press, 1936), pp. 70-71.

"pure," "precise," "energetic," have been used to describe individual styles or just "good" style with little attempt made to clarify the meaning of the terms. "Stylistics," says Milic, "has for most scholars still no method beyond the method of impressionistic descriptions and a vague use of rhetoric . . . when all these descriptions are placed side by side, they amount to little more than a glossary of adjectives."[14] The relationship between the "metaphysical adjectives" about which Milic is complaining and the nouns which I have called static abstractions is fairly clear: both sorts of approaches try to make terms that are essentially personal appear to be objective. Milic's metaphysical terms are analogous to SAs also in that they refer not to the activities of the writer but rather to the perceptions of the reader. "Though *clear* may seem to refer specifically to the writing process," says Milic, "it actually describes the response of the reader. What is clear to one may not be to another." Similarly, what is Emphatic to one may not be to another.

In critiquing metaphysical style criticism, Milic has also put his finger on the weakness of SAs as elements of writing-process pedagogy: both metaphysical terms and static abstractions are essentially based in the subjective responses of the reader rather than being in the affective domain of the writer. Trying to keep terms like "effectiveness" and "variety" in mind while composing is simply not possible—or at least is not productive. Even the master SAs, Unity, Coherence, and Emphasis, are simply abstract terms that cannot be mechanically gridded over the complex process of writing. Real writers do not process their intentions through such abstract and isolated terms, as the protocol analyses of Linda Flower and John Hayes show.

Admitting that SAs cannot be useful as part of the composing process, what about their use in editing already completed drafts? Here again we encounter the problem of an informed sensibility and a context for such terms. Here also, I believe, we can find the reason why textbook SAs have almost completely disappeared in the last six or seven years.

As Lathrop suggested in 1920, the most common student use of SAs was as editorial terms; students were encouraged to " 'apply' them successively to their work," checking their writing against the teacher's list of terms. That this process always had problems in practice there is little doubt, since it was condemned as early as 1910. But as long as students had any sort of informed context for SA terms like Unity and Coherence, their use was at least plausible, conceivable. They might not have worked well, but throughout most of this century they had *some* meaning for students, and their methodological problems could thus be ignored. So long as most students had read a few books and could vaguely grasp what Unity or Coherence looked like in practice, SAs were at least defensible.

After 1970, however, no teacher of composition could take for granted that any of his or her students had read anything more complex than *Jaws* or "Peanuts." With open admissions, teachers were suddenly confronted with students for whom the term "coherence" was not merely a crude generalization but a complete mystery. If SAs were made minimally useful only by an informed context for their use, it is not surprising that without such knowledge, SAs became for students in the seventies mere words,

14. Louis Tonko Milic, "Metaphysical Criticism of Style," in *New Rhetorics*, ed. Martin Steinmann (New York: Scribners, 1967), p. 164.

with no recognizable meaning at all. How can a student be expected to recognize "proportion" or "clarity" in writing when his or her reading has been confined to high-school textbooks and *TV Guide?*

Thus, the use of SAs has declined in textbooks until today there are relatively few texts that use such terms in any important way. UCE do live on in certain more traditional (and usually older) books, but the doughty trinity seems to be on the way out even there. After almost 90 years our discipline has finally grown beyond the facile use of "comparisons, analogies, and similarly crude approximations," as Milic puts it. We are at last coming to realize that all truly useful advice to students must be text-specific; such convenient generalities as SAs just don't help students compose or edit their own work. It took a genuine literacy crisis to give the quietus to SAs, but now that we are seeing the myths of current-traditional rhetoric dissolve we can at last begin to work toward replacing them with effective methods. If static abstractions teach us anything, it is that we as teachers must always be wary of neat, comprehensive-sounding conceptual schemes that are easy to teach but that have no real contact with what students need to learn.

Winston Weathers

Teaching Style: A Possible Anatomy

A general approach to the teaching of style can embrace any number of pedagogical tasks and obligations. There are three tasks, however, that seem obligatory: (1) making the teaching of style significant and relevant for our students, (2) revealing style as a measurable and viable subject matter, and (3) making style believable and real as a result of our own stylistic practices. These are all *sine qua non,* and to neglect them, one or all, is to do our discipline a disservice. They are not the only tasks involved in teaching style, of course, but they are the underlying concerns in all our particular classroom procedures. A discussion of these three tasks—the questing for relevance, viability, and credibility—may possibly serve as a kind of mapping of our pedagogical territory.

First, making style significant for our students. To teach style well, to reach the final goals we have in mind for the written page, we must confront our students not only with the discipline of style itself but with its justification. F. L. Lucas, the Cambridge professor, said that after forty years of trying, he had come to the conclusion that it was impossible to teach students to write well. "To write well," Professor Lucas said, "is a gift inborn; those who have it teach themselves; one can only try to help and hasten the process." But surely one of the best ways to "hasten the process" is to make it seem important. It is difficult to imagine any successful technical approach being made in teaching style if students are not aware of the great values involved. Surely many a student needs, at least in the context of freshman English, relevance pointed out to him, for otherwise he may think of style as a kind of aesthetic luxury, if not beyond his grasp at least beyond his interests.

I fear, though, that we often neglect to explain the significance of the discipline. In teaching *literature* we seem much more inclined to indicate relevance; in teaching *language,* once we have made the pitch about better communication we have a tendency, don't we, to drop the task of relevance altogether.

I think we should confirm for our students that style has something to do with better communication, adding as it does a certain technicolor to otherwise black-and-white language. But going beyond this "better communication" approach, we should also say that style is the proof of a human being's individuality; that style is a writer's revelation of himself; that through style, attitudes and values are communicated; that indeed our manner is a part of our message. We can remind students of Aristotle's observation, "character is the making of choices," and point out that since style, by its very nature, is the art of selection, how we choose says something about who we are.

From *College Composition and Communication* 21 (May 1970): 144-49.

In addition, we can tell students that style is a gesture of personal freedom against inflexible states of mind; that in a very real way—because it is the art of choice and option—style has something to do with freedom; that as systems—rhetorical or political—become rigid and dictatorial, style is reduced, unable as it is to exist in totalitarian environments. We can reveal to students the connection between democracy and style, saying that the study of style is a part of our democratic and free experience. And finally we can point out that with the acquisition of a plurality of styles (and we are after pluralities, aren't we? not *just* the plain style?) the student is equipping himself for a more adaptive way of life within a society increasingly complex and multifaceted.

To some, the "publicizing" task may seem beside the point in our discussion of approaches to style. Yet if we perform this task well, no student of ours will ever assume that we are teaching some dainty humanistic pastime; our students will know that we are playing the game for real. And I am convinced that it's this preparatory task that makes any other approach meaningful. Many students write poorly and with deplorable styles simply because they do not care; their failures are less the result of incapacity than the lack of will.

Now if this first approach, a la propaganda, can be successfully made, we can move on to the task of revealing style as a viable subject matter. Certainly we must keep rescuing style from what Professor Louis Milic has called the metaphysical approach—elevated descriptions that finally prove terribly nebulous—for if we find style unteachable because students see no relevance, we can also find style unteachable because students never get their fingers on it, never see it in measurable, quantitative terms.

To make style viable, we must teach students some rather specific skills—(1) how to recognize stylistic material, (2) how to master this stylistic material and make it a part of a compositional technique, (3) how to combine stylistic materials into particular stylistic modes, and (4) how to adapt particular stylistic modes to particular rhetorical situations. In teaching these four "how to's," we are providing students with a *modus operandi* for learning style and an overall strategy for using it. It is in these ways that style becomes a reality, a true discipline and a true art.

To begin, we do well to emphasize the concept of stylistic material; to explain to students that in the art of choosing, one *can* and must choose *from something*. We need to explain that certain real materials exist in style—measurable, identifiable, describable: Demetrius's "phrases, members, and periods," or Professor Josephine Miles' "linear units" (the terms do not really matter): but real material that serves as the substantive foundation of style, this material being of three general kinds: individual words; collections of words into phrases, sentences, and paragraphs; and larger architectural units of composition.

A certain amount of stylistic material is already a part of the student's repertoire when he comes to college, of course; the simple sentence, after all, is an ingredient of stylistic material, and any given word in a vocabulary is an element of style. But now, in college, the student must enlarge his collection of usable stylistic materials. The student learned a compound sentence in high school; he will now learn a periodic sentence. He learned a simile in high school; he will now learn reification. And it becomes our task to lead students to the storehouses of material from which they can make acquisition, to help students encounter the sources of stylistic materials and to draw from them.

There is certainly the traditional source—the established schemes and tropes, the established arrangement and procedures of writing. A metaphor, an oxymoron, an inductive paragraph. The student can draw upon the wealth of materials in classical and subsequent rhetorics. In addition, the student can draw upon materials that are not a part of tradition, but are the results of current achievements in the study of style. In the past decade or so, the great interest in rhetoric and style has effected new identifications of materials—such elements as serial sentences, patterned paragraphs, and the like are being analyzed and described in our professional literature. And finally—through the creative analysis of literary texts—the student can himself discover new materials. If, for instance, a student observes a writer habitually using a construction of ''opening prepositional phrase, a subject, a compound predicate, a closing prepositional phrase'' the student may note that sequence as a usable stylistic element. Admittedly, the student may be discovering material already discovered by Longinus—but that's fine; the student has the pleasure of confirming established knowledge. And if the student makes a discovery of material not heretofore identified, so much the better. Let the student name it: the D. H. Lawrence construction, the Hemingway verb, the Faulkner paragraph. The ingredients of style are that much more a reality. Indeed, this seems to me—this inductive approach—one of the great values in the stylistic analysis of a text; it is a chance to make discoveries about style that have not, amazingly enough, already been made.

Teaching the recognition of stylistic material, old or new, brings us to an interesting juncture, however, for it is easy to assume that with recognition and identification of an adequate supply of material, the student somehow has mastered style itself. But recognition of stylistic material is not the same as the practice of it; the knowledge must be converted into performance.

One widespread approach to the task of moving stylistic material from the depot of the student's mind to the front line of his writing fingers is, alas, the contemplative approach. If we use the contemplative approach, we tell students that by looking at style long enough they will finally find themselves practicing style—by a kind of osmosis. We say ''read a lot of good literature, make a lot of good stylistic analyses, and someday you'll wake up a writer.'' But can one learn to drive a car simply by taking a lot of car rides as a passenger? Surely one needs, in addition to contemplation, a great deal of involvement. One learns metaphor, not just by analysis, but by writing metaphors.

Some of us would advocate, therefore, a definite exercise system. We would advocate setting up recognized stylistic material as models; the models to be copied until the student can create similar but original versions of his own. It is a process of creative imitation that works like this: If we are going to teach a tricolon, an established bit of stylistic material, we first locate a tricolon in a text and point it out to our student; let us say the tricolon ''of the people, by the people, for the people.'' The student learns to identify and recognize the tricolon. But our second step is to isolate the tricolon sentence from the text, set it up as a model, and ask the student to make an exact copy of it in his own hand—word for word, comma for comma. After the student has made his perfect copy and we have checked it for accuracy, we then ask the student to discuss with us—or at least learn *from* us—the nature of the model tricolon, its use in the text in which it occurred, and the use of such tricola in general. Finally, we ask the student to compose a sentence of his own containing a tricolon—on some subject far removed from that which Lincoln was discussing. We ask the student to write a sentence or a topic of his own choosing, but following the ''model'' he has just studied.

In this process, the student is asked to *recognize, copy, understand,* and *imitate creatively.* And this process can be used to master all possible stylistic materials: from the use of particular words to the more complex combinations of materials found in long passages. Creative imitation or generative copying is not new, of course; we all know the famous essay by Rollo Walter Brown on "How the French Boy Learns to Write" and certainly Professor Edward P. J. Corbett has given great support to this method in his *Classical Rhetoric for the Modern Student.*

Teaching viability of style does not end here, however. The student must not be left in the lurch at this point either; he must not be left with the ability to recognize and imitate stylistic materials without having a rationale for using them. A student has the legitimate right to ask, "Now what do I do? I know how to recognize and compose a tricolon, but what do I do with one?"

What students are actually asking for at this point is a strategy of style—and we can establish such a strategy by doing two primary things: (1) identifying the categories of style, and (2) describing the constituency of those categories in terms of stylistic material.

First, a word about categories. The categories of style we choose to identify will depend upon our own individual way of seeing things: some of us may still use the four levels of style acknowledged by Demetrius; some of us may use the fairly conventional levels of usage—formal, informal, colloquial; some of us may prefer more elaborate categories combining both levels and intensities of language into a complex of rhetorical profiles; some of us may prefer such new categories as the styles of certitude, judiciousness, emotion, and absurdity, or "tough, sweet, and stuffy"; or some of us may even be so reductive as to prefer a simple two-category system of plain and literary style. But whatever our preferences, we must identify some *set of categories,* some *system* of categories, to serve as a framework in which various styles can be achieved.

Second, the constituency of these categories. Having established a system of diverse styles, we must establish recipes for achieving individual styles within the system. We must teach our students that certain stylistic material goes here; certain material goes there; that a certain combination and sequence of stylistic material creates one style; a certain combination and sequence creates another. Though given enough time a student might, by induction, discover the constituency of given categories himself, the burden of the description falls upon the teacher who is obligated to list the observable characteristics of the various styles. Indeed, I suspect that a good deal of our homework as teachers is, or should be, spent in discovering, in as great a detail as possible, these characteristics and pointing them out to our classes.

If we are able to effect for our classes these primary conditions of strategy—identification and description of categories—then we can exercise our students in the following ways:

Exercise One—We call for a student to write down all the possible verbalizations he can think of for any given message. How many ways can he say, "It's a beautiful day"? How many ways can he say, "Space exploration is too expensive"? Having made a list of all the possible ways, the student is then asked to allocate the various verbalizations on his list, placing them in the categories of style we have taught him. That is, given the recipes for a number of different styles—which verbalizations go where. Practicing this exercise over a period of time, the student—under our guidance—

comes to realize that nearly any verbalization he can think of per given message can play its part in a total system of style; he stops seeking one eternally correct verbalization, but seeks rather to place all verbalizations in their appropriate communities.

Exercise Two—We ask a student to write a paragraph—on any topic—and to identify the particular style he has used in that paragraph. Having done so, the student next transforms the paragraph into another style. If he has written about campus revolution in a militant style, we ask him to transform his composition—with the same facts, observations, data, and opinions—into the judicious style. If he has written about his flower garden in an elegant style, we ask him now to write about it in a plain style. If in a colloquial style, now in a formal style. The point of the exercise is to teach the student how to add or subtract or substitute particular stylistic materials so as to change one style to another. Ultimately by means of this transformational exercise, the student will be able to decline—as it were—any sentence, paragraph, or essay through all possible styles.

Finally, of course, after the exercises, we ask students to write complete compositions. Though in teaching viability perhaps we err too often by beginning with whole compositions, by plunging the student into the middle of stylistic performance without making it truly viable for him, step by step, we do not err by asking the student to make the final effort of demonstrating all that we have been talking about—to demonstrate a knowledge of stylistic material, piece by piece, and a capacity for its strategic incorporation into stylistic wholeness.

The third task in our general approach to style—after the tasks of relevance and viability—is that of making the practice of style tremendously believable as the result of our performances in front of students. Robert Zoellner recently wrote in *College English* that he had "never, repeat never, seen a composition instructor, whether full professor or graduate student, walk into a composition classroom cold-turkey, with no preparation, ask the class for a specific theme-topic . . . and then—off the top of his head—actually compose a paragraph which illustrates the rhetorical principles that are the current concern of the class."

Professor Zoellner was surely exaggerating with the "never, repeat never" but I suspect that in general his charge is valid. We are an amazing lot of piano players refusing to play the piano. Yet should not the student's most significant model, so far as style is concerned, be the teacher himself? Isocrates, that ancient member of the profession, did not, as Werner Jaeger points out in *Paideia*. "Merely discuss the technique of language and composition—the final inspiration was derived from the art of the master himself." And surely this is so: what the teacher writes on the blackboard in front of the student, or even what the teacher writes outside of class and brings to read to his students, is the teacher's commitment to the style he is urging his students to learn. Perhaps some of the difficulties in teaching style arise because of teacher failure: not failure in sincerity or industry or knowledge, but failure in demonstrating an art and a skill. Teacher failure ever to write and perform as a master stylist creates an amazing credibility gap.

I would propose a definite incorporation of teacher performance into our approach to style. Such a program would entail original composition by the teacher, at the blackboard, at least three to five minutes each class—or at least a five-to-ten minute performance once a week. We are limited by the physical circumstances of the classroom and by the pressures of time, but every blackboard is large enough for five or six sentences or a short paragraph, and every class period is long enough for a few minutes of teacher composition. Even if our demonstration of style is faulty and less than excellent, the fact that the teacher "did something" *for all to see* is noteworthy. And I have found that students actually learn a great deal from watching a teacher put in a word, take out a word, rewrite a sentence, even misspell, and then correct a spelling, ponder over the use of a comma or a semicolon. Believe me: the teacher's struggle amidst the chalk dust can become the student's education.

And to prove to our class that we are not conning them, we can have one student call out a noun, another call out a verb—then using the noun and verb, we can write without prearrangement what needs to be written that day: a balanced sentence, a serial sentence, a circular paragraph. We may be reluctant to do this sort of impromptu writing—yet we are obligated. We are supposed to be professionals, and we should know enough about style to do a passing job, if not a brilliant job, and do the job "on call."

Teacher performance can go beyond the blackboard even. I think a certain amount of talk—modest and judicious, but enthusiastic—about the writing we do outside the class is important in the teaching of style. And I wonder if we shouldn't write some of the essays we ask our students to write—or write something comparable at least—and on occasion read to our classes what we have written. We could risk offending students with our vanity in an attempt to convince them that composition, rhetoric, and style are things we really do, that they are a part of our lives, that *we* are involved.

Such are the three obligations that must be met, three important tasks that must be performed in a general approach to teaching style. Our decisions how best to make style relevant, viable, and credible in the classroom may indeed vary; you and I may not agree about the details; we may use different syllabi and different textbooks. But I hope we will agree concerning the obligations themselves. At least I offer this anatomy for teaching as a possibility.

Elizabeth D. Rankin

Revitalizing Style: Toward a New Theory and Pedagogy

In the past several years, we have seen various attempts to define and explain the dramatic changes occurring in the field of composition studies. In particular, Richard Young, James Berlin and Robert Inkster, and Maxine Hairston have spoken of a ''paradigm shift,'' a change in the basic assumptions and attitudes that underlie all our theory and pedagogy.[1] One result of this paradigm shift has been a noticeable decline in the status of style as a pedagogical concept. By this I mean that the teaching of style no longer enjoys a prominent place in our discipline.

Some see this as a change for the good. For too long, they argue, style dominated our pedagogy—almost to the exclusion of other concerns. And of course they are right. But what I fear may be happening now is an over-correction of sorts. Style hasn't just stepped back to take a less dominant role in our teaching—style is out of style.

Before I go any further, it is probably wise to stop and explain what I mean by style in this context. To do so is no easy task, for style has a protean identity over the years. ''Historically,'' says Linda Woodson, ''style has been interpreted both narrowly, as referring only to those figures that ornament discourse, and broadly, as representing a manifestation of the person speaking.''[2] Most often, however, those who write about style in texts and in journals define style, implicitly or explicitly, in terms such as these:

> The style of a piece of writing is the pattern of choices the writer makes in developing his or her purpose. If the choices are consistent, they create a harmony of tone and language that constitutes the style of the work. A description of the style of any piece of writing is therefore an explanation of the means by which the writer achieved his or her purpose.[3]

In practice, discussions of style nearly all deal with particular linguistic choices: with diction, syntax, and tone. So I'll start out my own discussion with this definition in mind, though the definition itself, if regarded narrowly, may well be a factor in style's decline.

From *Freshman English News* 14 (Spring 1985): 8-13. Reprinted by permission of the publisher.

1. Richard E. Young, ''Paradigms and Problems: Needed Research in Rhetorical Invention,'' in *Research in Composing*, ed. Charles R. Cooper and Lee Odell (Urbana: NCTE, 1978), pp. 29-47; James A. Berlin and Robert P. Inkster, ''Current-Traditional Rhetoric: Paradigm and Practice,'' *Freshman English News* 8 (Winter 1980), pp. 1-4, 13-14; Maxine Hairston, ''The Winds of Change: Thomas Kuhn and the Revolution in the Teaching of Writing,'' *CCC*, 33 (February 1982), pp. 76-88.
2. Linda Woodson, *A Handbook of Modern Rhetorical Terms* (Urbana: NCTE, 1979), p. 58.
3. James M. McCrimmon, *Writing With a Purpose* (Eighth Edition), (Boston: Houghton Mifflin, 1984), p. 311.

What, then, are the reasons for that decline? I think there are two. One has to do with the claims of competing concerns. As modern composition studies rediscovered its roots in classical rhetoric, it discovered whole areas of study that had been neglected for years. One of these areas, invention, has attracted much attention of late, and as rhetoricians rush in to fill the gap in research, interest in style has declined. The same could be said for other new areas of interest, such as the composing process. Although there is really no need to see these new fields as competing with the old, the politics of our profession and the rhetoric that it gives rise to have created such an effect.

Of course this is not to say that no one is interested in style these days. Style does have a place in the New Rhetoric—or rather, it has a place in both branches of the New Rhetoric that Richard Young has identified.[4] And this is the second reason for style's decline. For so far, neither branch of the New Rhetoric has offered us a sound, complete, and adequate theory of style. Until such a theory is formulated—or at least until such a formulation seems *possible*—we are left with a good deal of skepticism and confusion.

Is such confusion resolvable? Is there any way to reclaim for style the respect which it has lost? I believe there is, and a little later in this essay I will make some preliminary suggestions toward that end. But first, I must go back and fill in the substance of the argument I have just outlined.

A good place to begin, I think, is with the whole notion of a "paradigm shift," or, more specifically, with the concept of competing paradigms. When we look at the work of the scholars I mentioned earlier—Young, Berlin and Inkster, and Hairston—one thing we see them all doing is trying to define the characteristics of the old paradigm, the "current-traditional" paradigm, as it has come to be called. In doing so, all take note of the prominence given to style. Here is what Richard Young has observed:

> The overt features [of the current-traditional paradigm] are obvious enough: the emphasis on the composed product rather than the composing process, the analysis of discourse into words, sentences, and paragraphs, the classification of discourse into description, narration, exposition, and argument; *the strong concern with usage* (syntax, spelling, punctuation) *and with style* (economy, clarity, emphasis); the preoccupation with the informal essay and the research paper; and so on. ("Paradigms," 31; emphasis mine)

Picking up where Young left off, Berlin and Inkster trace the rhetorical history of the old paradigm, exposing the philosophical and epistemological assumptions on which it is based. As they do so, they also note the importance of style in the paradigm. The most salient assumption of current-traditional rhetoric, say Berlin and Inkster, is the notion that reality is fixed, knowable, and rational, and that discourse, to be valid, need only conform to that reality:

> One may ask how one piece of discourse is to be distinctive from any other discourse, given the powerful impetus for conformity that grows from the epistemology of the

4. Richard Young, "Arts, Crafts, Gifts and Knacks: Some Disharmonies in the New Rhetoric," in *Reinventing the Rhetorical Tradition,* ed. Aviva Freedman and Ian Pringle (Ottawa: CCTE, 1980), pp. 53-60.

> current-traditional paradigm. The answer lies in the concern for style, for here is the one avenue by which one may write distinctive prose, given the assumptions behind the paradigm. Hence, the elevation of style in the texts. (5)

What Young and Berlin and Inkster have said here is most assuredly true: we need only look at the more traditional handbooks and texts in our field to see how important the concept of style has been for teachers of composition. And I must point out that nowhere do these authors suggest that style is an inappropriate concern. If they see a problem with the old paradigm, the problem isn't with style *per se,* but with an over-emphasis on style. Maxine Hairston is particularly careful in this regard. She closes her essay with the caution that "it is important for us to preserve the best parts of earlier methods for teaching writing: the concern for style and the preservation of high standards for the written product" (88).

Alas, though, intentions are sometimes beside the point, and in this case I fear that the otherwise useful concept of competing paradigms has had some unfortunate side effects. For one thing, it sets up false dichotomies. The process/product opposition is one that many have been uncomfortable with. Less noticed, but just as unfortunate, is a similar implied opposition between invention and style. For instance, consider this prominent item in Hairston's proposed "new paradigm," a list of twelve tenets of current composition theory: "It teaches strategies for invention and discovery; instructors help students to generate content and discover purpose" (86). Here invention, the first of the five canons of classical rhetoric, gets mentioned by name; in contrast, neither style nor arrangement appears on the twelve-item list. Earlier in Hairston's essay, however, style was twice mentioned explicitly in association with the old paradigm. Thus, though no opposition between style and invention is stated, one does get conveyed in subtle ways nonetheless.

In Berlin and Inkster, too, invention seems to get counterpoised with style. Recounting the factors that led Hugh Blair "to reject the heuristic procedures of classical invention as mechanical algorithms," the authors go on to say, "One need not search far in modern texts to find that this legacy is still with us: 'The stylistic side of writing is, in fact, the only side that can be analyzed and learned" (3). The point that Berlin and Inkster make is a valid one: for modern rhetoricians and textbook authors since Blair, invention has not until recently been regarded as a teachable art. But to go on and quote an author who says that only style can be taught is a little misleading. After all, arrangement—that second canon of classical rhetoric—has also been a concern of the current-traditionalists. When it is left out of discussions like these, a false opposition of style and invention inadvertently gets implied.

Just as damaging to style's reputation as this implied opposition is the power of connotation. Let us look once again at Richard Young's definition of the old paradigm. Certain features he mentions are connotatively neutral, including "the strong concern . . . with style." Others, however, are negative, and they have a kind of rub-off effect on their less judgmental neighbors. Take, for example, "the emphasis on the composed product rather than the composing process." It's hard to say at this point when "composed product" acquired the negative connotations it holds for us today, but I would argue that the term itself gave us very little choice. Given our characteristic humanist antipathy toward the market-place, "product" will generally always have negative associations. (If you don't agree, try substituting a term like "finished discourse" and compare the effect. The latter term is more neutral, but it doesn't have that alliterative antithesis that "product/process" has.) The last item in Young's definition has an even stronger negative bias. Current traditionalists, says Young, have a "preoccupation with the informal essay and the research paper." Here the negative associations of "preoccupation" are obvious. (For an interesting contrast, compare this item in Hairston's proposed new paradigm: "It includes a variety of writing modes, expressive as well as expository.") Finally, consider the rhetorical effect of mentioning style in the same breath with usage. Usage has fallen on bad times of late—largely because it has become the domain of prescriptive grammarians and linguistic reactionaries like Edwin Newman and William Safire. What this casual yoking of terms may suggest is that style belongs in that same domain—a suggestion which produces the effect of guilt by association.

If the rhetoric of competing paradigms has had, in certain respects, a chilling effect on our attitudes towards style, so has the rhetoric of the composing process. In the past few years, considerable research has been devoted to investigating that process (or those processes, to be more accurate). Much of it remains descriptive, while some results in the formulation of models. Probably the most well-known and comprehensive model is the cognitive process model constructed by Linda Flower and John Hayes. In this model, "a writer uses a goal to generate ideas, then consolidates those ideas and uses them to revise or regenerate new, more complex goals."[5] As the examples that Flower and Hayes use indicate, many of the goals that writers set and revise are stylistic goals, having to do with such matters as word choice, syntax, and tone. Thus, their model could well be used to explore the complexities of style. But in actuality, the work that derives from the cognitive process model rarely pertains to style. Why is this? I suspect it's because of the rhetoric of the model itself.

One salient feature of the Flower-Hayes model is its hierarchical structure. Writing processes are "hierarchically organized" and writers "create a hierarchical network of goals." Within this hierarchical structure, "low level goals" (such as stylistic goals) are embedded within or subsumed to "higher level" goals involving, for instance, content, organization, or audience adaptation. As writers work, they "not only create a hierarchical network of guiding goals, but, as they compose, they continually return or 'pop' back up to their high-level goals. And these higher-level goals give direction and coherence to their next move" (379).

What Flower and Hayes say here makes perfect sense and rings true to our intuitive notion of how writers work. But almost inevitably, its effect has been to focus our interest—as teachers and researchers—on the "top level" or "middle-range" goals of the writer. Partially, this is because those "higher level" goals are so crucial. They do "give breadth and coherence to local decisions about what to say next" (379). But consider the connotations involved. When "narrow," "local," stylistic decisions occupy a "low-level" position in the "hierarchy," it's hard to see them as very important. Thus, though the Flower-Hayes model is broad and inclusive enough to account for stylistic decisions at all levels and stages in the writing process, the rhetoric of the

---

5. Linda Flower and John R. Hayes, "A Cognitive Process Theory of Writing," *CCC*, 32 (December 1981), p. 386.

model—like the rhetoric of competing paradigms—to some extent undermines the significance of style.

This same effect can be seen in some of the recent work on revision, particularly that of Donald Murray and Nancy Sommers. In a 1978 essay entitled "Internal Revision: A Process of Discovery," Murray divides the revision process into "two principal and quite separate editorial acts": internal and external revision. In setting up this division, Murray is not attempting to be rigorously scientific, as later researchers have been. Rather, he has a rhetorical purpose in mind. He wants us to see that revision includes not only matters of "form and language, mechanics and style" (external revision), but "everything writers do to discover and develop what they have to say, beginning with the reading of a completed first draft" (internal revision). It is this latter kind of revision that Murray is most interested in and goes on to discuss in his essay.[6]

If we look at the essay closely, we see that Murray clearly intends no denigration of style. For one thing, he concedes that "external revision has not been explored adequately or imaginatively." For another, as he goes on to discuss "internal revision," he is in fact discussing matters of style:

> language itself leads writers to meaning. During the process of internal revision [writers] reject words, choose new words, bring words together, switch their order around to discover what they are saying. 'I work with language,' says Bernard Malamud, 'I love the flowers of afterthought.'
>
> Finally, I believe there is [another] area, quite separate from content, form, or language, which is harder to define but may be as important as the other sources of discovery. That is what we call *voice*. I think voice, the way in which writers hear what they have to say, hear their point of view toward the subject, their authority, their distance from the subject, is an extremely significant form of internal revision. (44)

What Murray is saying here is important, and I will return to it later. For now, though, I will only point out one thing. Although Murray is clearly discussing style as an aspect of internal revision, he never actually uses the word there. He does mention style, however, when discussing external revision, and when he does, he inadvertently packs it with all the slightly disparaging connotations of words like "conventions" and "mechanics." Here again, as in delineations of the old paradigm, we have the effect of guilt by association. Very subtly, the pedagogy of style gets devalued.

A similar phenomenon occurs, I think, in the work of Nancy Sommers. In "Revision Strategies of Student Writers and Experienced Adult Writers," Sommers writes:

> Experienced writers see their revision process as a recursive process—a process with significant recurring activities—with different levels of attention and different agenda for each cycle. During the first revision cycle their attention is primarily directed towards narrowing the topic and delimiting their ideas. At this point, they are not as concerned as they are later about vocabulary and style. [However,] during the later

6. Donald Murray, "Internal Revision: A Process of Discovery," in *Research in Composing*, ed. Charles R. Cooper and Lee Odell (Urbana: NCTE, 1978), p. 91.

cycles, . . . the experienced writers' primary attention is focused upon stylistic concerns.[7]

Throughout this section of her essay, Sommers speaks of ideas, form, and style as separate "objectives." Because these objectives cannot be attended to all at once, she says, the writer must prioritize one or two. In her research, Sommers found that experienced writers tend to prioritize idea and form, leaving "stylistic concerns" for the later cycles of the process. Student writers, on the other hand, seem to get hung up on lexical concerns, perhaps because they "do not have strategies for handling the whole essay" (383).

The problem with Sommers' rhetoric here is that, as Murray has suggested, style and stylistic concerns are often tied in with other aspects of writing. It's true that narrow stylistic concerns can and do impede the composing process for many of our students; but likewise narrow ideas or narrow notions of form can impede the process. Although Sommers' research has broadened (and thereby strengthened) our concept of revision, she is still employing a fairly traditional and limited concept of style—and this in itself is part of a larger problem.

The problem, ultimately, has to do with our failure, as a profession, to come up with a unified theory of style. In his essay, "Theories of Style and Their Implications for the Teaching of Composition," Louis Milic distinguishes three different philosophies of style: (1) rhetorical dualism, which sees language and thought as separate entities, with style being the "dress" of ideas; (2) aesthetic monism, which sees style (and form) as an inevitable consequence of content; and (3) psychological monism, which regards style as the expression of a unique personality ("Style is the man").[8] In Milic's terms, Sommers would be a dualist. For her, by implication at least, style can be separated from idea and form, and when it is, it can be attended to in meaningful ways. In this, Sommers resembles a wide range of modern composition theorists and pedagogues, a heterogeneous group that Richard Young has labeled the "new classicists."

For the new classicists—including such people as Edward Corbett, Winston Weathers, Francis Christensen, and Joseph Williams, style is a teachable art. It is taught by first determining what expert stylists do, then constructing heuristics that will help beginners master—and eventually internalize—their techniques (Young, "Arts," 57-58). On the one hand, this view of style is liberating. It takes style out of the domain of the handbook rhetoricians and makes it a serious object of research and investigation. It encourages computer-based stylistic analysis and results in the formulation of such innovative heuristics as sentence-combining and generative sentence-building, as well as in the revitalization of classical concepts like imitation. On the other hand, though, the new classicists' pedagogy of style is problematic. Despite protests to the contrary, it often tends to be prescriptive. (One thinks immediately of people like Williams here, but Christensen too was accused of promoting a particular kind of style.)[9] In addition,

---

7. Nancy Sommers, "Revision Strategies of Student Writers and Experienced Adult Writers," *CCC*, 31 (December 1980), pp. 386-87.

8. Louis Milic, "Theories of Style and Their Implications for the Teaching of Composition," *CCC*, 16 (1965), pp. 66-69, 126.

9. Richard Lanham's "Paramedic Method" in *Revising Prose* is also frankly prescriptive; however, his broader theory of style, which I will refer to later, is more complex and inclusive.

the heuristics proposed by the new classicists are often hard to implement. Imitation exercises, sentence-combining, stylistic analysis—all of these take enormous amounts of time. For teachers committed to a process-centered approach to composition, such time takes away from the already limited time their students have for drafting and revising, and in their view the benefits may not be worth the costs.

More problematic than practical limitations, however, is one shaky assumption that underlies new classical pedagogy. This is the assumption that most stylistic decisions are conscious decisions—or can be made conscious with careful attention to the revision process. Ian Pringle has already raised objections to this assumption, in a review of Joseph Williams' text *Style: Ten Lessons in Clarity and Grace.* Calling attention to the work of second language researcher Stephen Krashen, which makes a distinction between conscious *learning* of language and unconscious *acquisition,* Pringle argues the following:

> If students are to produce literate writing, they will do so primarily on the basis of what is acquired, and what they acquire will come to them not through the explicit study of style (or for that matter grammar), but from 'comprehensible input,' from reading good, relevant models.[10]

Of course Pringle's point of view is also problematic. In the first place, it assumes that the model implies that stylistic decisions are "primarily" acquired, not learned. At this point, though, Krashen's model is purely hypothetical. It may *suggest* directions for further research, but as yet it can tell us little about what specific aspects of language may be consciously learned and which must be unconsciously acquired. And secondly, we must remember that Krashen's model is not only a construct, it's a construct that may be more field-specific than we realize. Its adequacy as a descriptive model of the acquisition of *written* language has yet to be fully explored.

Nevertheless, what Pringle is voicing here is a reservation that many would share. It's the same reservation implied by Milic in his essay "Rhetorical Choice and Stylistic Options." In that essay, Milic considers the very question that Pringle (via Krashen) has raised: to what extent are stylistic decisions *conscious* decisions? Citing personal experience, observations, and anecdotes about professional writers, Milic concludes the following:

> Without for a moment denying the possibility that some part of a writer's style is conscious artistry or craftsmanship, I am convinced that most writers, even some of the greatest, knew very little about what they were doing when they wrote and had much less conscious control over the final product than is commonly supposed.[11]

With some exceptions, he seems to corroborate Krashen.

What all this suggests is that there are complex aspects of style that the dualist theory of the new classicists is ill-equipped to deal with. These aspects, the unconscious aspects of style, are of interest to the "new romantics."

---

10. Ian Pringle, "Why Teach Style? A Review Essay," *CCC,* 34 (February 1983), p. 94. For an explication of Krashen's model see "On the Acquisition of Planned Discourse: Written English as a Second Dialect," in *Proceedings of the Claremont Reading Conference* (1978), pp. 173-185.

11. Louis Milic, "Rhetorical Choice and Stylistic Option," in *Literary Style: A Symposium,* ed. Seymour Chatman (London: Oxford University Press, 1971), p. 87.

The new romantics, says Richard Young, are committed to the notion that creativity is a mysterious process—a process that is ultimately unavailable to conscious, rational analysis and control. Thus, the creative process (which includes the composing process) cannot be taught in the conventional sense—but it can be facilitated by a teacher who acts as helper or guide (Young, "Arts," 55). For new romantics like William Coles, Peter Elbow, James Miller, Gordon Rohman, and others, style is also a matter cloaked in mystery. But what the romantics mean by style is radically different from what new classicists or current-traditionalists mean. Take Peter Elbow, for instance. For Elbow, style is virtually synonymous with voice. Thus, it isn't so much a skill to be learned as a capacity to be realized. Style is something the writer discovers within him or herself, and the teacher's role is not to teach it but to encourage and facilitate its expression.

Like the new classicists' approach, this new romantic attitude has had a mixed effect on the status of style in our pedagogy. On the one hand, it inflates the concept's importance. For Elbow, it is voice more than anything that gives writing *power*. On the other hand, though, just because voice is so important, the concept becomes all-pervasive and tends to diffuse. As it does so, it carries off with it the language we need to describe it, and what we are left with is just the kind of vague, subjective abstraction that Louis Milic has decried in traditional studies of style.[12] Here, for example, is Elbow's reaction to a sample student passage:

> This writing has the lively sound of speech. It has good timing. The words seem to issue naturally from a stance and personality. But what strikes me is how little I can feel the reality of any person in these words. I experience this as a lack of any deeper resonance. These words don't give off a solid thump that I can trust.[13]

If style as a pedagogical concept has suffered at the hands of the new romantics, it may be because of the necessary subjectivity of their approach.

At this point it becomes clear that the problematic assumptions underlying the new romantics' approach to style are in fact the opposite of equally problematic assumptions held by the new classicists. If the classicists embrace an essentially dualistic notion of style, the romantics are committed to the view that says "Style is the man." If the classicists view style as a conscious art to be mastered, the romantics see it as an unconscious voice to be discovered. The question is this: to what extent are these two theories of style mutually exclusive? Is there any way of reconciling them, and in doing so revitalizing the teaching of style?

For Louis Milic, the theories *are* reconcilable. He has no problem seeing that real writers "write in a certain way because they select the most effective artifices of expression, but also because they are unconsciously bound to the requirements of individual personality." Still, Milic thinks that "For teaching, a dualistic theory seems to be essential, at least in the early stages." He has no faith in new romantic approaches to teaching style ("Theories," 126). In part, Richard Young shares Milic's view: "The durability of these two fundamental conceptions of rhetorical art and the effectiveness of the pedagogical methods based on them suggest that in some sense both are true—in spite of the fact that they seem incompatible." Though he doesn't propose it himself, Young

---

12. Louis Milic, "Metaphysics in the Criticism of Style," *CCC*, 17 (1966), pp. 124-29.
13. Peter Elbow, *Writing With Power* (New York: Oxford University Press, 1981), p. 292.

believes that "there may be a more adequate conception of rhetorical art that does not lead us to affirm the importance of certain psychological powers at the cost of denying the importance of others" ("Arts," 60).

From what Young and Milic say, it is clear that they see the essential differences in theories of style as psychological differences. It could be, though, that the conflict goes even deeper. According to John Gage, for instance, different theories of style are manifestations of more deeply rooted philosophical assumptions about the nature of language and reality.[14] If these assumptions can be reconciled in a theory of hermeneutics—and Gage's remarks about rhetorical communities suggest that they can—then we may not have to wait long for a unified, comprehensive theory of style. What such a theory will look like we do not yet know. But in light of what I've been saying here, I would expect it to meet the following criteria:

1. *A new theory of style would offer a broad yet workable definition of style.* This definition would be specific enough to distinguish stylistic considerations from other concerns of the writing process, such as content and formal matters. At the same time, it would be broad and inclusive enough to account for overlap between style and invention, say, or stylistic decisions and audience adaptive techniques.

If we look once again at the textbook definition of style given earlier, we see that it contains the seeds of such a definition. On the one hand, it describes style as a "pattern," a "harmony," a set of consistent choices. On the other hand, it also sees style as "a means by which the writer achieve[s] his or her purpose." What the definition implies is that style is both "product" and "process," both a set of observable features of a finished text and a way of discovering what that text will become. Such a flexible definition is crucial if we are to develop style's full potential as a rhetorical and pedagogical concept.

2. *A new theory of style would take into account the wide range of psychological operations that go into the making of stylistic decisions.* That is, it would account for not only the conscious, rational decisions that Milic calls "rhetorical choices," but for the murkier matter of the formation of stores of unconscious "stylistic options" as well. In particular, it would provide some means of distinguishing when such operations are *acquired,* when they are *learned,* and if/when they might be effectively taught.

This is not to say, of course, that there is a clear line of demarcation between conscious and unconscious choices. Still, a complete definition of style will recognize that some of those choices are more indirect than others: they are influenced by social background, by linguistic experience, and by intellectual capacity; by deep psychological factors and momentary situational constraints. In short, they are complex and fascinating—a fertile subject for further research.

3. *A new theory of style would be grounded in sound and consistent philosophical/epistemological assumptions about the nature of language and reality.* Ideally, such a theory would mediate between those who see language (and style) as representative of a fixed, orderly reality, and those who see it instead, in John Gage's terms, as "a useful but imperfect . . . manifestation of a reality maybe orderly, maybe not" (617). As I

14. John T. Gage, "Philosophies of Style and Their Implications for Composition," *College English,* 41 (February 1980), pp. 615-22.

mentioned earlier, Gage himself has laid the foundation for such a theory. What it relies on ultimately is the concept of the rhetorical community, a community within which linguistic positivists and relativists can meet and, to some extent, coexist. In this community, which Stanley Fish has called the interpretive community, the essentially arbitrary and self-referential nature of language is acknowledged, yet a flexible set of acceptable meanings and standards is agreed upon. The fact that these meanings and standards change is not taken as cause for concern but as natural and expected—the result of a continuous dialogue in which meaning is negotiated.

In terms of style, such a concept *seems* to undercut monistic theories because it regards both the self (the individual personality) and meaning as relativistic constructs—entities that are themselves created by language (style). But if these constructs are seen as existing within a rhetorical (or interpretive) community, they can be provisionally regarded as determinant, even while their relativistic nature is acknowledged. Thus, the theory accommodates both dualistic and monistic concepts of style.

Richard Lanham outlines another way of resolving the problem of conflicting theories in his essay "*At* and *Through:* The Opaque Style and Its Uses." For Lanham, conflicting theories of art (and style) can be resolved if seen as points on a matrix constructed around a single variable: "degree of self-consciousness." Although he argues in different terms, Lanham seems to agree with Gage when he says that "each theory, if adequately rigorous and comprehensive, creates the object it criticizes."[15] In doing so, he implies, it also creates its audience, its rhetorical community, by whose standards and expectations its adequacy and rigor are judged.

In more practical terms, a new theory of style would have certain pedagogical consequences. By broadening our narrow definitions of style, it would force us to reconsider our notions of when, where, and how style can be taught in the process-centered classroom. Is it best to encourage students to prioritize content and form as they go through the writing process—or can style sometimes be profitably focused on even in earliest drafts? Is it enough to concentrate on those aspects of style that are most accessible to conscious control—or are there ways of reaching and shaping the less conscious processes too? And what about style as voice? Is it something the writer discovers within his or her unique individual self—or is it an interpretive construct the writer creates as he/she goes along?

All these are provocative questions, and they merit our serious attention. If we begin to address them in the context of a comprehensive theory—avoiding the negative rhetoric we hear in our discipline lately—we may be able to bring style back into style.

---

15. Richard Lanham, "*At* and *Through:* The Opaque Style and Its Uses," *Literacy and the Survival of Humanism* (New Haven: Yale University Press, 1983), pp. 58-86.

Richard Ohmann

Use Definite, Specific, Concrete Language

My title is Rule 12 from Strunk and White's *The Elements of Style,* and it probably comes as close as any precept to claiming the unanimous endorsement of writing teachers. After E. D. Hirsch, Jr., in *The Philosophy of Composition,* develops his principle of "relative readability," grounding it in historical and psychological evidence, he turns for support to "the accumulated wisdom of the handbooks." (The ones he chooses are Strunk and White, Crews, McCrimmon, Lucas, Gowers.) He reduces their wisdom to ten or a dozen maxims each: there is much overlap from book to book, but only two maxims appear in nearly the same form in all five books, and my title is also one of those two.[1]

Does anyone besides me feel uneasy when Strunk and White begin exemplifying this reasonable advice? For "A period of unfavorable weather set in," they substitute "It rained every day for a week." The rewrite is indeed more definite, specific, and concrete, and less pompous to boot. But it doesn't say the same thing, and in that difference there is a loss as well as a gain, especially if the writer means to relate the weather to some undertaking rather than just describing it. The original conveys—however inadequately—a more complex idea. The same is true when "He showed satisfaction as he took possession of his well-earned reward" becomes for Strunk and White "He grinned as he pocketed the coin."

In this essay I want to look at the way some authors of textbooks show students how to be definite, specific, and concrete. The questions I have in mind as I do so are whether in teaching a skill like this we may inadvertently suggest to students that they be less inquiring and less intelligent than they are capable of being, and whether the teaching of basic skills is an ideological activity. To bring suspense down to a tolerable level, let me reveal now that my answer to both questions is Yes.

I will examine just three textbooks, chosen not as bad examples—they seem to me lively, serious, and honest—but for these reasons: They are current (1978). They are second editions, an indication of acceptance in the market. Their authors teach in a large city university, a community college in a large northern city, and two community colleges in a southern town and a southern city; such institutions are close to the center of the freshmen composition industry. All three textbooks give unusually ample attention to style, and in particular to the matters I am concerned with here.

I will look first at the recently published second edition of David Skwire and Frances

From *College English* 41 (December 1979) 390-97.

1. The other is, avoid padding. *The Philosophy of Composition* (Chicago: University of Chicago Press, 1977), pp. 148-49.

Chitwood, *Student's Book of College English* (Glencoe Press), and refer to a section in it on "Specific Details" (pp. 347-349). Skwire and Chitwood introduce the section by saying "The use of specific details is the most direct way to avoid abstract writing." (And students *should* avoid it, since "abstract writing is the main cause of bored readers" [p. 346].) Detail is a plus. In fact, "within reason, the more specific the details, the better the writing." "Within reason" means that the detail must be relevant and neither obvious nor trivial. To illustrate, they offer three passages, labeled "Abstract (weak)," "More Specific (better)," and "Still More Specific (much better)." Here are the first and third.

1. Abstract (weak)
The telephone is a great scientific achievement, but it can also be a great inconvenience. Who could begin to count the number of times that phone calls have come from unwelcome people or on unwelcome occasions? Telephones make me nervous.

3. Still More Specific (much better)
The telephone is a great scientific achievement, but it can also be a great big headache. More often than not, that cheery ringing in my ear brings messages from the Ace Bill Collecting Agency, my mother (who is feeling snubbed for the fourth time that week), salesmen of encyclopedias and magazines, solicitors for the Policemen's Ball and Disease of the Month Foundation, and neighbors complaining about my dog. That's not to mention frequent wrong numbers—usually for someone named "Arnie." The calls always seem to come at the worst times, too. They've interrupted steak dinners, hot tubs, Friday night parties, and Saturday morning sleep-ins. There's no escape. Sometimes I wonder if there are any telephones in padded cells. (pp. 348-349)

Consider now how revision has transformed the style of the first passage. Most obviously, one generalization—"unwelcome people"—disappears entirely, to be replaced by a list of eight people or types of people the writer doesn't want to hear from; and another generalization—"unwelcome occasions"—is changed to "worst times," then amplified in another list. Seriation has become the main principle of structure. When items are placed in a series, the writer implies that they are alike in some respect. But in what respect? Here the angry neighbors and possessive mother are placed on par with salesmen and others connected to the writer only through the cash nexus. Are the callers unwelcome because the writer does not get along with his or her family and neighbors, or for a less personal reason: that businesses and other organizations in pursuit of money use the phone as a means of access to it? The answer may be both, of course, but in expanding the idea of "unwelcome people," Skwire and Chitwood add no insight to it. The specific details close off analysis.

The same holds for their treatment of "unwelcome occasions." An occasion is a time that is socially defined and structured: a party or a steak dinner, yes, but sleep and a bath are more private activities, hardly occasions. Of course a phone call is usually as unwelcome in the middle of a bath as during a party. My point is that in changing ''occasions'' to ''times'' and letting detail do the work, the germ of an idea has been lost: the idea that we like to control our own social time, and that the telephone allows other people to intervene and impose *their* structure. What the details communicate instead is a loose feeling of harassment—easier to visualize, more specific, but certainly not more precise in thought.

Other changes have a similar effect. ''Headache'' is more sensory than ''inconvenience,'' but less exact, and personal rather than social. The phrase, ''cheery ringing in my ears,'' is a distraction, from the perspective of developing an idea: the point is not the sound, but the fact, of the intrusions and their content, the social relations they put the writer into or take him or her out of. And where the final sentence of the original implicitly raised a fruitful question (why ''nervous''?), the new conclusion—''Sometimes I wonder if there are any telephones in padded cells''—closes off inquiry with a joke and points up the writer's idiosyncrasy rather than the social matter that is under discussion.

On the level of speech acts, too, the rewrite personalizes, moves away from social analysis. In the first passage, emphasis falls on the general claims made about phones and the people who use them and are used by them. The rewrite buries those claims in a heap of reports of ''my'' experience, reports for which only the writer need vouch. The speaker of an assertion must be in a position to make it, or it isn't ''felicitous,'' in Austin's terms. (I cannot felicitously assert that there is life in the next galaxy.) The writer of the second passage *risks* less by moving quickly from generalizations that require support from history and social analysis, to those that stand on private terrain. This reduction in scope accords well with the impression given by the rewrite of a person incapable of coping with events, victimized by others, fragmented, distracted—a kind of likable schlemiel. He or she may be a less ''boring'' writer, but also a less venturesome and more isolated person, the sort who chatters on in a harmless gossipy way without much purpose or consequence: a *character*.

If a student showed me the first passage (as the outline of a composition or the beginning of a draft), I would want to say that it expresses an interesting idea, inadequately handled to be sure, but begging for a kind of development that amplification by detail, alone, can never supply. The contradiction with which it begins is familiar but perplexing: How is it that so many of our scientists' ''achievements,'' with all their promise of efficiency and ease, turn out to be inconvenient or worse in the long run? Why does an invention designed to give people control over their lives make many of us feel so often in the control of others? Why does a device for bringing people together (as its proprietors are constantly telling us in commercials) in fact so often serve as the carrier of frictions and antagonisms?

To make any headway with such questions, it is necessary to stay with the abstractions a while, penetrate them, get at the center of the contradictions they express, not throw them out in favor of lists of details. ''Achievement'': By whom? Who calls ''science'' into being and engineers its discoveries into commodities? The telephone as we have it is a hundred-year-long achievement, of patent lawyers and corporate planners more than of Alexander Graham Bell. ''Inconvenience'': For whom? Not for the salesmen and bill collectors, presumably. And certainly not for executives barricaded behind secretaries making sure the boss talks only with people he wants to talk with, and at a time of his choosing. The telephone represents a network of social relations embedded in history. In order to gain any leverage on the badly expressed contradiction of the first passage, it is necessary to unpack some of those relations. Piling on the details, as in the rewrite, may create a kind of superficial interest, but no gain in insight. The strategy, as exemplified here, is a strategy for sacrificing thought to feckless merriment.

Skwire and Chitwood are concerned with added detail. In the section I wish to consider from Winston Weathers and Otis Winchester's *The New Strategy of Style* (McGraw-Hill), the authors show how to make detail more specific. They do this under the heading of "Texture" (pp. 135-144), explaining that different subjects call for different textures: the simpler the subject, the more elaborate the texture. The maxim begs the question to an extent, since whether or not a given subject is simple or complex depends partly upon the diction used in exploring it. But apparently the first passage below is about a simple subject, since as the authors take it through four revisions their instructions all advise elaboration of texture. *("Make your nouns more specific." "Make your adjectives more specific.")* Passage 2 is the second of the rewrites.

1. The country store was an interesting place to visit. In the very heart of the city, it had the air of a small town grocery store combined with a feed and hardware supply house. There were flower seeds and milk churns, coal buckets and saddle blankets, all mixed together. Walking down the crowded aisles, you felt you had gone back to the past—to the time of pot-belly stoves and kerosene lamps and giant pickle jars. You could smell the grain, you could touch the harnesses, you could even sit down in the old wooden chair. When you finally left the store and were once more in the activity of the city, you felt as you sometimes do when you come out of an old movie into the bright light of reality.

2. Charlie's Country Store was a *spell-binding* emporium. In the very heart of Minneapolis, Charlie's had the *dubious* charm of a smalltown grocery combined with a feed and hardware supply house. There were *zinnia* seeds and milk churns, *shiny* coal buckets and *garish* saddle blankets, all mixed together. Walking down its *quaint* passageways—*narrow, poorly lighted, but nevertheless immaculate*—you felt you had gone back to nineteenth-century America—to the lost years, the *faintly remembered* days of *squat* pot-belly stoves and *sturdy* kerosene lamps and *rotund, ceramic* crocks—meant for pickles or pastries. You could smell the cornmeal, you could touch the *leather* harnesses, you could even sit in the *stern* wooden rocker. And when you finally left this anachronism—and were once more in the bustle of the city—you felt as you sometimes do when you come out of an old cinema into the *blinding* glare of a *rocket-age* reality. (pp. 135-138: emphasis in original)

Passage 2 is the result of making nouns and adjectives more specific and also (though Weathers and Winchester don't say so) of adding adjectives. Setting aside some words that might be criticized as elegant variations (e.g., "emporium," which suggests a grander establishment than is implied by the rest of the passage), consider the ways the description has become more specific.

1. The scene is particularized. The store is now Charlie's; the city, Minneapolis; the past, nineteenth-century America. Note that this change blurs the two main contrasts in

which the description is grounded, country versus city and present versus past. For the sharpening of these contrasts, it does not matter whose store it is or in what city, or whether the visitor travels back in imagination to America or England. Some of the other specifics are equally irrelevant: zinnia seeds, pastries, cornmeal. The point, I take it, is not the kind of flowers people used to grow, but that they had gardens; not what kind of grain they used, but that they did more of the work of preparing their own food, and that the arts of preserving, packaging, and marketing were in a primitive state of development compared to our present attainments as represented by freezers full of TV dinners and by the Golden Arches.

2. The writer's own impressions and values are foregrounded, most often adjectivally. The store is now *spell-binding,* its passageways *quaint,* and so on through "dubious charm" "faintly remembered," "stern," and "blinding." The writer has become much more of a presence, reacting, exercising taste, judging. These responses seem to issue from no particular perspective; for instance, what's "dubious" about the store's charm? They scarcely relate to the content of the original passage, certainly not to the ideas latent in it. They seem like the reflexes of a dilettantish tourist whose fugitive sensations and values clutter the picture and block analysis.

3. Similarly, the adjectives highlight the thinginess of things, their physical appearance, rather than what they are, what they meant, how life might have been organized around them. "Shiny," "garish," "poorly lighted," "squat," "sturdy," "rotund," "ceramic." The picture turns into a kind of still life, crowded with visual detail apparently valuable in itself. Such emphasis on visible surfaces, along with the esthetic perspective, draws attention to a detached present experience, dissipating the image of an earlier kind of civilization in which most people lived on farms, the family was the main productive unit, few of people's needs were commercialized, and technology was manageable and local.

Like the telephone, the objects in the country store embody social relations. And even more clearly than the initial passage about telephones, Weathers and Winchester's original version supports a sense of history, of a society that has been utterly transformed so that most of the things in the store have lost their usefulness. The society these artifacts imply—in which local people grew the grain, harvested it, ground it into flour, and baked it into bread—has given way to one in which almost all of our labor is sold in the market and controlled by employers rather than expended at our own pace and to our own plans; and almost all of our consumption takes place through markets organized not by a village merchant but by distant corporations.

Of course the first passage doesn't say what I have just said, even by implication. But in the way it sets up contrasts and in the details it presents, it provides the ground and even the need for such analysis. The student who takes Weathers and Winchester's guidance in making the passage "richer," more "vivid," and more "intense" will lose the thread of *any* analysis in a barrage of sensory impressions, irrelevant details, and personalized or random responses. Once again, the rhetorical strategy scatters thought.

With my final example, I turn to the injunction to use concrete language. The textbook is the second edition of *Composition: Skills and Models,* by Sidney T. Stovall, Virginia B. Mathis, Linda C. Elliot, G. Mitchell Hagler, Jr., and Mary A. Poole (Houghton Mifflin). Here are two of the passages they present for comparison in their chapter on forming a style, the first from Fielding's *Tom Jones,* and the second from Nevil Shute's *On the Beach:*

1. The charms of Sophia had not made the least impression on Blifil; not that his heart was pre-engaged, neither was he totally insensible of beauty, or had any aversion to women; but his appetites were by nature so moderate that he was easily able by philosophy, or by study, or by some other method to subdue them; and as to that passion which we have treated of in the first chapter of this book, he had not the least tincture of it in his whole composition.

But though he was so entirely free from that mixed passion of which we there treated, and of which the virtues and beauty of Sophia formed so notable an object, yet was he altogether as well furnished with some other passions that promised themselves very full gratification in the young lady's fortune. Such were avarice and ambition, which divided the dominion of his mind between them. He had more than once considered the possession of this fortune as a very desirable thing, and had entertained some distant views concerning it, but his own youth and that of the young lady, and indeed, principally a reflection that Mr. Weston might marry again and have more children, had restrained him from too hasty or too eager a pursuit.

2. He went back to bed. Tomorrow would be an anxious, trying day; he must get his sleep. In the privacy of his little curtained cabin he unlocked the safe that held the confidential books and took out the bracelet; it glowed in the synthetic light. She would love it. He put it carefully in the breast pocket of his uniform suit. Then he went to bed again, his hand upon the fishing rod, and slept.

They surfaced again at four in the morning, just before dawn, a little to the north of Gray Harbor. No lights were visible on the shore, but as there were no towns and few roads in the district that evidence was inconclusive. They went down to periscope depth and carried on. When Dwight came to the control room at six o'clock the day was bright through the periscope and the crew off duty were taking turns to look at the desolate shore. He went to breakfast and then stood smoking at the chart table, studying the minefield chart that he already knew so well, and the well-remembered entrance to the Juan de Fuca Strait. (p. 390)

The authors have couched their discussion of style in historically relative terms. Styles change, and students will want to choose from among styles suited to contemporary life. Eighteenth-century readers could "idle" over "long sentences"; "leisure is at a premium" now. Stovall and his colleagues do not absolutely value Shute's style over Fielding's, but since they say that the earlier style would strike the modern reader as awkward, stilted, colorless, complex, plodding, tedious and wordy, their counsel to the student is reasonably plain.

They direct their judgment partly against Fielding's long and complex sentences, partly against the quality of his diction. The latter is my concern. Stovall et. al. object to phrases like "entertained some distant view" and "had not the least tincture," and especially to Fielding's dependence on the big abstractions, "passion," "virtues," "avarice," "ambition," words which "elicit no emotional response from the reader." They praise Shute for "concrete words" that give life to the passage, citing "curtained cabin," "glowed in the synthetic light," "surfaced," and "desolate shore." Later in the chapter they urge the student to "Strive for the concrete word" (pp. 390-391).

Abstract nouns refer to the world in a way quite different from concrete nouns. They do not point to a set of particulars—all curtained cabins—or to any one cabin. They are relational. For instance, in speaking of Blifil's "avarice," Fielding calls up at least the relation of a series of acts to one another (a single act of acquiring or hoarding is not enough); of Blifil's feelings to these actions and to the wealth that is their goal; of those acts and feelings to a scale of values that is socially established (avarice is a sin, and so related to salvation and damnation); and of Blifil to other people who make such judgments, as well as to people whose wealth he might covet and who would become poorer were he to become richer. The term also evokes a temporal relationship: an avaricious person like Blifil seeks to become wealthy over time, and it is this future goal that informs his conduct. Abstract nouns that characterize people do so through bundles of relationships like these.

In short, one need not adopt an eighteenth-century faculty psychology, or expect Nevil Shute to adopt it, to see that Fielding's abstract nouns give a rich social setting to Blifil's sordid intentions. This setting is made more rich as, in context, Fielding humorously brings avarice into parity with love, under the higher-level abstraction of passion. (Herein another relationship, that of the narrator to his subject and his reader.) Abstractions are for Fielding a speculative and interpretive grid against which he can examine the events of the novel, and which themselves are constantly tested and modified by those events.

Shute's language in this passage, by contrast, sets his hero's actions against a background mainly of objects and of other people treated more or less as objects. The moral implications of the passage will have to be supplied by the reader. And there is no way for the narrator, given his style, to place that moral content in a dynamic relationship with social values, at least within the passage cited. This may be appropriate enough in a story from which society has literally disappeared; I do not mean to disparage Shute's diction, only to question the wisdom of commending it to students as plainly superior (for the twentieth-century reader) to Fielding's. Some important kinds of thinking can be done only with the help of abstractions.

In sum, as this textbook teaches the skill of using definite, specific, concrete language, it joins the other two in preferring realia to more abstract inquiry about realia, and to the effort to connect them. In doing so, it seems to me, the authors convey a fairly well-defined ideological picture to students. I would characterize this picture in these terms:

1. Ahistoricism. The preferred style focuses on a truncated present moment. Things and events are frozen in an image, or they pass on the wing, coming from nowhere.

2. Empiricism. The style favors sensory news, from the surfaces of things.

3. Fragmentation. An object is just what it is, disconnected from the rest of the world. The style obscures the social relations and the relations of people to nature that are embedded in all things.

4. Solipsism. The style foregrounds the writer's own perceptions: This is what I saw and felt.

5. Denial of conflict. The style typically pictures a world in which the telephone has the same meaning for all classes of people, a world whose "rocket-age reality" is

just mysteriously *there,* outside the country store, a world where avarice is a superfluous and tedious concept.

Furthermore—and I think this, too, a matter of the ideology of style—the injunctions to use detail, be specific, be concrete, as applied in these books, push the student writer always toward the language that most nearly reproduces the immediate experience and away from the language that might be used to understand it, transform it, and relate it to everything else. The authors privilege a kind of revising and expanding that leaves the words themselves unexamined and untransformed. Susan Wells has suggested that Christensen's rhetoric does not open "to investigation the relations among language, vision, and their objects,"[2] but takes those relations for granted. Her comment applies well to the use of detail recommended in these textbooks.

In an epoch when so much of the language students hear or read comes from distant sources, via the media, and when so much of it is shaped by advertisers and other corporate experts to channel their thoughts and feelings and needs, I think it a special pity if English teachers are turning students away from critical scrutiny of the words in their heads, especially from those that are most heavily laden with ideology. When in the cause of clarity and liveliness we urge them toward detail, surfaces, the sensory, as mere *expansion* of ideas or even as a *substitute* for abstraction, we encourage them to accept the empirical fragmentation of consciousness that passes for common sense in our society, and hence to accept the society itself as just what it most superficially seems to be.

Yes, it is good to keep readers interested, bad to bore them. Like Hirsch's principle of readability, the injunction to be interesting is on one level a bit of self-evident practical wisdom, not to mention kindness. Whatever you are trying to accomplish through a piece of writing, you won't achieve it if the reader quits on you, or plods on in resentful tedium. But mechanically applied, the principles of interest and readability entail accepting the reader exactly as he or she is. The reader's most casual values, interests, and capacities become an inflexible measure of what to write and how to write it, a Nielsen rating for prose. What happens to the possibility of challenging or even changing the reader? If keeping readers' attention is elevated to the prime goal of our teaching, the strategies we teach may well lead toward triviality and evasion.

Yes, I also realize that most students don't handle abstractions and generalizations well. I know that they often write badly when they try, and how depressing an experience it can be to read a batch of compositions on free will or alienation or capital punishment. And I am aware of the pressure many English teachers now feel to teach basic skills, whatever they are, rather than critical inquiry.[3] But I can't believe that the best response to this pressure is valorizing the concrete, fragmented, and inconsistent world-views that many of our students bring to college with them. Jeffrey Youdelman refers to colleagues he has heard say, "They can't handle abstraction . . . and therefore

---

2. "Classroom Heuristics and Empiricism," *College English,* 39 (1977), 471.

3. Obviously critical inquiry requires both abstractions and details, and a fluid exchange between them. I hope not to be taken as merely inverting the values I have criticized and recommending the abstract and general over the concrete.

I always give them topics like 'describe your favorite room.' '' Youdelman continues: ''Already stuck in a world of daily detail, with limited horizons and stunted consciousness, students are forced deeper into their solipsistic prison.''[4] Like him, I am concerned that in the cause of improving their skills we may end up increasing their powerlessness.

---

4. ''Limiting Students: Remedial Writing and the Death of Open Admissions,'' *College English*, 39 (1978), 563-64. Anyone interested in the politics of rhetoric and composition should read this excellent article and that of Susan Wells, cited earlier. I consider the present essay a supplement to theirs.

BASIC WRITING

Mina P. Shaughnessy

Diving In: An Introduction to Basic Writing

Basic writing, alias remedial, developmental, pre-baccalaureate, or even handicapped English, is commonly thought of as a writing course for young men and women who have many things wrong with them. Not only do medical metaphors dominate the pedagogy (*remedial, clinic, lab, diagnosis,* and so on), but teachers and administrators tend to discuss basic-writing students much as doctors tend to discuss their patients, without being tinged by mortality themselves and with certainly no expectations that questions will be raised about the state of *their* health.

Yet such is the nature of instruction in writing that teachers and students cannot easily escape one another's maladies. Unlike other courses, where exchanges between teacher and student can be reduced to as little as one or two objective tests a semester, the writing course requires students to write things down regularly, usually once a week, and requires teachers to read what is written and then write things back and every so often even talk directly with individual students about the way they write.

This system of exchange between teacher and student has so far yielded much more information about what is wrong with students than about what is wrong with teachers, reinforcing the notion that students, not teachers, are the people in education who must do the changing. The phrase "catching up," so often used to describe the progress of BW students, is illuminating here, suggesting as it does that the only person who must move in the teaching situation is the student. As a result of this view, we are much more likely in talking about teaching to talk about students, to theorize about *their* needs and attitudes or to chart *their* development and ignore the possibility that teachers also change in response to students, that there may in fact be important connections between the changes teachers undergo and the progress of their students.

I would like, at any rate, to suggest that this is so, and since it is common these days to "place" students on developmental scales, saying they are eighth-graders or fifth-graders when they read and even younger when they write or that they are stalled some place on Piaget's scale without formal propositions, I would further like to propose a developmental scale for teachers, admittedly an impressionistic one, but one that fits the observations I have made over the years as I have watched traditionally prepared English teachers, including myself, learning to teach in the open-admissions classroom.

My scale has four stages, each of which I will name with a familiar metaphor intended to suggest what lies at the center of the teacher's emotional energy during

From *College Composition and Communication* 27 (October 1976): 234-39.

that stage. Thus I have chosen to name the first stage of my developmental scale GUARDING THE TOWER, because during this stage the teacher is in one way or another concentrating on protecting the academy (including himself) from the outsiders, those who do not seem to belong in the community of learners. The grounds for exclusion are various. The mores of the times inhibit anyone's openly ascribing the exclusion to genetic inferiority, but a few teachers doubtless still hold to this view.

More often, however, the teacher comes to the basic-writing class with every intention of preparing his students to write for college courses, only to discover, with the first batch of essays, that the students are so alarmingly and incredibly behind any students he has taught before that the idea of their ever learning to write acceptably for college, let alone learning to do so in one or two semesters, seems utterly pretentious. Whatever the sources of their incompetence—whether rooted in the limits they were born with or those that were imposed upon them by the world they grew up in—the fact seems stunningly, depressingly obvious: they will never "make it" in college unless someone radically lowers the standards.

The first pedagogical question the teacher asks at this stage is therefore not "How do I teach these students?" but "What are the consequences of flunking an entire class?" It is a question that threatens to turn the class into a contest, a peculiar and demoralizing contest for both student and teacher, since neither expects to win. The student, already conditioned to the idea that there is something wrong with his English and that writing is a device for magnifying and exposing this deficiency, risks as little as possible on the page, often straining with what he does write to approximate the academic style and producing in the process what might better be called "written Anguish" rather than English—sentences whose subjects are crowded out by such phrases as "it is my conviction that" or "on the contrary to my opinion," inflections that belong to no variety of English, standard or non-standard, but grow out of the writer's attempt to be correct, or words whose idiosyncratic spellings reveal not simply an increase in the number of conventional misspellings but new orders of difficulty with the correspondences between spoken and written English. Meanwhile, the teacher assumes that he must not only hold out for the same product he held out for in the past but teach unflinchingly in the same way as before, as if any pedagogical adjustment to the needs of students were a kind of cheating. Obliged because of the exigencies brought on by open admissions to serve his time in the defense of the academy, he does if not his best, at least his duty, setting forth the material to be mastered, as if he expected students to learn it, but feeling grateful when a national holiday happens to fall on a basic-writing day and looking always for ways of evading conscription next semester.

But gradually, student and teacher are drawn into closer range. They are obliged, like emissaries from opposing camps, to send messages back and forth. They meet to consider each other's words and separate to study them in private. Slowly, the teacher's preconceptions of his students begin to give way here and there. It now appears that, in some instances at least, their writing, with its rudimentary errors and labored style has belied their intelligence and individuality. Examined at a closer range, the class now appears to have at least some members in it who might, with hard work, eventually "catch up." And it is the intent of reaching these students that moves the teacher into the second stage of development—which I will name CONVERTING THE NATIVES.

As the image suggests, the teacher has now admitted at least some to the community of the educable. These learners are perceived, however, as empty vessels, ready to be filled with new knowledge. Learning is thought of not so much as a constant and often troubling reformulation of the world so as to encompass new knowledge but as a steady flow of truth into a void. Whether the truth is delivered in lectures or modules, cassettes or computers, circles or squares, the teacher's purpose is the same: to carry the technology of advanced literacy to the inhabitants of an underdeveloped country. And so confident is he of the reasonableness and allure of what he is presenting, it does not occur to him to consider the competing logics and values and habits that may be influencing his students, often in ways that they themselves are unaware of.

Sensing no need to relate what he is teaching to what his students know, to stop to explore the contexts within which the conventions of academic discourse have developed, and to view these conventions in patterns large enough to encompass what students do know about language already, the teacher becomes a mechanic of the sentence, the paragraph, and the essay. Drawing usually upon the rules and formulas that were part of his training in composition, he conscientiously presents to his students flawless schemes for achieving order and grammaticality and anatomizes model passages of English prose to uncover, beneath brilliant, unique surfaces, the skeletons of ordinary paragraphs.

Yet too often the schemes, however well meant, do not seem to work. Like other simplistic prescriptions, they illuminate for the moment and then disappear in the melee of real situations, where paradigms frequently break down and thoughts will not be regimented. S's keep reappearing or disappearing in the wrong places; regular verbs shed their inflections and irregular verbs acquire them; tenses collide; sentences derail; and whole essays idle at one level of generalization.

Baffled, the teacher asks, "How is it that these young men and women whom I have personally admitted to the community of learners cannot learn these simple things?" Until one day, it occurs to him that perhaps these simple things—so transparent and compelling to him—are not in fact simple at all, that they only appear simple to those who already know them, that the grammar and rhetoric of formal written English have been shaped by the irrationalities of history and habit and by the peculiar restrictions and rituals that come from putting words on paper instead of into the air, that the sense and nonsense of written English must often collide with the spoken English that has been serving students in their negotiations with the world for many years. The insight leads our teacher to the third stage of his development, which I will name SOUNDING THE DEPTHS, for he turns now to the careful observation not only of his students and their writing but of himself as writer and teacher, seeking a deeper understanding of the behavior called writing and of the special difficulties his students have in mastering the skill. Let us imagine, for the sake of instruction, that the teacher now begins to look more carefully at two common problems among basic writers—the problem of grammatical errors and the problem of undeveloped paragraphs.

Should he begin in his exploration of error not only to count and name errors but to search for patterns and pose hypotheses that might explain them, he will begin to see that while his lessons in the past may have been "simple," the sources of the error he was trying to correct were often complex. The insight leads not inevitably or finally to a rejection of all rules and standards, but to a more careful look at error, to the formulation of what might be called a "logic" of errors that serves to mark a pedagogical path for teacher and student to follow.

Let us consider in this connection the "simple" *s* inflection on the verb, the source of a variety of grammatical errors in BW papers. It is, first, an alien form to many students whose mother tongues inflect the verb differently or not at all. Uniformly called for, however, in all verbs in the third person singular present indicative of standard English, it would seem to be a highly predictable or stable form and therefore one easily remembered. But note the grammatical concepts the student must grasp before he can apply the rule: the concepts of person, tense, number, and mood. Note that the *s* inflection is an atypical inflection within the modern English verb system. Note too how often it must seem to the student that he hears the stem form of the verb after third person singular subjects in what sounds like the present, as he does for example whenever he hears questions like "Does *she want* to go?" or "Can the *subway stop?*" In such sentences, the standard language itself reinforces the student's own resistance to the inflection.

And then, beyond these apparent unpredictabilities within the standard system, there is the influence of the student's own language or dialect, which urges him to ignore a troublesome form that brings no commensurate increase in meaning. Indeed, the very *s* he struggles with here may shift in a moment to signify plurality simply by being attached to a noun instead of a verb. No wonder then that students of formal English throughout the world find this inflection difficult, not because they lack intelligence or care but because they think analogically and are linguistically efficient. The issue is not the capacity of students finally to master this and the many other forms of written English that go against the grain of their instincts and experience but the priority this kind of problem ought to have in the larger scheme of learning to write and the willingness of students to mobilize themselves to master such forms at the initial stages of instruction.

Somewhere between the folly of pretending that errors don't matter and the rigidity of insisting that they matter more than anything, the teacher must find his answer, searching always under pressure for short cuts that will not ultimately restrict the intellectual power of his students. But as yet, we lack models for the maturation of the writing skill among young, native-born adults and can only theorize about the adaptability of other models for these students. We cannot say with certainty just what progress in writing ought to look like for basic-writing students, and more particularly how the elimination of error is related to their over-all improvement.

Should the teacher then turn from problems of error to his students' difficulties with the paragraphs of academic essays, new complexities emerge. Why, he wonders, do they reach such instant closure on their ideas, seldom moving into even one subordinate level of qualification but either moving on to a new topic sentence or drifting off into reverie and anecdote until the point of the essay has been dissolved? Where is that attitude of "suspended conclusion" that Dewey called thinking, and what can one infer about their intellectual competence from such behavior?

Before consigning his students to some earlier stage of mental development, the teacher at this stage begins to look more closely at the task he is asking students to perform. Are they aware, for example, after years of right/wrong testing, after the ACT'S and the GED'S and the OAT'S, after straining to memorize what they read but never learning to doubt it, after "psyching out" answers rather than discovering them, are they aware that the rules have changed and that the rewards now go to those who can sustain a play of mind upon ideas—teasing out the contradictions and ambiguities and frailties of statements?

Or again, are the students sensitive to the ways in which the conventions of talk differ from those of academic discourse? Committed to extending the boundaries of what is known, the scholar proposes generalizations that cover the greatest possible number of instances and then sets about supporting his case according to the rules of evidence and sound reasoning that governs his subject. The spoken language, looping back and forth between speakers, offering chances for groping and backing up and even hiding, leaving room for the language of hands and faces, of pitch and pauses, is by comparison generous and inviting. The speaker is not responsible for the advancement of formal learning. He is free to assert opinions without a display of evidence or recount experiences without explaining what they "mean." His movements from one level of generality to another are more often brought on by shifts in the winds of conversation rather than by some decision of his to be more specific or to sum things up. For him the injunction to "be more specific" is difficult to carry out because the conditions that lead to specificity are usually missing. He may not have acquired the habit of questioning his propositions, as a listener might, in order to locate the points that require amplification or evidence. Or he may be marooned with a proposition he cannot defend for lack of information or for want of practice in retrieving the history of an idea as it developed in his own mind.

Similarly, the query "What is your point?" may be difficult to answer because the conditions under which the student is writing have not allowed for the slow generation of an orienting conviction, that underlying sense of the direction he wants his thinking to take. Yet without this conviction, he cannot judge the relevance of what comes to his mind, as one sentence branches out into another or one idea engenders another, gradually crowding from his memory the direction he initially set for himself.

Or finally, the writer may lack the vocabulary that would enable him to move more easily up the ladder of abstraction and must instead forge out of a nonanalytical vocabulary a way of discussing thoughts about thoughts, a task so formidable as to discourage him, as travelers in a foreign land are discouraged, from venturing far beyond bread-and-butter matters.

From such soundings, our teacher begins to see that teaching at the remedial level is not a matter of being simpler but of being more profound, of not only starting from "scratch" but also determining where "scratch" is. The experience of studenthood is the experience of being just so far over one's head that it is both realistic and essential to work at surviving. But by underestimating the sophistication of our students and by ignoring the complexity of the tasks we set before them, we have failed to locate in precise ways where to begin and what follows what.

But I have created a fourth stage in my developmental scheme, which I am calling DIVING IN in order to suggest that the teacher who has come this far must now make a decision that demands professional courage—the decision to remediate himself, to become a student of new disciplines and of his students themselves in order to perceive both their difficulties and their incipient excellence. "Always assume," wrote Leo Strauss, to the teacher, "that there is one silent student in your class who is by far superior to you in head and in heart." This assumption, as I have been trying to suggest, does not come easily or naturally when the teacher is a college teacher and the young men and women in his class are labeled remedial. But as we come to know these students better, we begin to see that the greatest barrier to our work with them is our ignorance of them and of the very subject we have contracted to teach. We see that we must grope our ways into the turbulent disciplines of semantics and linguistics for fuller, more accurate data about words and sentences; we must pursue more rigorously the design of developmental models, basing our schemes less upon loose comparisons with children and more open case studies and developmental research of the sort that produced William Perry's impressive study of the intellectual development of Harvard students; we need finally to examine more closely the nature of speaking and writing and divine the subtle ways in which these forms of language both support and undo each other.

The work is waiting for us. And so irrevocable now is the tide that brings the new students into the nation's college classrooms that it is no longer within our power, as perhaps it once was, to refuse to accept them into the community of the educable. They are here. DIVING IN is simply deciding that teaching them to write well is not only suitable but challenging work for those who would be teachers and scholars in a democracy.

Min-zhan Lu

Redefining the Legacy of Mina Shaughnessy: A Critique of the Politics of Linguistic Innocence

The aim of this paper is to critique an essentialist assumption about language that is dominant in the teaching of basic writing. This assumption holds that the essence of meaning precedes and is independent of language, which serves merely as a vehicle to communicate that essence. According to this assumption, differences in discourse conventions have no effect on the essential meaning communicated. Using Mina Shaughnessy's *Errors and Expectations* as an example, I examine the ways in which such an assumption leads to pedagogies which promote what I call a politics of linguistic innocence: that is, a politics which preempts teachers' attention from the political dimensions of the linguistic choices students make in their writing.

My critique is motivated by my alignment with various Marxist and poststructuralist theories of language.[1] In one way or another, these theories have argued that language is best understood not as a neutral vehicle of communication but as a site of struggle among competing discourses. Each discourse puts specific constraints on the construction of one's stance—how one makes sense of oneself and gives meaning to the world. Through one's gender; family; work; religious, educational, or recreational life; each individual gains access to a range of competing discourses which offer competing views of oneself, the world, and one's relation with the world. Each time one writes, even and especially when one is attempting to use one of these discourses, one experiences the need to respond to the dissonance among the various discourses of one's daily life. Because different discourses do not enjoy equal political power in current-day America, decisions on how to respond to such dissonance are never politically innocent.

From the perspective of such a view of language, Shaughnessy's stated goal for her basic writers—the mastery of written English and the "ultimate freedom of deciding how and when and where" to use which language (11)—should involve at least three

1. My view of language has been informed by Louis Althusser's notion of ideology, Antonio Gramsci's analysis of hegemony, Jacques Derrida's critique of the metaphysics of presence, Michel Foucault's theory of discourse and power, and the distinction Raymond Williams makes between practical and official consciousness.

challenges for student writers. First, the students need to become familiar with the conventions or "the stock of words, routines, and rituals that make up" academic discourse (198). Second, they need to gain confidence as learners and writers. Third, they need to decide how to respond to the potential dissonance between academic discourse and their home discourses. These decisions involve changes in how they think and how they use language. Yet, most pedagogies informed by the kind of essentialist assumption I defined earlier, including the one Shaughnessy presents in *Errors and Expectations,* tend to focus attention on only the first two of these challenges.

I choose *Errors and Expectations* as an example of such pedagogies because, following Robert Lyons, I interpret the operative word in that book to be "tasks" rather than "achievements." As Lyons cogently points out, Shaughnessy's work "resists closure; instead, it looks to the future, emphasizing what needs to be learned and done" (186). The legacy of Shaughnessy, I believe, is the set of tasks she maps out for composition teachers. To honor this legacy, we need to examine the pedagogical advice she gives in *Errors and Expectations* as tasks which point to the future—to what needs to be learned and done—rather than as providing closure to our pedagogical inquiry. One of the first tasks Shaughnessy establishes for composition teachers is that of "remediating" ourselves ("Diving In" 238). She urges us to become "students" of our students and of new disciplines. Reading *Errors and Expectations* in light of current theories of language is one way of continuing that "remediation." Shaughnessy also argues that a good composition teacher should inculcate interest in and respect for linguistic variety and help students attain discursive option, freedom, and choice. She thus maps out one more task for us: to carry out some democratic aspirations in the teaching of basic writing.[2] Another task she maps out for composition teachers is the need to "sound the depths" of the students' difficulties as well as their intelligence ("Diving In" 236). If, as I will argue, some of her own pedagogical advice indicates that an essentialist view of language could impede rather than enhance one's effort to fulfill these tasks, then the only way we can fully benefit from the legacy of Shaughnessy is to take the essentialist view of language itself to task.

In *Errors and Expectations,* Shaughnessy argues that language "is variously shaped by situations and bound by conventions, none of which is inferior to the others but none of which, also, can substitute for the others" (121). Using such a view of language, she makes several arguments key to her pedagogy. For example, she uses such a view to argue for the "systematic nature" of her students' home discourses, the students' "quasi-foreign relationship" with academic discourse and, thus, the logic of some of their errors. She also uses this view of language to call attention to basic writers' existing mastery of at least one variety of English and thus, their "intelligence and linguistic aptitudes" (292). She is then able to increase the confidence of both teachers and students in the students' ability to master a new variety of English—academic English.

Shaughnessy's view of language indicates her willingness to "remediate" herself by studying and exploring the implications which contemporary linguistic theories have for the teaching of basic writing.[3] However, in looking to these fields for "fresh insights

2. For discussion of Shaughnessy's pedagogy in relation to her democratic aspirations, see Robert Lyons and rebuttals to Rouse's "The Politics of Shaughnessy" by Michael Allen, Gerald Graff, and William Lawlor.

3. In arguing for the need to show "interest in and respect for language variety," Shaughnessy cites William Labov's analysis of the inner logic, grammar, and ritual forms in Black English Vernacular (17,

and new data,'' Shaughnessy seems to have also adopted an essentialist assumption which dominates these theories of language: that linguistic codes can be taught in isolation from the production of meaning and from the dynamic power struggle within and among diverse discourses.[4]

We see this assumption operating in Shaughnessy's description of a writer's ''consciousness (or conviction) of what [he] means'':

> It seems to exist at some subterranean level of language—but yet to need words to coax it to the surface, where it is communicable, not only to others but, in a different sense, to the writer himself. (80)

The image of someone using words to coax meaning ''to the surface'' suggests that meaning exists separately from and ''at some subterranean level of language.'' Meaning is thus seen as a kind of essence which the writer carries *in* his or her mind prior to writing, although the writer might not always be fully conscious of it. Writing merely serves to make this essence communicable to oneself and others. As David Bartholomae puts it, Shaughnessy implies that ''writing is in service of 'personal thoughts and styles' '' (83). Shaughnessy does recognize that writing is ''a deliberate process whereby meaning is crafted, stage by stage'' (81), even that ''the act of articulation refines and changes [thought]'' (82). But the pedagogy she advocates seldom attends to the changes which occur in that act. Instead, it presents writing primarily as getting ''as close a fit as possible between what [the writer] means and what he says on paper,'' or as ''testing the words that come to mind against the thought one has in mind'' (79, 204). That is, ''meaning is crafted'' only to match what is already in the writer's mind (81-82).

Such a view of the relationship between words and meaning overlooks the possibility that different ways of using words—different discourses—might exercise different constraints on how one ''crafts'' the meaning ''one has in mind.'' This is probably why the pedagogical advice Shaughnessy offers in *Errors and Expectations* seldom considers the possibility that the meaning one ''has in mind'' might undergo substantial change as one tries to ''coax'' it and ''communicate'' it in different discourses. In the following section, I use Shaughnessy's responses to three student writings to examine this tendency in her pedagogy. I argue that such a tendency might keep her pedagogy from achieving all the goals it envisions. That is, it might teach students to ''write something in formal English'' and ''have something to say'' but can help students obtain only a very limited ''freedom of deciding *how* and when and where'' to ''use which language'' (11, emphasis mine).

The following is a sentence written by one of Shaughnessy's students:

> In my opinion I believe that you there is no field that cannot be effected some sort of advancement that one maybe need a college degree to make it. (62)

Shaughnessy approaches the sentence ''grammatically,'' as an example of her students' tendency to use ''fillers'' such as ''I think that . . .'' and ''It is my opinion that . . .''

237, 304). Shaughnessy also cites theories in contrastive analysis (156), first-language interference (93), and transformational grammar (77-78) to support her speculations on the logic of basic writers' error.

4. For a critique of the way modern linguistics of language, code, and competence (such as Labov's study of Black English Vernacular) tend to treat discourses as discrete and autonomous entities, see Mary Louise Pratt's ''Linguistic Utopias.''

(62). She argues that these ''fillers'' keep the writers from ''making a strong start with a *real subject*'' and make them lose their *''bearings''* (62, my emphasis). The distinction between a ''real subject'' and ''fillers'' suggests that in getting rid of the ''fillers,'' the teacher is merely helping the writer to retrieve the real subject or bearings he has in mind. I believe Shaughnessy assumes this to be the case because she sees meaning as existing ''at some subterranean level of language.'' Yet, in assuming that, her attention seems to have been occluded from the possibility that as the writer gets rid of the ''fillers,'' he might also be qualifying the subject or bearing he originally has in mind.

For instance, Shaughnessy follows the student's original sentence with a consolidated sentence: ''A person with a college degree has a better chance for advancement in any field'' (63). Shaughnessy does not indicate whether this is the student's revised sentence or the model the teacher might pose for the student. In either case, the revised sentence articulates a much stronger confidence than the original in the belief that education entails advancement. For we might read some of the phrases in the original sentence, such as ''in my opinion,'' ''I believe that you,'' ''some sort of,'' and ''one maybe need,'' as indications not only of the writer's inability to produce a grammatically correct sentence but also of the writer's attempt to articulate his uncertainty or skepticism towards the belief that education entails advancement. In learning ''consolidation,'' this student is also consolidating his attitude towards that belief. Furthermore, this consolidation could involve important changes in the writer's political alignment. For one can well imagine that people of different economic, racial, ethnic, or gender groups would have different feelings about the degree to which education entails one's advancement.

In a footnote to this passage, Shaughnessy acknowledges that ''some would argue'' that what she calls ''fillers'' are ''indices of involvement'' which convey a stance or point of view (62 n. 4). But her analysis in the main text suggests that the sentence is to be tackled ''grammatically,'' without consideration to stance or point of view. I think the teacher should do both. The teacher should deliberately call the student's attention to the relationship between ''grammar'' and ''stance'' when teaching ''consolidation.'' For example, the teacher might ask the student to consider if a change in meaning has occurred between the original sentence and the grammatically correct one. The advantage of such an approach is that the student would realize that decisions on what are ''fillers'' and what is one's ''real subject'' are not merely ''grammatical'' but also political: they could involve a change in one's social alignment. The writer would also perceive deliberation over one's stance or point of view as a normal aspect of learning to master grammatical conventions. Moreover, the writer would be given the opportunity to reach a self-conscious decision. Without practice in this type of decision making, the kind of discursive options, freedom, or choice the student could obtain through education is likely to be very limited.

Attention to this type of deliberation seems just as necessary if the teacher is to help the student who wrote the following paper achieve the style of ''weav[ing] personal experience into analytical discourse'' which Shaughnessy admires in ''mature and gifted writers'' (198):

> It can be said that my parents have led useful live but that usefulness seems to deteriorate when they found themselves constantly being manipulated for the benefit of one and not for the benefit of the community. If they were able to realize that were being manipulate successful advancements could of been gained but being that they had no

strong political awareness their energies were consumed by the politicians who saw personal advancements at the expenses of dedicated community workers. And now that my parents have taken a leave of absence from community involvement, comes my term to participate on worthwhile community activities which well bring about positive results and to maintain a level of consciousness in the community so that they will know what policies affect them, and if they don't quite like the results of the policies I'll make sure, if its possible, to abolish the ones which hinder progress to ones which well present the correct shift in establishing correct legislation or enactments. In order to establish myself and my life to revolve around the community I must maintain a level of awareness to make sure that I can bring about positive actions and to keep an open mind to the problems of the community and to the possible manipulation machinery which is always on the watch when progressive leaders or members of the community try to build effective activities for the people to participate. (197)

Shaughnessy suggests that the reason this writer has not yet "mastered the style" is because he has just "begun to advance into the complexity of the new language" and "is almost certain to sound and feel alien with the stock of words, routines, and rituals that make up that language" (198). The "delicate task" of the teacher in such a situation, Shaughnessy points out, is to "encourag[e] the enterprise and confidence of the student" while "improving his judgment about both the forms and meanings of the words he chooses" (198).

I believe that there is another dimension to the teacher's task. As Shaughnessy points out, this writer might be "struggling to develop a language that will enable him to talk analytically, with strangers, about the oppression of his parents and his own resolve to work against that oppression" (197). If what Shaughnessy says of most of her basic writers is true of this writer—that he too has "grown up in one of New York's ethnic or racial enclaves" (3)—then the "strangers" for whom he writes and whose analytical discourse he is struggling to use are "strangers" both in the political and linguistic sense. To this writer, these "strangers" are people who already belong to what Shaughnessy calls the world of "public transactions—educational, civic, and professional" (125), a world which has traditionally excluded people like the writer and his parents. These "strangers" enjoy power relationships with the very "politicians" and "manipulation machinery" against whom this writer is resolved to fight. In trying to "talk analytically," this writer is also learning the "strangers' " way of perceiving people like his parents, such as viewing the oppression of his parents and his resolution to work against that oppression with the "curiosity and sentimentality of strangers" (197-198). Thus, their "style" might put different constraints than the student's home discourse on how this writer re-views "the experiences he has in mind" (197). If all of this is so, the teacher ought to acknowledge that possibility to the students.

Let me use the writings of another of Shaughnessy's students to illustrate why attention to a potential change in point of view might benefit students. The following are two passages written by one of Shaughnessy's students at the beginning and the end of a semester:

Essay written at beginning of semester

Harlem taught me that light skin Black people was better look, the best to suceed, the best off fanicially etc this whole that I trying to say, that I was brainwashed and

> people aliked. I couldn't understand why people (Black and white) couldn't get alone. So as time went along I began learned more about myself and the establishment.
>
> *Essay written at end of semester*
> In the midst of this decay there are children between the ages of five and ten playing with plenty of vitality. As they toss the football around, their bodies full of energy, their clothes look like rainbows. The colors mix together and one is given the impression of being in a psychedelic dream, beautiful, active, and alive with unity. They yell to eachother increasing their morale. They have the sound of an organized alto section. At the sidelines are the girls who are shy, with the shyness that belongs to the very young. They are embarrassed when their dresses are raised by the wind. As their feet rise above pavement, they cheer for their boy friends. In the midst of the decay, children will continue to play. (278)

In the first passage, the writer approaches the "people" through their racial and economic differences and the subject of childhood through racial rift and contention. In the second paper, he approaches the "children" through the differences in their age, sex, and the color of their clothes. And he approaches the subject of childhood through the "unity" among children. The second passage indicates a change in how this writer makes sense of the world around him: the writer has appeased his anger and rebellion against a world which "brainwashed" children with discriminatory perceptions of Blacks and Whites. Compared to the earlier and more labored struggle to puzzle out "why people (Black and white) couldn't get alone *[sic],*" the almost lyrical celebration of the children's ability to "continue to play" "in the midst of the decay" seems a much more "literary" and evasive form of confronting the world of "decay."

Shaughnessy characterizes this writer as a student who "discovered early in the semester that writing gave him *access* to thoughts and feelings he had not *reached* any other way" (278, my emphasis). She uses these essays to illustrate "the measure of his improvement in one semester." By that, I take Shaughnessy to have in mind the changes in length and style. By the end of the semester, the student is clearly not only finding more to say on the subject but also demonstrating better control over the formal English taught in the classroom. This change in length and style certainly illustrates the effectiveness of the kind of pedagogical advice Shaughnessy gives.

Yet, these two passages also indicate that the change in the length and style of the student's writing can be accompanied by a change in thinking—in the way one perceives the world around one and relates to it. This latter change is often political as well as stylistic. I think that Shaughnessy's responses to these student writings overlook this potential change in thinking because she believes that language will only help the writers "reach" but not change how they think and feel about a certain subject or experience. Thus, attention to a potential change in one's point of view or political stance seems superfluous.

If mastery of academic discourse is often accompanied by a change in one's point of view, as my reading of these three student writings suggests, then it ought to be the teacher's task to acknowledge to the students this aspect of their learning. However, teachers may hesitate to do so because they are worried that doing so might confirm the students' fear that education will distance them from their home discourses or communities and, as a result, slow down their learning. As Shaughnessy cogently points out, her students are already feeling overwhelmed by their sense of the competition between home and college:

> Neglected by the dominant society, [basic writers] have nonetheless had their own worlds to grow up in and they arrive on our campuses as young adults, with opinions and languages and plans already in their minds. College both beckons and threatens them, offering to teach them useful ways of thinking and talking about the world, promising even to improve the quality of their lives, but threatening at the same time to take from them their distinctive ways of interpreting the world, to assimilate them into the culture of academia without acknowledging their experience as outsiders. (292)

Again and again, Shaughnessy reminds us of her students' fear that college may distance them from "their own worlds" and take away from them the point of view they have developed through "their experience as outsiders." She argues that this fear causes her students to mistrust and psychologically resist learning to write (125). Accordingly, she suggests several methods which she believes will help students assuage that fear.

For example, when discussing her students' difficulty in developing an "academic vocabulary," Shaughnessy points out that they might resist a new meaning for a familiar word because accepting it would be like consenting to a "linguistic betrayal that threatens to wipe out not just a word but the reality that the word refers to" (212). She then goes on to suggest that "if we consider the formal (rather than the contextual) ways in which words can be made to shift meaning we are closer to the kind of practical information about words BW students need" (212). This seems to be her rationale: if a "formal" approach (in this case, teaching students to pay attention to prefixes and suffixes) can help students learn that words can be made to shift meaning, then why not avoid the "contextual" approach, especially since the "contextual" approach will only activate their sense of being pressured to "wipe out not just a word but the reality that the word refers to"?

But taking this "formal" approach only circumvents the students' attention to the potential change in their thinking and their relationship with home and school. It delays but cannot eliminate their need to deal with that possibility. As a result, students are likely to realize the change only after it has already become a fact. At the same time, because the classroom has suggested that learning academic discourse will not affect how they think, feel, or relate to home, students are also likely to perceive their "betrayal" of home in purely personal terms, the result of purely personal choices. The sense of guilt and confusion resulting from such a perception is best illustrated in Richard Rodriguez's narrative of his own educational experience, *Hunger of Memory*. Rodriguez's narrative also suggests that the best way for students to cope constructively with their sense of having consented to a "betrayal" is to perceive it in relation to the politics of education and language. The long, lonely, and painful deliberation it takes for Rodriguez to contextualize that "betrayal" suggests that teachers might better help students anticipate and cope with their sense of "betrayal" if they take the "contextual" as well as the "formal" approach when teaching the conventions of academic discourse. In fact, doing both might even help students to minimize that "betrayal." When students are encouraged to pay attention to the ways in which diverse discourses constrain one's alignments with different points of view and social groups, they have a better chance to deliberate over how they might resist various pressures academic discourse exercises on their existing points of view. As Shaughnessy points out, "English has been robustly inventing itself for centuries—stretching and reshaping and enriching itself with every language and dialect it has encountered" (13). If the teacher acknowledges that all practitioners of academic discourse, including those who are learning to master it as well as those who have already mastered it, can participate in this process of reshaping, then students might be less passive in coping with the constraints that academic discourse puts on their alignments with their home discourses.

In preempting Shaughnessy's attention from the political decisions involved in her students' formal or linguistic decisions, the essentialist view of language also seems to have kept her from noticing her own privileging of academic discourse. Shaughnessy calls formal written English "the language of public transactions—educational, civic, and professional"—and the students' home discourse the language one uses with one's family and friends (125). Shaughnessy insists that no variety of English can "substitute for the others" (121). She reassures her students that their home discourses cannot be substituted by academic discourse, but neither can their home discourses substitute for academic discourse. Thus, she suggests that academic discourse is a "necessary" and "advantageous" language for *all* language users because it *is* the language of public transaction (125, 293). This insistence on the nonsubstitutive nature of language implies that academic discourse has been, is, and will inevitably be the language of public transaction. And it may very well lead students to see the function of formal English as a timeless linguistic law which they must respect, adapt to, and perpetuate rather than as a specific existing circumstance resulting from the historically unequal distribution of social power, and as a condition which they must recognize but can also call into question and change.

Further, she differentiates the function of academic discourse from that of the students' home discourses through the way she characterizes the degree to which each discourse mobilizes one's language learning faculty. She presents the students' efforts to seek patterns and to discriminate or apply rules "*self-sustaining* activities" (127, emphasis mine). She argues that the search for causes, like the ability to compare, is "a constant and deep *urge* among people of *all* cultures and ages" and "part of an *unfolding intellective power* that begins with infancy and continues, at least in the lives of some, until death" (263, emphasis mine). Academic discourse and the students' home discourses, Shaughnessy suggests, unfold their "intellective power" differently. The home discourses of basic writers are seen as allowing such power to remain "largely intuitive," "simplistic," and "unreasoned" (263), while the conventions of written English are seen as demanding that such power be "more thoroughly developed," "more consciously organized" (261). Thus, academic discourse is endowed with the power to bring the "native intelligence" or the "constant and deep urge" in *all* language learners to a higher and more self-conscious level.

This type of depiction suggests that learning academic discourse is not a violation but a cultivation of what basic writers or "people of all cultures and ages" have in and of themselves. Shaughnessy thus suggests basic writers are being asked to learn academic discourse because of its distinctive ability to utilize a "human" resource. Hence, her pedagogy provides the need to learn academic discourse with a "human," and hence with yet another seemingly politically innocent, justification. It teaches students to see discursive decisions made from the point of view of academic culture as "human" and therefore "innocent" decisions made absolutely free from the pressures of specific social and historical circumstances. If it is the student's concern to align himself or herself with minority economic and ethnic groups in the very act of learning academic discourse, the politics of "linguistic" innocence can only pacify rather than activate such a concern.

Shaughnessy's desire to propose a pedagogy which inculcates respect for discursive diversity and freedom of discursive choice articulates her dissatisfaction with and reaction to the unequal social power and prestige of diverse discourses in current day America. It also demonstrates her belief that education can and should attempt to change these prevailing unequal conditions. However, the essentialist view of language which underlies her pedagogy seems also to have led her to believe that a vision of language which insists on the equality and nonsubstitutive nature of linguistic variety, and an ideal writing classroom which promotes such a view, can stand in pure opposition to society, adjusting existing social inequality and the human costs of such inequality from somewhere "outside" the socio-historical space which it is trying to transform. As a result, her pedagogy enacts a systematic denial of the political context of students' linguistic decisions.

The need to critique the essentialist view of language and the politics of linguistic innocence is urgent when viewed in the context of the popular success of E. D. Hirsch, Jr.'s proposals for educational "reforms." Hirsch argues for the "validity" of his "vocabulary" by claiming its political neutrality. Hirsch argues that "it is used to support *all* conflicting values that arise in public discourse" and "to communicate *any* point of view effectively" or "in *whatever* direction one wishes to be effective" (*Cultural Literacy* 23, 102, 103: my emphasis). Hirsch thus implies that the "vocabulary" one uses is separate from one's "values," "point of view," or "direction." Like Shaughnessy, he assumes an essence in the individual—a body of values, points of view, a sense of direction—which exists prior to the act of "communication" and outside of the "means of communication" (*Cultural Literacy* 23).

Like Shaughnessy, Hirsch also argues for the need for *everyone* to learn the "literate" language by presenting it as existing "beyond the narrow spheres of family, neighborhood, and region" (*Cultural Literacy* 21). Furthermore, he assumes that there can be only one cause of one's failure to gain "literacy": one's unfamiliarity with "the background information and the linguistic conventions that are needed to read, write, speak effectively" in America (*Cultural Literacy* 22, "Primal Scene" 31). Thus, Hirsch also denies the students' need to deal with cultural differences and to negotiate the competing claims of multiple ways of using language when writing. He thereby both simplifies and depoliticizes the challenges facing the student writer.

Hirsch self-consciously invokes a continuity between Shaughnessy's pedagogy and his "educational reforms" ("Culture and Literacy" 27; *Cultural Literacy* 10). He legitimizes his New Right rhetoric by reminding us that Shaughnessy had approved of his work. For those of us concerned with examining writing in relation to the politics of gender, race, nationality, and class, the best way to forestall Hirsch's use of Shaughnessy is to point out that the continuity resides only in the essentialist view of language underlying both pedagogies and the politics of linguistic innocence it promotes. Critiquing the essentialist view of language and the politics of linguistic innocence in Shaughnessy's work contributes to existing criticism of Hirsch's New Right rhetoric (see Armstrong, Bizzell, Moglen, Scholes, and Sledd). It makes clear that if, as Hirsch self-consciously maintains, there is a continuity between Shaughnessy's work and Hirsch's ("Culture and Literacy" 27; *Cultural Literacy* 10); the continuity resides only in the most limiting aspect of Shaughnessy's pedagogy. Recognition of some of the limitations of Shaughnessy's pedagogy can also be politically constructive for the field of composition by helping us appreciate Shaughnessy's legacy. Most of the lessons she taught us in *Errors and Expectations,* such as students' "quasi-foreign relationship" with academic discourse, their lack of confidence as learners and writers, their desire to participate in academic work, and their intelligence and language-learning aptitudes, continue to be central to the teaching of basic writing. The tasks she delineates for us remain urgent for those of us concerned with the politics of the teaching of writing. Recognizing the negative effects that an essentialist view of language have on Shaughnessy's own efforts to execute these tasks can only help us identify issues that need to be addressed if we are to carry on her legacy: a fuller recognition of the social dimensions of students' linguistic decisions.[5]

Works Cited

Allen, Michael. "Writing Away from Fear: Mina Shaughnessy and the Uses of Authority." *College English* 41 (1980): 857-867.

Armstrong, Paul B. "Pluralistic Literacy." *Profession* 88: 29-32.

Bartholomae, David. "Released into Language: Errors, Expectations, and the Legacy of Mina Shaughnessy." *The Territory of Language: Linguistics, Stylistics, and the Teaching of Composition.* Ed. Donald A. McQuade. Carbondale, IL: U of Southern Illinois P, 1986. 65-88.

Bizzell, Patricia. "Arguing About Literacy." *College English* 50 (1988): 141-153.

Graff, Gerald. "The Politics of Composition: A Reply to John Rouse." *College English* 41 (1980): 851-856.

Hirsch, E. D., Jr. *Cultural Literacy: What Every American Needs to Know.* Boston: Houghton, 1987.

———. "Culture and Literacy." *Journal of Basic Writing* 3.1 (Fall/Winter 1980): 27-35.

———. "The Primal Scene of Education." *New York Review of Books* 2 Mar. 1989: 29-35.

Lawlor, William. "The Politics of Rouse." *College English* 42 (1980): 195-199.

Lyons, Robert. "Mina Shaughnessy." *Traditions of Inquiry.* Ed. John Brereton. New York: Oxford UP, 1985, 171-189.

Moglen, Helene. "Allan Bloom and E. D. Hirsch: Educational Reform as Tragedy and Farce." *Profession* 88: 59-64.

5. Material from this essay is drawn from my dissertation, directed by David Bartholomae at the University of Pittsburgh. I would like to thank my teachers and colleagues at the University of Pittsburgh and Drake University, especially David Bartholomae and Joseph Harris, for their criticism and support. I want to acknowledge particularly Bruce Horner's contributions to the conception and revisions of this essay.

Pratt, Mary Louise. ''Linguistic Utopias.'' *The Linguistics of Writing: Arguments Between Language and Literature*. Ed. Nigel Fabb, Derek Attridge, Alan Durant, and Colin MacCabe. New York: Methuen, 1987. 48-66.

Rodriguez, Richard. *Hunger of Memory*. New York: Bantam, 1982.

Rouse, John. ''The Politics of Composition.'' *College English* 41 (1979): 1-12.

Scholes, Robert. ''Three Views of Education: Nostalgia, History, and Voodoo.'' *College English* 50 (1988): 323-332.

Shaughnessy, Mina. *Errors and Expectations: A Guide for the Teacher of Basic Writing*. New York: Oxford UP, 1977.

———. ''Diving In: An Introduction to Basic Writing.'' *College Composition and Communication* 27 (1976): 234-239.

Sledd, Andrew, and James Sledd. ''Hirsch's Use of His Sources in Cultural Literacy: A Critique.'' *Profession* 88: 33-39.

David Bartholomae

The Study of Error

It is curious, I think, that with all the current interest in "Basic Writing," little attention has been paid to the most basic question: What is it? What is "basic writing," that is, if the term is to refer to a phenomenon, an activity, something a writer does or has done, rather than to a course of instruction? We know that across the country students take tests of one sort or another and are placed in courses that bear the title, "Basic Writing." But all we know is that there are students taking courses. We know little about their performance as writers, beyond the bald fact that they fail to do what other, conventionally successful, writers do. We don't, then, have an adequate description of the variety of writing we call "basic."

On the other hand, we have considerable knowledge of what Basic Writing courses are like around the country, the texts that are used, the approaches taken. For some time now, "specialists" have been devising and refining the technology of basic or developmental instruction. But these technicians are tinkering with pedagogies based on what? At best on models of how successful writers write. At worst, on old text-book models that disregard what writers actually do or how they could be said to learn, and break writing conveniently into constituent skills like "word power," "sentence power," and "paragraph power." Neither pedagogy is built on the results of any systematic inquiry into what basic writers do when they write or into the way writing skills develop for beginning adult writers. Such basic research has barely begun. Mina Shaughnessy argued the case this way:

> Those pedagogies that served the profession for years seem no longer appropriate to large numbers of students, and their inappropriateness lies largely in the fact that many of our students . . . are adult beginners and depend as students did not depend in the past upon the classroom and the teacher for the acquisition of the skill of writing.

If the profession is going to accept responsibility for teaching this kind of student, she concludes, "We are committed to research of a very ambitious sort."[1]

Where might such research begin, and how might it proceed? We must begin by studying basic writing—the phenomenon, not the course of instruction. If we begin here, we will recognize at once that "basic" does not mean simple or childlike. These

From *College Composition and Communication* 31 (October 1980): 253-69.

1. Mina Shaughnessy, "Some Needed Research on Writing," *CCC,* 28 (December, 1977), 317, 388.

are beginning writers, to be sure, but they are not writers who need to learn to use language. They are writers who need to learn to command a particular variety of language—the language of a written, academic discourse—and a particular variety of language use—writing itself. The writing of a basic writer can be shown to be an approximation of conventional written discourse; it is a peculiar and idiosyncratic version of a highly conventional type, but the relation between the approximate and the conventional forms is not the same as the relation between the writing, say, of a 7th grader and the writing of a college freshman.

Basic writing, I want to argue, is a variety of writing, not writing with fewer parts or more rudimentary constituents. It is not evidence of arrested cognitive development, arrested language development, or unruly or unpredictable language use. The writer of this sentence, for example, could not be said to be writing an "immature" sentence, in any sense of the term, if we grant her credit for the sentence she intended to write:

> The time of my life when I learned something, and which resulted in a change in which I look upon life things. This would be the period of my life when I graduated from Elementary school to High school.

When we have used conventional T-analysis, and included in our tabulations figures on words/clause, words/T-unit and clauses/T-unit that were drawn from "intended T-units" as well as actual T-units, we have found that basic writers do not, in general, write "immature" sentences. They are not, that is, 13th graders writing 7th grade sentences. In fact, they often attempt syntax whose surface is more complex than that of more successful freshman writers. They get into trouble by getting in over their heads, not only attempting to do more than they can, but imagining as their target a syntax that is *more* complex than convention requires. The failed sentences, then, could be taken as stages of learning rather than the failure to learn, but also as evidence that these writers are using writing as an occasion to learn.

It is possible to extend the concept of "intentional structures" to the analysis of complete essays in order to determine the "grammar" that governs the idiosyncratic discourse of writers imagining the language and conventions of academic discourse in unconventional ways. This method of analysis is certainly available to English teachers, since it requires a form of close reading, paying attention to the language of a text in order to determine not only what a writer says, but how he locates and articulates meaning. When a basic writer violates our expectations, however, there is a tendency to dismiss the text as non-writing, as meaningless or imperfect writing. We have not read as we have been trained to read, with a particular interest in the way an individual style confronts and violates convention. We have read, rather, as policemen, examiners, gate-keepers. The teacher who is unable to make sense out of a seemingly bizarre piece of student writing is often the same teacher who can give an elaborate explanation of the "meaning" of a story by Donald Barthelme or a poem by e. e. cummings. If we learn to treat the language of basic writing *as* language and assume, as we do when writers violate our expectations in more conventional ways, that the unconventional features in the writing are evidence of intention and that they are, therefore, meaningful, then we can chart systematic choices, individual strategies, and characteristic processes of thought. One can read Mina Shaughnessy's *Errors and Expectations* as the record of just such a close reading.[2]

There is a style, then, to the apparently bizarre and incoherent writing of a basic writer because it is, finally, evidence of an individual using language to make and transcribe meaning. This is one of the axioms of error analysis, whether it be applied to reading (as in "miscue analysis"), writing, or second-language learning. An error (and I would include errors beyond those in the decoding or encoding of sentences) can only be understood as evidence of intention. They are the only evidence we have of an individual's idiosyncratic way of using the language and articulating meaning, of imposing a style on common material. A writer's activity is linguistic and rhetorical activity; it can be different but never random. The task for both teacher and researcher, then, is to discover the grammar of *that* coherence, of the "idiosyncratic dialect" that belongs to a particular writer at a particular moment in the history of his attempts to imagine and reproduce the standard idiom of academic discourse.[3]

All writing, of course, could be said to only approximate conventional discourse; our writing is never either completely predictable or completely idiosyncratic. We speak our own language as well as the language of the tribe and, in doing so, make concessions to both ourselves and our culture. The distance between text and conventional expectation may be a sign of failure and it may be a sign of genius, depending on the level of control and intent we are willing to assign to the writer, and depending on the insight we acquire from seeing convention so transformed. For a basic writer the distance between text and convention is greater than it is for the run-of-the-mill freshman writer. It may be, however, that the more talented the freshman writer becomes, the more able she is to increase again the distance between text and convention. We are drawn to conclude that basic writers lack control, although it may be more precise to say that they lack choice and option, the power to make decisions about the idiosyncracy of their writing. Their writing is not, however, truly uncontrolled. About the actual distance from text to convention for the basic writer, we know very little. We know that it will take a long time to traverse—generally the greater the distance the greater the time and energy required to close the gap. We know almost nothing about the actual sequence of development—the natural sequence of learning—that moves a writer from basic writing to competent writing to good writing. The point, however, is that "basic writing" is something our students *do* or *produce;* it is not a kind of writing we teach to backward or unprepared students. We should not spend our time imagining simple or "basic" writing tasks, but studying the errors that emerge when beginning writers are faced with complex tasks.

The mode of analysis that seems most promising for the research we need on the writer's sequence of learning is error analysis. Error analysis provides the basic writing teacher with both a technique for analyzing errors in the production of discourse, a technique developed by linguists to study second language learning, and a theory of error, or, perhaps more properly, a perspective on error, where errors are seen as

---

2. Mina Shaughnessy, *Errors and Expectations: A Guide for the Teacher of Basic Writing* (New York: Oxford University Press, 1977).

3. The term "idiosyncratic dialect" is taken from S. P. Corder, "Idiosyncratic Dialects and Error Analysis," in Jack C. Richards, ed., *Error Analysis: Perspectives on Second Language Acquisition* (London: Longman, 1974), pp. 158-71.

(1) necessary stages of individual development and (2) data that provide insight into the idiosyncratic strategies of a particular language user at a particular point in his acquisition of a target language. Enough has been written lately about error analysis that I'll only give a brief summary of its perspective on second language or second dialect acquisition.[4] I want to go on to look closely at error analysis as a method, in order to point out its strengths and limits as a procedure for textual analysis.

George Steiner has argued that all acts of interpretation are acts of translation and are, therefore, subject to the constraints governing the passage from one language to another.[5] All our utterances are approximations, attempts to use the language of, say, Frank Kermode or the language, perhaps, of our other, smarter, wittier self. In this sense, the analogy that links developmental composition instruction with second language learning can be a useful one—useful that is, if the mode of learning (whatever the "second" language) is writing rather than speaking. (This distinction, I might add, is not generally made in the literature on error analysis, where writing and speech are taken as equivalent phenomena.) Error analysis begins with the recognition that errors, or the points where the actual text varies from a hypothetical "standard" text, will be either random or systematic. If they are systematic in the writing of an individual writer, then they are evidence of some idiosyncratic rule system—an idiosyncratic grammar or rhetoric, an "interlanguage" or "approximative system."[6] If the errors are systematic across all basic writers, then they would be evidence of generalized stages in the acquisition of fluent writing for beginning adult writers. This distinction between individual and general systems is an important one for both teaching and research. It is not one that Shaughnessy makes. We don't know whether the categories of error in *Errors and Expectations* hold across a group, and, if so, with what frequency and across a group of what size.

Shaughnessy did find, however, predictable patterns in the errors in the essays she studied. She demonstrated that even the most apparently incoherent writing, if we are sensitive to its intentional structure, is evidence of systematic, coherent, rule-governed behavior. Basic writers, she demonstrated, are not performing mechanically or randomly but making choices and forming strategies as they struggle to deal with the varied demands of a task, a language, and a rhetoric. The "systems" such writing exhibits provide evidence that basic writers *are* competent, mature language users. Their attempts at producing written language are not hit and miss, nor are they evidence of simple translation of speech into print. The approximate systems they produce are evidence that they can conceive of and manipulate written language as a structured, systematic code. They are "intermediate" systems in that they mark stages on route to mastery (or, more properly, on route to conventional fluency) of written, academic discourse.

This also, however, requires some qualification. They *may* be evidence of some tran-

---

4. Barry M. Kroll and John C. Schafer, "Error Analysis and the Teaching of Composition," *CCC*, 29 (October, 1978), 243-48. See also my review of *Errors and Expectations* in Donald McQuade, ed., *Linguistics, Stylistics and the Teaching of Composition* (Akron, Ohio: L & S Books, 1979), pp. 209-20.

5. George Steiner, *After Babel: Aspects of Language and Translation* (New York: Oxford University Press, 1975).

6. For the term "interlanguage," see L. Selinker, "Interlanguage," in Richards, ed., *Error Analysis*, pp. 31-55. For "approximate system," see William Nemser, "Approximate Systems of Foreign Language Learners," in Richards, ed., *Error Analysis*, pp. 55-64. These are more appropriate terms than "idiosyncratic dialect" for the study of error in written composition.

sitional stage. They may also, to use Selinker's term, be evidence of "stabilized variability," where a writer is stuck or searching rather than moving on toward more complete approximation of the target language.[7] A writer will stick with some intermediate system if he is convinced that the language he uses "works," or if he is unable to see errors *as* errors and form alternate hypotheses in response.

Error analysis begins with a theory of writing, a theory of language production and language development, that allows us to see errors as evidence of choice or strategy among a range of possible choices or strategies. They provide evidence of an individual style of using the language and making it work; they are not a simple record of what a writer failed to do because of incompetence or indifference. Errors, then, are stylistic features, information about *this* writer and *this* language; they are not necessarily "noise" in the system, accidents of composing, or malfunctions in the language process. Consequently, we cannot identify errors without identifying them in context, and the context is not the text, but the activity of composing that presented the erroneous form as a possible solution to the problem of making a meaningful statement. Shaughnessy's taxonomy of error, for example, identifies errors according to their source, not their type. A single type of error could be attributed to a variety of causes. Donald Freeman's research, for example, has shown that, "subject-verb agreement . . . is a host of errors, not one." One of his students analyzed a "large sample of real world sentences and concluded that there are at least eight different kinds, most of which have very little to do with one another."[8]

Error analysis allows us to place error in the context of composing and to interpret and classify systematic errors. The key concept is the concept of an "interlanguage" or an "intermediate system," an idiosyncratic grammar and rhetoric that is a writer's approximation of the standard idiom. Errors, while they can be given more precise classification, fall into three main categories: errors that are evidence of an intermediate system; errors that could truly be said to be accidents, or slips of the pen as a writer's mind rushes ahead faster than his hand; and, finally, errors of language transfer, or, more commonly, dialect interference, where in the attempt to produce the target language, the writer intrudes forms from the "first" or "native" language rather than inventing some intermediate form. For writers, this intrusion most often comes from a spoken dialect. The error analyst is primarily concerned, however, with errors that are evidence of some intermediate system. This kind of error occurs because the writer *is* an active, competent language user who uses his knowledge that language is rule-governed, and who uses his ability to predict and form analogies, to construct hypothe-

---

7. The term "stabilized variability" is quoted in Andrew D. Cohen and Margaret Robbins, "Toward Assessing Interlanguage Performance: The Relationship Between Selected Errors, Learner's Characteristics and Learner's Explanations," *Language Learning,* 26 (June, 1976), p. 59. Selinker uses the term "fossilization" to refer to single errors that recur across time, so that the interlanguage form is not evidence of a transitional stage. (See Selinker, "Interlanguage.") M. P. Jain distinguishes between "systematic," "asystematic" and "nonsystematic" errors. (See "Error Analysis: Source, Cause and Significance" in Richards, ed., *Error Analysis,* pp. 189-215.) Unsystematic errors are mistakes, "slips of the tongue." Systematic errors "seem to establish that in certain areas of language use the learner possesses construction rules." Asystematic errors lead one to the "inescapable conclusion" that "the learner's capacity to generalize must improve, for progress in learning a language is made by adopting generalizations and stretching them to match the facts of the language."

8. Donald C. Freeman, "Linguistics and Error Analysis: On Agency," in Donald McQuade, ed., *Linguistics, Stylistics and The Teaching of Composition* (Akron, Ohio: L & S Books, 1979), pp. 143-44.

ses that can make an irregular or unfamiliar language more manageable. The problem comes when the rule is incorrect or, more properly, when it is idiosyncratic, belonging only to the language of this writer. There is evidence of an idiosyncratic system, for example, when a student adds inflectional endings to infinitives, as in this sentence, "There was plenty the boy had to *learned* about birds." It also seems to be evident in a sentence like this: "This assignment calls on *choosing* one of my papers and making a last draft out of it." These errors can be further sub-divided into those that are in flux and mark a fully transitional stage, and those that, for one reason or another, become frozen and recur across time.

Kroll and Schafer, in a recent *CCC* article, argue that the value of error analysis for the composition teacher is the perspective it offers on the learner, since it allows us to see errors "as clues to inner processes, as windows into the mind."[9] If we investigate the pattern of error in the performance of an individual writer, we can better understand the nature of those errors and the way they "fit" in an individual writer's program for writing. As a consequence, rather than impose an inappropriate or even misleading syllabus on a learner, we can plan instruction to assist a writer's internal syllabus. If, for example, a writer puts standard inflections on irregular verbs or on verbs that are used in verbals (as in "I used to runned"), drill on verb endings will only reinforce the rule that, because the writer is over-generalizing, is the source of the error in the first place. By charting and analyzing a writer's errors, we can begin in our instruction with what a writer *does* rather than with what he fails to do. It makes no sense, for example, to impose lessons on the sentence on a student whose problems with syntax can be understood in more precise terms. It makes no sense to teach spelling to an individual who has trouble principally with words that contain vowel clusters. Error analysis, then, is a method of diagnosis.

Error analysis can assist instruction at another level. By having students share in the process of investigating and interpreting the patterns of error in their writing, we can help them begin to see those errors as evidence of hypotheses or strategies they have formed and, as a consequence, put them in a position to change, experiment, imagine other strategies. Studying their own writing puts students in a position to see themselves as language users, rather than as victims of a language that uses them.

This, then, is the perspective and the technique of error analysis. To interpret a student paper without this frame of reference is to misread, as for example when a teacher sees an incorrect verb form and concludes that the student doesn't understand the rules for indicating tense or number. I want, now, to examine error analysis as a procedure for the study of errors in written composition. It presents two problems. The first can be traced to the fact that error analysis was developed for studying errors in spoken performance.[10] It can be transferred to writing only to the degree that writing is like speech, and there are significant points of difference. It is generally acknowledged, for example, that written discourse is not just speech written down on paper. Adult written

9. Kroll and Schafer, "Error Analysis and the Teaching of Composition."

10. In the late 60's and early 70's, linguists began to study second language acquisition by systematically studying the actual performance of individual learners. What they studied, however, was the language a learner would speak. In the literature of error analysis, the reception and production of language is generally defined as the learner's ability to hear, learn, imitate, and independently produce *sounds*. Errors, then, are phonological substitutions, alterations, additions, and subtractions. Similarly, errors diagnosed as rooted in

discourse has a grammar and rhetoric that is different from speech. And clearly the activity of producing language is different for a writer than it is for a speaker.

The "second language" a basic writer must learn to master is formal, written discourse, a discourse whose lexicon, grammar, and rhetoric are learned not through speaking and listening but through reading and writing. The process of acquisition is visual not aural. Furthermore, basic writers do not necessarily produce writing by translating speech into print (the way children learning to write would); that is, they must draw on a memory for graphemes rather than phonemes. This is a different order of memory and production from that used in speech and gives rise to errors unique to writing.

Writing also, however, presents "interference" of a type never found in speech. Errors in writing may be caused by interference from the act of writing itself, from the difficulty of moving a pen across the page quickly enough to keep up with the words in the writer's mind, or from the difficulty of recalling and producing the conventions that are necessary for producing print rather than speech, conventions of spelling, orthography, punctuation, capitalization and so on. This is not, however, just a way of saying that writers make spelling errors and speakers do not. As Shaughnessy pointed out, errors of syntax can be traced to the gyrations of a writer trying to avoid a word that her sentence has led her to, but that she knows she cannot spell.

The second problem in applying error analysis to the composition classroom arises from special properties in the taxonomy of errors we chart in student writing. Listing varieties of errors is not like listing varieties of rocks or butterflies. What a reader finds depends to a large degree on her assumptions about the writer's intention. Any systematic attempt to chart a learner's errors is clouded by the difficulty of assigning intention through textual analysis. The analyst begins, then, by interpreting a text, not by describing features on a page. And interpretation is less than a precise science.

Let me turn to an example. This is part of a paper that a student, John, wrote in response to an assignment that asked him to go back to some papers he had written on significant moments in his life in order to write a paper that considered the general question of the way people change:

> This assignment call on chosing one of my incident making a last draft out of it. I found this very differcult because I like them all but you said I had to pick one so the Second incident was decide. Because this one had the most important insight to my life that I indeed learn from. This insight explain why adulthood mean that much as it dose to me because I think it alway influence me to change and my outlook on certain thing like my point-of-view I have one day and it might change the next week on the same issue. So in these frew words I going to write about the incident now. My exprience took place in my high school and the reason was out side of school but I will show you the connection. The sitution took place cause of the type of school I went too. Let me tell you about the sitution first of all what happen was that I got

---

the mode of production (rather than, for example, in an idiosyncratic grammar or interference from the first language) are errors caused by the difficulty a learner has hearing or making foreign sounds. When we are studying written composition, we are studying a different mode of production, where a learner must see, remember, and produce marks on a page. There may be some similarity between the grammar-based errors in the two modes, speech and writing (it would be interesting to know to what degree this is true), but there should be marked differences in the nature and frequency of performance-based errors.

> suspense from school. For thing that I fell was out of my control sometime, but it taught me alot about respondability of a growing man. The school suspense me for being late ten time. I had accummate ten dementic and had to bring my mother to school to talk to a conselor and Prinpicable of the school what when on at the meet took me out mentally period.

One could imagine a variety of responses to this. The first would be to form the wholesale conclusion that John can't write and to send him off to a workbook. Once he had learned how to write correct sentences, then he could go on to the business of actually writing. Let me call this the "old style" response to error. A second response, which I'll call the "investigative approach," would be to chart the patterns of error in this particular text. Of the approximately 40 errors in the first 200 words, the majority fall under four fairly specific categories: verb endings, noun plurals, syntax, and spelling. The value to pedagogy is obvious. One is no longer teaching a student to "write" but to deal with a limited number of very specific kinds of errors, each of which would suggest its own appropriate response. Furthermore, it is possible to refine the categories and to speculate on and organize them according to cause. The verb errors almost all involve "s" or "ed" endings, which could indicate dialect interference or a failure to learn the rules for indicating tense and number. It is possible to be even more precise. The passage contains 41 verbs; only 17 of them are used incorrectly. With the exception of four spelling errors, the errors are all errors of inflection and, furthermore, these errors come only with regular verbs. There are no errors with irregular verbs. This would suggest, then, that when John draws on memory for a verb form, he gets it right; but when John applies a rule to determine the ending, he gets it wrong.

The errors of syntax could be divided into those that might be called punctuation errors (or errors that indicate a difficulty perceiving the boundaries of the sentence), such as

> Let me tell you about the sitution first of all what happen was that I got suspense from school. For thing that I fell was out of my control sometime, but it taught me alot about respondability of a growing man.

and errors of syntax that would fall under Shaughnessy's category of consolidation errors,

> This insight explain why adulthood mean that much as it dose to me because I think it alway influence me to change and my outlook on certain thing like my point-of-view I have one day and it might change the next week on the same issue.

One would also want to note the difference between consistent errors, the substitution of "sitution" for "situation" or "suspense" for "suspended," and unstable ones, as, for example, when John writes "cause" in one place and "because" in another. In one case John could be said to have fixed on a rule; in the other he is searching for one. One would also want to distinguish between what might seem to be "accidental" errors, like substituting "frew" for "few" or "when" for "went," errors that might best be addressed by teaching a student to edit, and those whose causes are deeper and require time and experience, or some specific instructional strategy.

I'm not sure, however, that this analysis provides an accurate representation of John's writing. Consider what happens when John reads this paper out loud. I've been taping students reading their own papers, and I've developed a system of notation, like that used in miscue analysis,[11] that will allow me to record the points of variation between the writing that is on the page and the writing that is spoken, or, to use the terminology of miscue analysis, between the expected response (ER) and the observed response (OR). What I've found is that students will often, or in predictable instances, substitute correct forms for the incorrect forms on the page, even though they are generally unaware that such a substitution was made. This observation suggests the limits of conventional error analysis for the study of error in written composition.

I asked John to read his paper out loud, and to stop and correct or note any mistakes he found. Let me try to reproduce the transcript of that reading. I will underline any substitution or correction and offer some comments in parentheses. The reader might first go back and review the original. Here is what John read:

> This assignment calls on *choosing* one of my incident making a last draft out of it. I found this very diff*i*cult because I like them all but you said I *had* to pick one so the Second incident was decide*d on*. Because (John goes back and rereads, connecting up the subordinate clause.) So the second incident was decided on because this one had the most important insight to my life that I indeed learn*ed* from. This insight explains why adulthood *meant* that much as it dose to me because I think it always influence*s* me to change and my outlook on certain thing*s* like my point-of-view I have one day and it might change the next week on the same issue. (John goes back and rereads, beginning with "like my point of view," and he is puzzled but he makes no additional changes.) So in these *few* words *I'm* going to write about the incident now. My exp*e*rience took place *be*cause of the type of school I went to (John had written "too.") Let me tell you about the situation (John comes to a full stop.) first of all what happen*ed* was that I got *suspended* from school (no full stop) for thing*s* that I *felt* was out of my control sometime, but it taught me a lot about *responsibility* of a growing man. The school *suspended* me for being late ten time*s*. I had *accumulated* (for "accumate") ten *demerits* (for "dementic") and had to bring my mother to school to talk to a counselor and *the Principal* of the school (full stop) what *went* on at the meet*ing* took me out mentally (full stop) period (with brio).

I have chosen an extreme case to make my point, but what one sees here is the writer correcting almost every error as he reads the paper, even though he is not able to recognize that there *are* errors or that he has corrected them. The only errors John spotted (where he stopped, noted an error and corrected it) were the misspellings of "situation" and "Principal," and the substitution of "chosing" for "choosing." Even when he was asked to reread sentences to see if he could notice any difference between what he was saying and the words on the page, he could not. He could not, for example, see the error in "frew" or "dementic" or any of the other verb errors, and yet he spoke the correct form of every verb (with the exception of "was" after he had changed "thing" to "things" in "for thing*s* that I *felt* was out of my control") and he corrected every plural. His phrasing as he read produced correct syntax, except in the case of the

11. See Y. M. Goodman and C. L. Burke, *Reading Miscue Inventory: Procedure for Diagnosis and Evaluation* (New York: Macmillan, 1972).

consolidation error, which he puzzled over but did not correct. It's important to note, however, that John did not read that confused syntax as if no confusion were there. He sensed the difference between the phrasing called for by the meaning of the sentence and that which existed on the page. He did not read as though meaning didn't matter or as though the "meaning" coded on the page was complete. His problem cannot be simply a syntax problem, since the jumble is bound up with his struggle to articulate this particular meaning. And it is not simply a "thinking" problem—John doesn't write this way because he thinks this way—since he perceives that the statement as it is written is other than that which he intended.

When I asked John why the paper (which went on for two more pages) was written all as one paragraph, he replied, "It was all one idea. I didn't want to have to start all over again. I had a good idea and I didn't want to give it up." John doesn't need to be "taught" the paragraph, at least not as the paragraph is traditionally taught. His prose is orderly and proceeds through blocks of discourse. He tells the story of his experience at the school and concludes that through his experience he realized that he must accept responsibility for his tardiness, even though the tardiness was not his fault but the fault of the Philadelphia subway system. He concludes that with this realization he learned "the responsibility of a growing man." Furthermore John knows that the print code carries certain conventions for ordering and presenting discourse. His translation of the notion that "a paragraph develops a single idea" is peculiar but not illogical.

It could also be argued that John does not need to be "taught" to produce correct verb forms, or, again, at least not as such things are conventionally taught. Fifteen weeks of drill on verb endings might raise his test scores but they would not change the way he writes. He *knows* how to produce correct endings. He demonstrated that when he read, since he was reading in terms of his grammatical competence. His problem is a problem of performance, or fluency, not of competence. There is certainly no evidence that the verb errors are due to interference from his spoken language. And if the errors could be traced to some intermediate system, the system exists only in John's performance as a writer. It does not operate when he reads or, for that matter, when he speaks, if his oral reconstruction of his own text can be taken as a record of John "speaking" the idiom of academic discourse.[12]

John's case also highlights the tremendous difficulty such a student has with editing, where a failure to correct a paper is not evidence of laziness or inattention or a failure

---

12. Bruder and Hayden noticed a similar phenomenon. They assigned a group of students exercises in writing formal and informal dialogues. One student's informal dialogue contained the following:

> What going on?
> It been a long time . . .
> I about through . . .
> I be glad . . .

When the student read the dialogue aloud, however, these were spoken as

> What's going on?
> It's been a long time . . .
> I'm about through . . .
> I'll be glad . . .

See Mary Newton Bruder and Luddy Hayden, "Teaching Composition: A Report on a Bidialectal Approach," *Language Learning*, 23 (June, 1973), 1-15.

to know correct forms, but evidence of the tremendous difficulty such a student has objectifying language and seeing it as black and white marks on the page, where things can be wrong even though the meaning seems right.[13] One of the hardest errors for John to spot, after all my coaching, was the substitution of "frew" for "few," certainly not an error that calls into question John's competence as a writer. I can call this a "performance" error, but that term doesn't suggest the constraints on performance in writing. This is an important area for further study. Surely one constraint is the difficulty of moving the hand fast enough to translate meaning into print. The burden imposed on their patience and short term memory by the slow, awkward handwriting of many inexperienced writers is a very real one. But I think the constraints extend beyond the difficulty of forming words quickly with pen or pencil.

One of the most interesting results of the comparison of the spoken and written versions of John's text is his inability to *see* the difference between "frew" and "few" or "dementic" and "demerit." What this suggests is that John reads and writes from the "top down" rather than the "bottom up," to use a distinction made by cognitive psychologists in their study of reading.[14] John is not operating through the lower level process of translating orthographic information into sounds and sounds into meaning when he reads. And conversely, he is not working from meaning to sound to word when he is writing. He is, rather, retrieving lexical items directly, through a "higher level" process that by-passes the "lower level" operation of phonetic translation. When I put *frew* and *few* on the blackboard, John read them both as "few." The lexical item "few" is represented for John by either orthographic array. He is not, then, reading or writing phonetically, which is a sign, from one perspective, of a high level of fluency, since the activity is automatic and not mediated by the more primitive operation of translating speech into print or print into speech. When John was writing, he did not produce "frew" or "dementic" by searching for sound/letter correspondences. He drew directly upon his memory for the look and shape of those words; he was working from the top down rather than the bottom up. He went to stored print forms and did not take the slower route of translating speech into writing.

John, then, has reached a stage of fluency in writing where he directly and consistently retrieves print forms, like "dementic," that are meaningful to him, even though they are idiosyncratic. I'm not sure what all the implications of this might be, but we surely must see John's problem in a new light, since his problem can, in a sense, be attributed to his skill. To ask John to slow down his writing and sound out words would

13. See Patricia Laurence, "Error's Endless Train: Why Students Don't Perceive Errors," *Journal of Basic Writing*, 1 (Spring, 1975), 23-43, for a different explanation of this phenomenon.

14. See, for example, J. R. Frederiksen, "Component Skills in Reading" in R. R. Snow, P. A. Federico, and W. E. Montague, eds., *Aptitude, Learning, and Instruction* (Hillsdale, N.J.: Erlbaum, 1979); D. E. Rumelhart, "Toward an Interactive Model of Reading," in S. Dornic, ed., *Attention and Performance VI* (Hillsdale, N.J.: Erlbaum, 1977); and Joseph H. Denks and Gregory O. Hill, "Interactive Models of Lexical Assessment during Oral Reading," paper presented at Conference on Interactive Processes in Reading, Learning Research and Development Center, University of Pittsburgh, September 1979.

Patrick Hartwell argued that "apparent dialect interference in writing reveals partial or imperfect mastery of a neural coding system that underlies both reading and writing" in a paper, " 'Dialect Interference' in Writing: A Critical View," presented at CCCC, April 1979. This paper is available through ERIC. He predicts, in this paper, that "basic writing students, when asked to read their writing in a formal situation, . . . will make fewer errors in their reading than in their writing." I read Professor Hartwell's paper after this essay was completed, so I was unable to acknowledge his study as completely as I would have desired.

be disastrous. Perhaps the most we can do is to teach John the slowed down form of reading he will need in order to edit.

John's paper also calls into question our ability to identify accidental errors. I suspect that when John substitutes a word like "when" for "went," this is an accidental error, a slip of the pen. Since John spoke "wen*t*" when he read, I cannot conclude that he substituted "when" for "went" because he pronounces both as "wen." This, then, is not an error of dialect interference but an accidental error, the same order of error as the omission of "the" before "Principal." Both were errors John corrected while reading (even though he didn't identify them as errors).

What is surprising is that, with all the difficulty John had identifying errors, he immediately saw that he had written "chosing" rather than "choosing." While textual analysis would have led to the conclusion that he was applying a tense rule to a participial construction, or over-generalizing from a known rule, the ease with which it was identified would lead one to conclude that it was, in fact, a mistake, and not evidence of an approximative system. What would have been diagnosed as a deep error now appears to be only an accidental error, a "mistake" (or perhaps a spelling error).

In summary, this analysis of John's reading produces a healthy respect for the tremendous complexity of transcription, for the process of recording meaning in print as opposed to the process of generating meaning. It also points out the difficulty of charting a learner's "interlanguage" or "intermediate system," since we are working not only with a writer moving between a first and a second language, but a writer whose performance is subject to the interference of transcription, of producing meaning through the print code. We need, in general, to refine our understanding of performance-based errors, and we need to refine our teaching to take into account the high percentage of error in written composition that is rooted in the difficulty of performance rather than in problems of general linguistic competence.

Let me pause for a moment to put what I've said in the context of work in error analysis. Such analysis is textual analysis. It requires the reader to make assumptions about intention on the basis of information in the text. The writer's errors provide the most important information since they provide insight into the idiosyncratic systems the writer has developed. The regular but unconventional features in the writing will reveal the rules and strategies operating for the basic writer.

The basic procedure for such analysis could be outlined this way. First the reader must identify the idiosyncratic construction; he must determine what is an error. This is often difficult, as in the case of fragments, which are conventionally used for effect. Here is an example of a sentence whose syntax could clearly be said to be idiosyncratic:

> In high school you learn alot for example Kindergarten which I took in high school.[15]

The reader, then, must reconstruct that sentence based upon the most reasonable interpretation of the intention in the original, and this must be done *before* the error can be classified, since it will be classified according to its cause.[16] Here is Shaughnessy's reconstruction of the example given above: "In high school you learn a lot. For example, I took up the study of Kindergarten in high school." For any idiosyncratic sentence,

---

15. This example is taken from Shaughnessy, *Errors and Expectations,* p. 52.
16. Corder refers to "reconstructed sentences" in "Idiosyncratic Dialects and Error Analysis."

however, there are often a variety of possible reconstructions, depending on the reader's sense of the larger meaning of which this individual sentence is only a part, but also depending upon the reader's ability to predict how this writer puts sentences together, that is, on an understanding of this individual style. The text is being interpreted, not described. I've had graduate students who have reconstructed the following sentence, for example, in a variety of ways:

> Why do we have womens liberation and their fighting for Equal Rights ect. to be recognized not as a lady but as an Individual.

It could be read, "Why do we have women's liberation and why are they fighting for Equal Rights? In order that women may be recognized not as ladies but as individuals." And, "Why do we have women's liberation and their fight for equal rights, to be recognized not as a lady but as an individual?" There is an extensive literature on the question of interpretation and intention in prose, too extensive for the easy assumption that all a reader has to do is identify what the writer would have written if he wanted to "get it right the first time." The great genius of Shaughnessy's study, in fact, is the remarkable wisdom and sympathy of her interpretations of student texts.

Error analysis, then, involves more than just making lists of the errors in a student essay and looking for patterns to emerge. It begins with the double perspective of text and reconstructed text and seeks to explain the difference between the two on the basis of whatever can be inferred about the meaning of the text and the process of creating it. The reader/researcher brings to bear his general knowledge of how basic writers write, but also whatever is known about the linguistic and rhetorical constraints that govern an individual act of writing. In Shaughnessy's analysis of the "kindergarten" sentence, this discussion is contained in the section on "consolidation errors" in the chapter on "Syntax."[17] The key point, however, is that any such analysis must draw upon extra-textual information as well as close, stylistic analysis.

This paper has illustrated two methods for gathering information about how a text was created. A teacher can interview the student and ask him to explain his error. John wrote this sentence in another paper for my course:

> I would to write about my experience helping 1600 childrens have a happy christmas.

The missing word (I would *like* to write about . . .) he supplied when reading the sentence aloud. It is an accidental error and can be addressed by teaching editing. It is the same kind of error as his earlier substitution of "when" for "went." John used the phrase, "1600 childrens," throughout his paper, however. The conventional interpretation would have it that this is evidence of dialect interference. And yet, when John read the paper out loud, he consistently read "1600 children," even though he said he did not see any difference between the word he spoke and the word that was on the page. When I asked him to explain why he put an "s" on the end of "children," he replied, "Because there were 1600 of them." John had a rule for forming plurals that he used when he wrote but not when he spoke. Writing, as he rightly recognized, has its own peculiar rules and constraints. It is different from speech. The error is not due to inter-

17. Shaughnessy, *Errors and Expectations*, pp. 51-72.

ference from his spoken language but to his conception of the "code" of written discourse.

The other method for gathering information is having students read aloud their own writing, and having them provide an oral reconstruction of their written text. What I've presented in my analysis of John's essay is a method for recording the discrepancies between the written and spoken versions of a single text. The record of a writer reading provides a version of the "intended" text that can supplement the teacher's or researcher's own reconstruction and aid in the interpretation of errors, whether they be accidental, interlingual, or due to dialect interference. I had to read John's paper very differently once I had heard him read it.

More importantly, however, this method of analysis can provide access to an additional type of error. This is the error that can be attributed to the physical and conceptual demands of writing rather than speaking; it can be traced to the requirements of manipulating a pen and the requirements of manipulating the print code.[18]

In general, when writers read, and read in order to spot and correct errors, their responses will fall among the following categories:

1. overt corrections—errors a reader sees, acknowledges, and corrects;
2. spoken corrections—errors the writer does not acknowledge but corrects in reading;
3. no recognition—errors that are read as written;
4. overcorrection—correct forms made incorrect, or incorrect forms substituted for incorrect forms;
5. acknowledged error—errors a reader senses but cannot correct;
6. reader miscue—a conventional miscue, not linked to error in the text;
7. nonsense—In this case, the reader reads a non-sentence or a nonsense sentence as though it were correct and meaningful. No error or confusion is acknowledged. This applies to errors of syntax only.

Corrections, whether acknowledged or unacknowledged, would indicate performance-based errors. The other responses (with the exception of "reader miscues") would indicate deeper errors, errors that, when charted, would provide evidence of some idiosyncratic grammar or rhetoric.

John "miscues" by completing or correcting the text that he has written. When reading researchers have readers read out loud, they have them read someone else's writing, of course, and they are primarily concerned with the "quality" of the miscues.[19] All fluent readers will miscue; that is, they will not repeat verbatim the words on the page. Since fluent readers are reading for meaning, they are actively predicting what will come and processing large chunks of graphic information at a time. They do not read

18. For a discussion of the role of the "print code" in writer's errors, see Patrick Hartwell, " 'Dialect Interference' in Writing: A Critical View."

19. See Kenneth S. Goodman, "Miscues: Windows on the Reading Process," in Kenneth S. Goodman, ed., *Miscue Analysis: Applications to Reading Instruction* (Urbana, Illinois: ERIC, 1977), pp. 3-14.

individual words, and they miscue because they speak what they expect to see rather than what is actually on the page. One indication of a reader's proficiency, then, is that the miscues don't destroy the "sense" of the passage. Poor readers will produce miscues that jumble the meaning of a passage, as in

Text: Her wings were folded quietly at her sides.
Reader: Her wings were floated quickly at her sides.

or they will correct miscues that do not affect meaning in any significant way.[20]

The situation is different when a reader reads his own text, since this reader already knows what the passage means and attention is drawn, then, to the representation of that meaning. Reading also frees a writer from the constraints of transcription, which for many basic writers is an awkward, laborious process, putting excessive demands on both patience and short-term memory. John, like any reader, read what he expected to see, but with a low percentage of meaning-related miscues, since the meaning, for him, was set, and with a high percentage of code-related miscues, where a correct form was substituted for an incorrect form.

The value of studying students' oral reconstruction of their written texts is threefold. The first is as a diagnostic tool. I've illustrated in my analysis of John's paper how such a diagnosis might take place.

It is also a means of instruction. By having John read aloud and, at the same time, look for discrepancies between what he spoke and what was on the page, I was teaching him a form of reading. The most dramatic change in John's performance over the term was in the number of errors he could spot and correct while re-reading. This far exceeded the number of errors he was able to eliminate from his first drafts. I could teach John an editing procedure better than I could teach him to be correct at the point of transcription.

The third consequence of this form of analysis, or of conventional error analysis, has yet to be demonstrated, but the suggestions for research are clear. It seems evident that we can chart stages of growth in individual basic writers. The pressing question is whether we can chart a sequence of "natural" development for the class of writers we call basic writers. If all non-fluent adult writers proceed through a "natural" learning sequence, and if we can identify that sequence through some large, longitudinal study, then we will begin to understand what a basic writing course or text or syllabus might look like. There are studies of adult second language learners that suggest that there is a general, natural sequence of acquisition for adults learning a second language, one that is determined by the psychology of language production and language acquisition.[21] Before we can adapt these methods to a study of basic writers, however, we need to better understand the additional constraints of learning to transcribe and manipulate the "code" of written discourse. John's case illustrates where we might begin and what we must know.[22]

---

20. This example was taken from Yetta M. Goodman, "Miscue Analysis for In-Service Reading Teachers," in K. S. Goodman, ed., *Miscue Analysis,* p. 55.

21. Nathalie Bailey, Carolyn Madden, and Stephen D. Krashen, "Is There a 'Natural Sequence' in Adult Second Language Learning?" *Language Learning,* 24 (June, 1974), 235-43.

22. This paper was originally presented at CCCC, April 1979. The research for this study was funded by a research grant from the National Council of Teachers of English.

Mike Rose

Remedial Writing Courses: A Critique and a Proposal

Many of our attempts to help college remedial[1] writers, attempts that are often well-intentioned and seemingly commonsensical, may, in fact, be ineffective, even counterproductive, for these attempts reduce, fragment, and possibly misrepresent the composing process. I believe we may be limiting growth in writing in five not unrelated ways. 1) Our remedial courses are self-contained; that is, they have little conceptual or practical connection to the larger academic writing environment in which our students find themselves. 2) The writing topics assigned in these courses—while meant to be personally relevant and motivating and, in their simplicity, to assist in the removal of error—in fact might not motivate and might not contribute to the production of a correct academic prose. 3) The writing teacher's vigilance for error most likely conveys to students a very restricted model of the composing process. 4) Our notion of "basic skills" has become so narrow that we attempt to separate the intimately related processes of reading and thinking from writing. 5) In some of our attempts to reform staid curricula we have inadvertently undercut the expressive and exploratory possibilities of academic writing and have perceived fundamental discourse strategies and structures as restricting rather than enhancing the production and comprehension of prose.

At various places in my speculations I will offer potential solutions to the problems I pose. For the most part, these solutions come from programs I run at UCLA, though I should mention that some of these solutions were spawned during my days as a teacher of "developmental" writers in special programs for returning Vietnam veterans, parole aids, and newly released convicts. So, though this paper is primarily addressed to teachers of traditional college "remedial" writers vs. truly "basic" writers of the sort we saw and continue to see during periods of open admissions, many of the ideas I will present grew out of my work with students from both camps. I strongly believe, therefore, that what I am about to say has, with appropriate modification, broad applicability to that large, complex stratum of writers who have been labeled "substandard."

Remedial Courses Are Self-contained

Many remedial courses do not fit conceptually and practically into the larger writing environment in which students find themselves. Much of the writing we have our stu-

From *College English* 45 (February 1983): 109-28.

1. I will use the adjective "remedial" and occasionally the adjective "basic" throughout this essay. I should note, though, that I use them with some reservation, for they are often more pejorative than accurately descriptive.

dents do is distressingly like that Arthur Applebee found in his survey of secondary education—phrase to paragraph length fill-ins or brief responses, often in workbooks or on worksheets (*Writing in the Secondary School* [Urbana, Ill.: National Council of Teachers of English, 1981]). When fuller assignments are given, the topics are most often personal and simple. They are meant to be relevant and accessible but in fact are usually old-hat and unacademic—a unique artifact of the composition classroom. Furthermore, though some courses and programs give the student more than a class period to work on a paper, a number of others (shades of Applebee's report again) limit writing to twenty to fifty-minute in-class sprints.[2] The end result, of course, is that our students' papers are flawed, not only by the writers' current compositional inadequacies but also by the writers' very composing situations. These papers are then returned to their source: composition teachers. But how can we teachers honestly provide an engaged response to spurts of writing on topics like "Describe a favorite place or event" or "Give your opinion of X"—X being some broad, complex social issue about which students are usually ill-informed? Clearly these topics and these situations are not preparing students for their university lives. In all too many cases we have created a writing course that does not lead outward toward the intellectual community that contains it. And that's a pity, for a remedial writing curriculum must fit into the overall context of a university education: students must, early on, begin wrestling with academically oriented topics that help them develop into more critical thinkers, that provide them with some of the tools of the examined life, and that, practically, will assist them in the courses they take. I am sure that many of my colleagues would agree, but establishing such a context is easier said than done.

Let me suggest one way to lay groundwork for a meaningful context, and this is, in fact, the procedure we used in developing UCLA's Freshman Preparatory Program. (I should add that we thought it important to explain our procedure to our students. Freshman writers should know the origins of their curriculum.) Here is what we did: we wanted to find out what our students were being asked to do when they wrote for university classes, so we collected 445 essay and take-home examination questions as well as paper topics from seventeen departments and performed some relatively simple analyses. (We hope eventually to collect more assignments—and the essays resulting from them—from more campuses and conduct a more sophisticated analysis.) Our (rather predictable) findings:

1. We determined what discourse mode the questions and topics seemed to require. Most called for exposition—transaction in James Britton's scheme, reference in James Kinneavy's (*Language and Learning* [Harmondsworth: Penguin, 1970]; *A Theory of Discourse* [New York: Norton, 1980]). The balance required argument (Britton's conative mode, Kinneavy's persuasive), but a special kind of argument I will label academic argument, that is, not a series of emotionally charged appeals and exhortations, as one often finds in oratorical persuasion, but a calculated marshalling of information, a sort of exposition aimed at persuading.[3]

---

2. Many think this fifty-minute limit prepares students for essay exams. But not so. In those situations students already have a wealth of information to spew forth. In the remedial course they are expected to retrieve, give meaning to, and organize information that, in some cases, they simply do not have.

3. For some additional and informative discussion of argument in the academy, see Charles Kneupper, "Teaching Argument: An Introduction to the Toulmin Model." *College Composition and Communication*,

2. It was also obvious that, in their writing, students had to work with large bodies of information garnered from lectures and readings and often had to write from texts. There were simply no assignments calling for the student to narrate or describe personal experiences, to observe relatively immediate objects or events like the architecture of campus buildings, to express a general opinion on something not studied closely, to reflect on self.

3. While in-class essay examinations clearly required a quick, nearly regurgitative—albeit structured—response, the students' other assignments assumed the ability to reflect on a broad range of complex material, to select and order information, and to see and re-see data and events in various contexts.

4. Our surveys also suggested that various academic audiences write and read with an elaborate and—unfortunately for our students—often subtle, even tacit set of philosophical and methodological assumptions that determine what they will consider acceptable or unacceptable reasoning, presenting of evidence, and inferring.[4] For example, an individual's reflections on personal events are considered legitimate evidence in many areas of sociology and anthropology, but are considered much less legitimate by behavioristic psychologists. Developing a sensitivity to the plurality of these assumptive foundations and the conventions that arise from them is crucial, for they shape the complex rhetorical relationship between writer and reader in the academy.

5. In another, related, survey (Mike Rose, "When Faculty Talk About Writing," *College English,* 41 [1979-80], 272-279), we found that some of the stylistic information we had been giving students in English composition simply did not reflect any sort of broad consensus among even as limited an audience as academics. Now, while it was no big surprise to find out that our concerns about, say, writing in the first person and avoiding passive constructions were not fervently shared by other disciplines, it was sobering to hear remarks like the following—this one from a professor of management:

   > Students come to us writing over-academic, highly embroidered prose. We, in turn, have to retrain them to write simple, direct reports for companies, reports that someone will feel like reading. ("When Faculty Talk About Writing," p. 273)

   Felicitous techniques like the Christensen cumulative sentence (which Christensen primarily derived from studying fiction) or the topic sentence seductively placed at the end rather than the beginning of a paragraph, for some readers in some contexts, hampered rather than enhanced or enlivened communication.

Considering the complex discourse demands our surveys revealed, our students and our courses fell pitifully short. We needed a remedial program that slowly but steadily and systematically introduced remedial writers to transactional/expositional academic discourse; that relied on texts and bits of texts, preferably from a variety of disciplines

---

29 (1978), 237-241; Patricia Bizzell, "The Ethos of Academic Discourse," *CCC,* 29 (1978), 351-355; and Paul Bator, "Aristotelian and Rogerian Rhetoric," *CCC,* 31 (1980), 427-432.

4. A good general introduction to this complex assumptive plurality is Philip Phenix's *Realms of Meaning* (New York: McGraw-Hill, 1964).

so that students would learn how to work with data presented in social science exposition as well as with detail from a short story; that created full, rich assignments which, again slowly and systematically, encouraged the student to develop his or her structural, rhetorical, stylistic facility; that alerted students to stylistic/rhetorical variation within the university.

Simple Topics, Motivation, and the Elimination of Error

One of the aspects of traditional remedial curricula that I have questioned is the simple, personal topic. But clearly teachers do not assign topics like "Describe your favorite place" because they hold some deep affection for them. They have reasons. One very popular reason is best expressed thus: "I want to make the topic simple so if the student writes poorly, I'll know it had nothing to do with the strain of a complicated topic." Another popular reason concerns motivation: "If I can give them success experiences, they'll feel better about writing and want to write more." The reasoning behind assigning simple, personal topics, then, often lies in beliefs about cognitive interference and motivation. Let me deal with the issue of motivation first.

Certainly it is a sound motivation and learning principle to begin with the simple—let the student experience success—and then move toward the more complex. No argument. But we should not assume that the successful completion of an assignment the student might well perceive as being simple, even juvenile, is going to make him feel better about himself or his writing. Let me illustrate with a brief anecdote. My colleague at UCLA, George Gadda, is currently attempting to phase into our orientation program a more university-oriented set of diagnostic topics. He still gives a typical "Describe your favorite object" topic, but also gives a topic, with a brief academically oriented reading passage, that requires summarizing or analyzing skills. Though it is seemingly more difficult, a number of students reported via questionnaire that they preferred the academic topic because, as one student succinctly put it. "This is the real thing." Motivation to achieve in writing is much more complex than some composition theorists suppose. Personal topics are not necessarily more relevant than academic ones, and some of our students—particularly those from certain minority cultures—might not feel comfortable revealing highly personal experiences. Current work on achievement motivation has shown it to be a highly complex cognitive-affective phenomenon that includes such dimensions as perception of the difficulty of a task and perception of the role of luck or skill in completing the task.[5] Add to these perceptions psychodynamic variables such as the degree of comfort with the content of a task, and it is no longer clear that simple and personal topics are most motivating.

Now to the issue of cognitive overload and the concentration on error. It makes sense to assume that if we reduce the interfering strain of the challenging topic, or, put another way, if we give students a topic for which they already have information (e.g., "Describe an important person in your life"), they and we can more readily focus on grammatical problems. But this might be a case of common sense misleading us, for we have evidence to suggest that while a writer might eventually produce grammatically correct prose for one kind of assignment, that correctness might not hold when she faces

---

5. Bernard Weiner, "Achievement Motivation, Attribution Theory, and the Educational Process," *Review of Educational Research,* 42 (1972), 203-215; Bernard Weiner, ed., *Achievement Motivation and Attribution Theory* (Morristown, N.J.: General Learning Press, 1974).

other kinds of tasks. Brooke Nielson, for example, found that when her sample of traditional writers shifted registers from the informal (writing to peers) to the formal (writing to an academic audience), their proficiency fell apart ("Writing as a Second Language: Psycholinguistic Processes in Composing," Diss., University of California at San Diego, 1979). I suspect that similar difficulties arise when the student shifts from simpler discourse structures to more complex ones and from simpler to more complex topics. So we might guide a student to the point where she writes with few errors about her dorm room, but when she is asked, say, to compare and contrast two opinions on dormitory housing, not to mention two economic theories, the organizational demands of comparing and contrasting and the more syntactically complicated sentences often attending more complex exposition or argument[6] put such strain on her cognitive resources and linguistic repertoire that error might well reemerge. Error, in short, is not something that, once fixed in a simple and clean environment, will never emerge again. It is not a culture we can isolate and alter in a petri dish. What we must do, therefore, is carefully define and describe the kind of writing demanded of students in the academy (which—lest this suggestion seems mind-shackling—is also the kind of writing students would use to challenge the academy), and then focus on that kind of writing, scaling our assignments down and building slowly, but scaling and building within the same discourse domain. For we cannot assume a simple transfer of skills across broadly different discourse demands.

There is a related issue here. When we think that simple topics coupled with concentration on error will lead to the correction of error, we might be misleading ourselves. As any one who has read large numbers of remedial-level exams knows, there is a nagging doubt that one reason one sees, say, fewer comma splices or misspelled words in post-tests is that the students have simply stopped writing complex sentences or using tricky words. I don't know if the following study has ever been attempted, but I would wager that a careful examination of remedial students' pre- and post-tests would show that pre-tests evince more ambitious lexical to sentential attempts gone awry than would cleaner post-tests. We may be training students to be simple and safe rather than urging them toward the ambitious experimentation that will enhance their linguistic repertoire.

Creating simple topics to aid in the correction of error, then, might be a less successful strategy than we think—error cannot be isolated and removed; it can reemerge whenever a student moves onto a task that challenges him or her in new ways. Furthermore, we might be demoralizing our students by giving them the same kind of topic they have been writing on for so many years. I would suggest that we develop curriculums that offer academically oriented topics, the difficulty of each being systematically gradated so that the student is continually challenged in ways that don't overwhelm. (I will give examples of such assignments later.) Until students begin to develop some familiarity with such topics, we will not be able to help them in their attempts to write a relatively correct university prose.

Error Vigilance and Reductionistic Models of Composing

Our concern for error leads us to create overly simple topics, but I suspect it also results in something even more counterproductive. We might be unwittingly passing on an

6. Marion Crowhurst, "Syntactic Complexity and Teachers' Quality Ratings of Narrations and Arguments," *Research in the Teaching of English,* 14 (1980), 223-231.

extremely constricted notion of what composing is. This occurs on three broad levels: the process, the conceptual, the rhetorical. Many of our students come to see the writing process itself as a matter of framing a thought in correct language. The results of such perception are disastrous. Sondra Perl, for example, noted that the basic writers she studied wrote in halting spurts and produced extremely truncated products. They were, she discovered, so vigilant for error, so concerned with placing every bit of language in its correct place, that their writing processes were stymied—they could not get the flow of their thoughts onto paper (''Five Writers Writing: Case Studies of the Composing Processes of Unskilled College Writers,'' Diss., New York University, 1978). The possibilities in writing—even ''incorrect'' writing—for discovering, connecting, playing were lost. Conceptually, our students come to believe that what counts is not the thought they give to a topic but how correctly that thought is conveyed. The results? Clean but empty papers. Barbara Tomlinson (personal communication, 1982) reports that even though the remedial writers' papers she studied for her doctoral dissertation were relatively error-free, her independent evaluators were ''stunned'' by the vapidity of the contents of the essays. The papers said nothing. On the rhetorical level students may not grow beyond their limited notions of connecting thought to reader, not because they are—to cite one current misapplication of the Piagetian developmental framework—egocentric, but rather because the local to global semantic and syntactic devices that establish that connection have not been opened up to them and, perhaps worse, the social base of making meaning and conveying it has not been established in their writing classroom. Just about the only rhetorical connection the correctness model establishes is the negative sociolinguistic one: don't err lest ye be judged. That is sound advice, but not when it becomes the only rhetorical advice students get.

In closing this section, let me cite a chilling—though admittedly preliminary—study conducted by Patrick Hartwell (''Writers as Readers,'' paper presented at the Conference on College Composition and Communication, March, 1981). He asked elementary, secondary, and college students labeled by their teachers as weaker readers/writers and better readers/writers to respond to the question, ''What do people do when they write?'' Notice the model of writing implied in the typical responses of weaker elementary readers/writers (I corrected the few misspellings in their responses): ''They hold the pencil tightly.'' ''Move their fingers. And write neat.'' Of weaker secondary readers/writers: ''They put the point of the pencil to the paper and start making words and letters.'' Of weaker college readers/writers. ''People write through English grammar, punctuation, etc.'' ''First you pick your topic, then you make sure that you have enough information. Then you rewrite and check the spelling and copy it down.'' The responses of better readers/writers were qualitatively different. A few examples. The primary level: ''They think of what they are going to write. They ask a person if it sounds good.'' The secondary level: ''They get stuff across to other people.'' The college level: ''People explain their ideas, theories, stories and imagination to each other.'' Because these data are preliminary and because Hartwell has not yet shown whether these self-reports are linked to actual writing behaviors, I do not want to make too much of this. Though I must admit that because my own investigations of stymied writers revealed the power of rigid rules and inaccurate assumptions (''The Cognitive Dimension of Writer's Block: An Examination of University Students,'' Diss., UCLA, 1981), I find it hard to take these self-reports lightly. It seems safe to say that the impressions of these selected weaker student writers implies a very limited notion of what composing is, a notion based on simple behaviors, narrow linear steps, and shriveled rhetorical possibilities. This notion might reflect their limited skills (i.e., writers fixate on error because they keep erring), but it also stands as a barrier to their improving. If they continue to conceive of writing as holding a pencil tightly and using correct grammar, how will they grow beyond those constraints?

But let me be quick to point out that I am not trying to lay blame on the remedial writing teacher alone, if at all. For there are powerful reasons to explain why some teachers reduce the process, conceptual, and rhetorical possibilities of composing. The public, spurred by an often misconceived "back to basics" movement and the misinformed, but profitable, arrogance of "pop grammarians" like John Simon and Edwin Newman, make a teacher feel negligent and vulnerable if he or she does not attempt to clear up error. Furthermore, as Patricia Laurence points out in "Error's Endless Train: Why Students Don't Perceive Errors" (*Journal of Basic Writing* [Spring, 1975], 23-42), our scholars have not provided us with a comprehensive theory of error—a rich perceptual/cognitive/linguistic framework that will enable us to study error, see patterns in our students' errors, and provide guidelines on how to assist most effectively the student in understanding and remedying them. (And, I suspect, such a theory would also tell us when to ignore error.) Thus there is little for the conscientious teacher to do but keep marking. To do less in the absence of any other guidelines seems like shirking responsibility.

And there is a third reason for excusing us teachers. It seems to me that, in general, we have been offered pretty limited definitions of "writing skills." The reasons for the emergence of these narrow definitions are historically and sociologically complex. But one powerful reason lies in the energetic movement of the 1920s and 30s to insure mass education by reducing all learning to discrete steps, stages, bits of information and then holding teachers accountable for imparting fixed numbers of these steps, stages, bits during a given period of time. This "cult of efficiency," as Raymond Callahan called it (*Education and the Cult of Efficiency* [Chicago: University of Chicago Press, 1962]), found its theoretical constructs and evaluative devices in the storehouse of a burgeoning educational psychology that was rapidly devising methods and models for quantifying and structuring knowledge. In this milieu writing was reduced to text production and text to its most salient constituent parts. And though we currently hold to more advanced notions about composing and don't expect, for example, teachers to teach x number of grammar rules per week, we still are partially entrapped by reductionistic models and measures. Look, for example, at current evaluation procedures—for evaluation schemes reveal powerful assumptions about the object of evaluation.

Many of our evaluation schemes focus on product alone, do not incorporate issues of writer's intention and the actual playing out of that intention in the process of composing the essay, nor do they take account of a writer's relation to audience in any full way.[7] Furthermore, in all too many cases, crucial dimensions of any meaningful communica-

7. Lee Odell and Charles R. Cooper, "Procedures for Evaluating Writing: Assumptions and Needed Research," *College English,* 42 (1980), 35-43; Anne Ruggles Gere, "Written Composition: Toward a Theory of Evaluation," *CE,* 42 (1980), 44-58. See also Ruth Mitchell and Mary Taylor, "The Integrating Perspective: An Audience-Response Model for Writing," *CE,* 41 (1979), 247-271.

tion—like the accuracy of information and the legitimacy of the writer's reasoning with it—are put aside so that the felicity of the writing itself can be evaluated.[8] (Such separation, of course, only communicates to the student that cleanliness and charm matter. Accuracy and sense do not.) These procedures, for the most part, suggest that what counts most is a fixed and final and fairly limited product, a product containing or lacking certain countable features or broad structural relationships or specific connections to the writing task. Now this sounds like a limited and fragmented notion of writing skills to me—perhaps *conceptually* not far removed from earlier rigid product notions, the kinds of seriously limited and perhaps limiting notions that emerged in Hartwell's pilot study.

Clearly we need to rethink our definitions of writing skills and make special efforts to change the models of composing our students have internalized. It is possible, for example, that our remedial classes—at least some significant portion of them—should be very process-oriented. This does not simply mean that we would have our students freewrite daily, though there is certainly value in that. It also means that we would help our students experience the rich possibilities of the writing process. We could lead them to see the value of writing as an ordering and storing aid by making them amateur ethnographers and turning them loose on a campus event with pad and pencil in hand. We could break them of conceiving of their written texts as static by introducing exercises like our Preparatory Program's "Revision Scramble." In this exercise students are given a five to ten-minute lecture on an academic topic, must take notes, and then must write either a summary or a critical reaction. In the next class they are given a further brief lecture on the same topic but with new information included. They must revise their summary or reaction in ways that account for the new. Thus they come to see that texts can—even must—evolve. We get them to experience writing as making meaning for self and others by engineering group projects that necessitate collaborative research, composing, and editing, thus moving them through process to product. A nice illustration of this is provided by my Preparatory Program colleague Bill Creasy. He gives his entire class the task of preparing a brief guidebook for incoming Preparatory Program students. Teams of two to three students are each assigned aspects of campus life: dormitories, athletics, ethnic study centers. Throughout the quarter the class collects and selects information (by interviewing officials, reading pamphlets and brochures, distributing questionnaires to seniors), writes it up, puts it on a word processor, and edits it. Professor Creasy provides assistance as it is needed at each point along the way.

Sure, these assignments must be carefully structured, scaled down, and built upon, and, yes, for a while our students will struggle, and some will produce very flawed papers. But we will have to train ourselves to wait, to live with uncertainty and comma splices. Some students may leave our classes writing papers that aren't as clean as some of us would like them to be, but at least these students will hold conceptions of composing that will foster rather than limit growth in writing.[9]

---

8. Barbara Gross Davis, Michael Scriven, and Susan Thomas. *The Evaluation of Composition* (Inverness, Cal.: Edgepress, 1981).

9. I am not suggesting that teachers should completely turn their backs on error. There is little doubt that many academic readers and, as Maxine Hairston reminds us, readers outside the university react strongly to grammatical/mechanical errors ("Not All Errors Are Created Equal: Non-Academic Readers in the Professions Respond to Lapses in Usage," *CE*, 43 (1981), 794-806). We wouldn't want our students to blithely write

The Separation of Writing from Reading and Thinking

In our attempts to isolate and thereby more effectively treat "basic skills" we have not only reduced discourse complexity, we have separated writing from reading and thinking.

When I bring up "thinking skills" I sometimes get responses like the following: "Yes, we used to teach the syllogism, but it didn't carry over to the students' writing." Presenting students with intellectually worthwhile problems, assisting them as they work through them, offering them strategies with which to explore them, showing them how to represent and, when necessary, reduce them, seems to have been equated with formal Aristotelian logic and relegated to the philosophy department. What a disservice. We should help remedial writers become familiar with heuristic routines, too often saved for standard composition, that will enhance their interpretive powers. (Stripped of its theoretical baggage, the tagmemist's particle/wave/field heuristic comes to mind as a particularly useful technique.) And if certain of us question heuristics as being too gimmicky, there is good old-fashioned patient and careful guidance with assignments requiring classifying, comparing, analyzing. In any case—and this is why formal logic always failed in the composition classroom—"thinking skills" must not be taught as a set of abstract exercises (which, of course, they will be if they are not conceived of as being part of writing), but must be intimately connected to composition instruction. Otherwise students hear one more lecture on isolated mental arabesques.

When I bring up the need to incorporate reading into our basic courses, and particularly when I suggest we have our students work from a simple passage in writing their diagnostics (thereby making their diagnostics more equivalent to the academic writing tasks they will face), here is what I hear: "If you do that, you'll confound reading skills and writing skills. You'll never know why they make the mistakes they do. Is it because they can't read or they can't write?"[10] Yes, reading and writing are different processes, but it is simply not true that they are unconnected. Anthony Petrosky explains how current theories present reading as a kind of "composing process"; people construct meaning from text rather than passively internalize it. Teun Van Dijk points out that while we need to know the conventions, structures, and intentions of particular discourses to produce them, we likewise need such knowledge to comprehend them. And Stephen Krashen suggests that one's repertoire of discourse skills is built slowly and comprehensively through reading.[11] Reading and writing are intimately connected in ways we are only beginning to understand. Furthermore, even if they weren't, a major

---

their ways into the dens of the error-vigilant. What I am suggesting is that we might better serve our students if we free them from dulling and limiting notions of composing and then focus and refocus on correctness as an editing, not generating and producing, concern. If we run out of time—if some of our students still have not mastered points of mechanics and usage when they leave us—they will at least be open to writing as a discovering/ordering/communicating process. And if we taught the editing process well, they will know how to use dictionaries, handbooks, and a friend with a copy-editor's eye to help them clean up their final drafts.

10. Anthony Petrosky, "From Story to Essay: Reading and Writing," *CCC*, 33 (1982), 19-36. Petrosky reports similar artificial encapsulation of reading and writing skills. In his case evaluators were telling him that to have students write about their reading would muddy the assessment of their "reading" ability.

11. Petrosky, "From Story to Essay"; Van Dijk, *Macrostructures* (Hillsdale, N.J.: Erlbaum, 1980); Krashen, "The Role of Input (Reading) and Instruction in Developing Writing Ability," unpublished manuscript, Department of Linguistics, University of Southern California, 1981.

skill in academic writing is the complex ability to write from other texts—to summarize, to disambiguate key notions and useful facts and incorporate them in one's own writing, to react critically to prose. Few academic assignments (outside of composition) require a student to produce material ex nihilo; she is almost always writing about, from, or through others' materials.

It seems to me that we have no choice but to begin—and to urge the scholars who have sequestered themselves in segmented disciplines to begin—conceiving of composition as a highly complex thinking/learning/reading/writing skill that demands holistic, not neatly segmented and encapsulated, pedagogies.[12]

The Narrowing of Exploratory Discourse and the Misperception of Discourse Structures

But not all causes of a narrowed model of and curriculum for composing stem from an overzealous vigilance for error or from limited conceptualizations of writing skill. Ironically, one cause might inadvertently have come from some of the most serious critics of standard composition fare, for in their eagerness to torpedo staid and wrong-headed notions about composing, they sometimes polarize issues that in fact lie along a continuum. I suspect that when Ken Macrorie inveighs against "Engfish" and William Coles against "theme-writing," when politically conscious critics like John Rouse complain that teaching standard patterns of discourse socializes and thus regiments minds, when Stephen Judy says "the best student writing is motivated by personal feelings and experience," when Janet Emig distinguishes between extensive and reflexive writing[13]—when all these critics express their observations, they establish in the minds of some of their readers an essentially false set of dichotomies: to write in a voice other than one's most natural is to write inauthentically, to master and use strategies like comparing and contrasting is to sacrifice freedom, to write on academic topics that don't have deep personal associations is to be doomed to mechanical, lifeless composing, and to write expositional, extensive academic prose is to sabotage the possibility of reflexive exploration. Again, let me restate that Coles, Judy, et al. do not necessarily make distinctions this rigidly. Many folks who cite them do.

What does this polarization do to the remedial writer's curriculum? At least two things: The reflexive, exploratory possibilities of engaging in academic (vs. personal) topics are not exploited, and instruction in more complex patterns of discourse is delayed or soft-pedaled.

Reflexive, exploratory discourse has been too exclusively linked to "personal" writing, writing that deals with making sense of one's own feelings and experiences. In fact, making meaning for the self, ordering experience, establishing one's own relation to it is what informs any serious writing. A student writing a paper on Rousseau or on

---

12. Several other writers have recently called for variations of such integrated pedagogies: Charles Bazerman, "A Relationship between Reading and Writing: The Conversational Model," *CE,* 41 (1980), 656-661; Marilyn S. Sternglass, "Assessing Reading, Writing, and Reasoning," *CE,* 43 (1981), 269-275.

13. Macrorie, *Uptaught* (New York: Hayden, 1970); Coles, *Composing* (Rochell Park, N.J.: Hayden, 1974); Rouse, "Knowledge, Power and the Teaching of English," *CE,* 40 (1979), 473-491; Judy, "The Experiential Approach: Inner Worlds to Outer Worlds," in *Eight Approaches to Teaching Composition,* ed. Timothy R. Donovan and Ben W. McClelland (Urbana, Ill.: National Council of Teachers of English, 1980), pp. 37-51; Emig. *The Composing Processes of Twelfth Graders* (Urbana, Ill.: NCTE, 1971).

operant conditioning could, and should be encouraged to, engage in a good deal of self-referenced writing to make sense of difficult notions and, possibly, to weigh these notions against other readings, personal experience, and values. If we don't see all the possibilities for exploration inherent in academic writing, we won't encourage our writers to talk to each other and to us, to plumb their own thoughts, to freely explore the conceptual intricacies of a topic. Again, it is not Emig who would disagree. But in my experience many teachers who read—or misread—her would.

When we delay or insufficiently emphasize the writing of complex discourse, we deprive writers of a chance to learn what coherent, extended texts should look like, what shape or structure they will take. And until they possess a sense of the form of such texts, they cannot write them. Textlinguists like Van Dijk and Robert de Beaugrande have begun to demonstrate the reality of global discourse structures. Margaret Atwell has shown the central role discourse structures can play in the production of coherent text. Frank D'Angelo has recently suggested that these structures, or "paradigms" as he calls them, could well be related to Aristotelian topoi, the classical orator's repository of generating and structuring aids.[14]

The sad truth is that many of our students, particularly remedial students, do not get that much opportunity to read or write extended academic discourse before reaching us[15] and thus are not afforded the chance to develop a wide repertoire of discourse structures or schemata, as they are called by cognitive psychologists. (Lately there has been a proliferation of labels for these discourse representations: superstructures, paradigms, frames, plans, schemata. For consistency's sake, I will stick to schemata, but because I am interested both in the investigative, interpretive capability of schemata as cognitive strategies as well as the role of schemata in the production of written discourse, in certain contexts I will interchange "strategies" with "structures" or "patterns.") The lack of appropriate schemata, of course, will have disastrous results as remedial writers are asked to produce structurally complex prose by readers who, according to Sarah Freedman in "Why Do Teachers Give the Grades They Do?" (*CCC*, 30 [1979], 161-164), evaluate student writing with a good deal of emphasis on organization, a product feature resulting from appropriate discourse schemata.[16] We have little

---

14. Van Dijk, *Macrostructures;* de Beaugrande, *Text, Discourse, and Process* (Norwood, N.J.: Ablex, 1980); Atwell, "The Evolution of Text: The Inter-relationship of Reading and Writing in the Composing Process," Diss., Indiana University, 1981; D'Angelo, "Paradigms as Structural Counterparts of Topoi," in *Linguistics, Stylistics and the Teaching of Composition,* ed. Donald McQuade (Akron, Ohio: University of Akron, 1979), pp. 41-51; D'Angelo, "Topoi, Paradigms, and Psychological Schemata," in *Proceedings of the Inaugural Conference of the University of Maryland Junior Writing Program,* University of Maryland, 1981, pp. 9-23.

15. Applebee, *Writing in the Secondary School;* John Mellon, "Language Competence," in *The Nature and Measurement of Competency in English,* ed. Charles Cooper (Urbana, Ill.: NCTE, 1981), pp. 21-64.

16. Relying on an analysis of variance, Freedman found that her evaluators were most affected by the content of an essay, then its organization. Mechanics ranked third. Interestingly, mechanics were most influential when organization was strong. To my mind, these findings lend further credence to a point I made earlier: clean but empty (or directionless) papers count for little. Correctness begins strongly to affect a reader once he or she has substantial and sensible prose to read. Let me close this note with a quote from Freedman's article:

> if society values content and organization as much as the teachers in this project did, then according to the definition of content and organization I used in this study, a pedagogy for teaching writing should aim first to help students develop their ideas logically. . . . Then it should focus on teaching students to organize the developed ideas so that they would be easily understood and favorably evalu-

choice, then, but to teach these schemata. And I should stress here that we have no reason to believe that a student has to have every pronoun and antecedent correctly in place before he can learn discourse structures. Sentence level mechanics and discourse structures are not developmentally lockstepped.

Now some would agree that students must master these complex schemata and add that a sensible remedial approach would be to begin with "simpler" patterns like narration and then eventually shift to higher order discourse. But, as I suggested earlier, we have reason to doubt that work on narration or on description will build in students a repertoire of more abstract and complex schemata, schemata, that is, that are not based on chronological sequences or spatial arrangements.[17] Academic expositional discourse seems to be more cognitively demanding than simple narration or description;[18] it seems to embody global syntactic and semantic structures that are different from those found in narration and description;[19] and it requires kinds of sentences that, on the average, tend to be syntactically different from those found in narration and description.[20] The studies that support the aforementioned propositions only represent beginnings in the exploration of the cognitive demands and structural features of different kinds of discourse, but though beginnings they should make us wary of assuming that mastering, say, narrative structure will enable students to construct analytic essays.

I suggest, therefore, that we determine the organizational patterns required of our students in academic discourse, and slowly and systematically teach these patterns. They should not be conceived of or taught as "modes" of discourse or as rigid frameworks but, simultaneously, as strategies by which one explores information and structures by which one organizes it. It would not restrict students' freedom to learn these strategies/ structures; in fact, such learning would enhance their freedom, afford them more discourse options. Without these options much academic discourse (the very kind of discourse written by critics like Rouse) will be beyond these writers. For that fact they are essential to the making and conveying of meaning in our culture. The question is, how should they be taught? The two most natural ways to assimilate or learn these patterns are by reading a good deal of discourse containing them (see Krashen's essay cited in footnote 11, and Mellon's cited in footnote 15), and experiencing the need for them as one encounters barriers while writing. The trouble is, of course, that our remedial writers don't have much time. They are enrolled in other classes, some of which demand

---

> ated. . . . It seems today that many college-level curricula begin with a focus on helping students correct mechanical and syntactic problems rather than with the more fundamental aspects of the discourse. (pp. 163-164)

17. Writers often resort to narrative frameworks when they are unable to execute more complex or abstract discourse schema. Narration becomes a substitute rather than a building block: see Suzanne E. Jacobs and Adela B. Karliner, "Helping Writers to Think," *CE*, 38 (1976-77), 484-505; Linda Flower, "Writer-based Prose: A Cognitive Basis for Problems in Writing," *CE*, 41 (1979-80), 19-37.

18. Jim Williams and Micky Riggs, "Subvocalization During Writing," unpublished manuscript, Department of English, University of Southern California, 1981; Ann Matsuhashi, "Producing Written Discourse: A Theory-based Description of the Temporal Characteristics of Three Discourse Types from Four Competent Grade 12 Writers," Diss., State University of New York at Buffalo, 1979.

19. Van Dijk, *Macrostructures;* C. Cooper and A. Matsuhasi, "A Theory of the Writing Process," in Elizabeth Martlew, ed., *The Psychology of Writing* (New York: Wiley, in press).

20. Crowhurst, "Syntactic Complexity"; Joseph Williams, "Defining Complexity," *CE*, 40 (1978-79), 595-609; Sandra Thompson, "Grammar and Discourse: The English Detached Participial Clause," in Flora Klein, ed., *Discourse Approaches to Syntax* (Norwood, N.J.: Ablex, in press).

structurally complex written responses to complex assignments. And while we can suspend concern about, say, comma splices, we can't wait for students to assimilate these global structures—they are essential to reader comprehension. Students need them now. Yet we don't want to put our students through one more lifeless and disembodied drill on the "compare/contrast mode." What to do? I propose a four-tiered plan:

A. *Make sure that the patterns/strategies are real.* That is, that they are derived not from the theorist's speculations on how discourse ought to be taxonomized, but from writing situations the students face daily. In the aforementioned surveys of university assignments, I found calls or cues for a number of global discourse strategies or patterns. These were either explicitly requested or seemed to be the best approaches to apply to purposely ambiguous or to simply poorly written questions. I will list the most salient of them now and illustrate the teaching of several of them momentarily: definition, seriation, classification, summary, compare/contrast, analysis, academic argument. With the exception of analysis and academic argument—which I define in special ways—these patterns are too familiar for comfort. We see some combination of them in our rhetorics and readers, and they have, in all too many hands, come to represent much that stifles our pedagogy. But—and here I'll look for support to D'Angelo's speculation about the connection of schemata to topoi—the problem is not that these patterns are inherently rigid or unuseful, but that they have not been taught as thinking strategies as well as discourse structures. (I am tempted, therefore, to write out definition, classification, etc., with participial endings to stress their actively strategic nature.) I would speculate—and at this stage, it *is* speculation—that the reason these strategies/structures appeared so frequently in our surveys of academic writing situations (and, my colleague Ruth Mitchell tells me, in her surveys of business and professional writing as well) is that they are so central to the way we explore, order, and present information when we are engaged in transactional/referential discourse.[21]

B. *Create a meaningful context for their use.* Here I would like to raise again the issues of the context of writing and the motivation to write. It is usually assumed that for writing to be meaningful, it must generate from issues and experiences that centrally involve the student. This assumption, as I suggested earlier, can result in some pretty superficial curricula, but can also lead to challenging pedagogies like Coles', that leads students to examine their own writing as ways of knowing and becoming (see footnote 13), or like David Bartholomae's, that weaves academic writing into students' reflections on, for example, their coming of age and requires them to turn an analytic eye onto their own and others' autobiographies ("Teaching Basic Writing: An Alternative to Basic Skills," *Journal of Basic Writing* [Spring/Summer 1979], pp.

21. For compatible viewpoints presenting somewhat different lists of strategies, see Chapter Seven of Mina Shaughnessy, *Errors and Expectations* (New York: Oxford University Press, 1977) and Anne Ruggles Gere and Eugene Smith, *Writing and Learning,* 2nd ed. (New York: Macmillan, 1988).

85-109). I believe, though, that another meaningful context for student writing is the very academic environment in which students find themselves; it is a strange, complicated place, at times feared in its newness, at others, appreciated, a place of promise and a place of limitation, sometimes cynically apprehended, sometimes enjoyed. If the discourse schemata I have listed are, on one hand, pragmatically offered as being central to success in the university and, on the other, offered as investigative tools necessary for examining reality, for examining the academic environment itself, then it seems to me we have established a motivating and meaningful context for their use.

C. *Teach the schemata as strategies as well as structures.* Classifying or comparing/contrasting can be taught in cookbook formulaic fashion, and with writers who hold few complex schemata in their repertoire, this patterned instruction might be necessary. At first. We would then want to show students other uses, other shapes, to wean them from learning an inflexible discourse pattern. But we also want to teach comparing and contrasting as a cognitive strategy, as a way to explore things, events, phenomena as well as a way to organize what we discover about them. Note the difference from usual textbook treatment—no static modes, but, borrowing Linda Flower's words, "ways to think systematically about complex topics" (*Problem-Solving Strategies for Writers* [New York: Harcourt Brace Jovanovich, 1981], p. 74).

D. *Sequence the schemata appropriately.* In teaching these structures/strategies, we need to keep the importance of proper sequencing in mind. Educational theorists like Benjamin Bloom and Robert Gagné[22] repeatedly show the benefits of arranging tasks in ways that allow subsequent tasks to build on previously learned ones, and in our own field Britton and his colleagues have noted:

> Our experience suggests that there is likely to be a hierarchy of kinds of writing which is shaped by the thinking problems with which the writer is confronted. (James Britton, Tony Burgess, Nancy Martin, Alex McLeod, Harold Rosen, *The Development of Writing Abilities (11-18)* [London: Macmillan, 1975], p. 52)

Considering the schemata I presented, this discussion of sequencing suggests that we offer students tasks that require, for example, the presenting of steps in a series before confronting them with tasks requiring summarizing before assigning tasks involving argument. And, as frequently as possible, a previously learned structure/strategy should be incorporated into the preliminary stages of a new assignment (e.g., have the student summarize a theory that she will need for an assignment requiring analysis) or should become part of the assignment itself. It is possible to sequence assignments and the schemata they embody so that they lead to highly complex discourse like analyzing and arguing.

---

22. Bloom, ed., *Taxonomy of Educational Objectives. Handbook I: Cognitive Domain* (New York: David McKay, 1956); Gagné, *The Conditions of Learning,* 2nd ed. (New York: Holt, Rinehart, and Winston, 1970).

Teaching these schemata allows students to begin writing academic discourse early on. They are able, even crudely, to construct a complicated essay into which they can weave their reading. They are participating in an academic context.[23]

Let me illustrate the above discussion by providing three kinds of assignments from our Freshman Preparatory Program.

> *Seriation.* For one seriation assignment the teacher reads a brief account of a burn patient's daily routine of hydrotherapy, medication, meals, etc. Late in the day the patient experiences chills, nausea, and dizziness. The students take notes and then, in groups, write a brief paper retelling the day's events. In a somewhat related assignment students take notes while the teacher reads a description of the reduplication process of an intestinal virus. Then they are given a list of the steps in the reduplication process, but the steps are scrambled. In groups, students have to order the steps correctly and write a brief paper detailing the accurate sequence of process steps. Another seriation assignment has the teacher and a colleague extemporaneously act out five minutes of a psychotherapy session. Students become clinical observers and take notes on the interaction. Again, in groups, the students write an essay presenting the stages of the interaction. For all assignments students must underline and orally or in footnotes explain the function of each connective and transition they use. In this way they become sensitive to the differences among simple concatenation vs. correlation vs. causality. For example, can they say the medication "caused" the patient's reaction, or must they be cautious and say the reaction "followed" the medication, physical therapy, and meals?
>
> *Classification.* One classification assignment calls for students to view twenty slides of the human form—Raphael to Rivera. They are shown the slides two or three times and are told to observe as keenly as possible, jotting down notes on whatever characteristics of each painting strike them. They are not told anything about era, school, or painter. Then they are told to decide, with the aid of their notes, whether or not any two or more paintings could be grouped under the same characteristic. (For teachers so inclined, this can be done in small groups.) At first, students tend to offer very general observations: "Paintings #1, 3, 10, 19 are all very bright." "2, 8, 20 don't look like human beings." The teacher does not

---

23. I realize I am asking that we teach discourse structures often assumed to be beyond the grasp of remedial writers. One way to teach such complicated structures has been suggested throughout this fifth section. The approach simply entails a scaling down of potentially complex tasks and a gradual building of skill through carefully sequenced, increasingly complex assignments. But it would also prove helpful to have some idea of the written discourse sophistication students possess when they enter our classes. The following procedure could provide such information: During the first few days of class, the teacher gives an assignment that requires students to bring into play a discourse structure that will be dealt with at some later point in the course. In reading the students' responses, the teacher would suspend concern with sentence-level error and attempt to estimate the discourse sophistication of each student's essay. If, for example, the task called for comparing and contrasting, did a student attempt to organize the essay in a way that indicates comparison? How adequate was the attempt? Did the student rely on simpler—inadequate—structures like narration? Certainly this procedure will not provide clear entry to the complexities of a student's discourse repertoire, but—especially if repeated with one further assignment—it can offer some suggestions as to the level of particular students' discourse sophistication. Thus the teacher will have an indication of where and how he or she needs to begin instruction on discourse frameworks.

criticize and puts all characteristics on the board. What she does begin doing is asking questions to point out similarity or generality in the categories. Students are led to refine, collapse, subdivide categories. What the students finally arrive at is their own classificational system. They are then asked to write a brief essay that proposes the system and illustrates it with specific paintings. In subsequent exercises they work with items from a personality inventory, a list of definitions of genius, and a collection of first paragraphs from novels, textbooks, and articles. In this way students actually experience the process of classification as well as learn how to present its results.

Analysis. Textbook authors usually define analysis as a careful exploration that breaks an artifact, event, or phenomenon down to its constituent parts. What they don't mention is that no analyzer operates without some belief or value system, without some exploratory framework.[24] Toward the end of our program we introduce students to this complex nature of analysis. We give them a list of raw data, or a description of an event, or a scene or story, or we show them a film. To enhance pedagogical effect, the data, descriptions, or episodes are in some way puzzling or incomplete. We encourage discussion about them. Then we provide a theoretical framework of some sort and ask them to assume it and analyze and explain the data, event, film, etc. Thus their analysis is informed by another's perspective. Finally, we ask them to criticize the perspective, to consider ways it might be lacking in accuracy, explanatory power, or comprehensibility. In these ways we introduce students to the fact that analyses are always founded on assumptions and orientations—from personal analyses of moral issues to statistical analyses of biological data—and that any given analysis is open to investigation once its informing framework is made specific. Some examples: Students are given a newspaper account of a suicide. They discuss it, providing their own interpretations. They are then offered an explanation of depression and "learned helplessness." They investigate the suicide with the aid of the latter clinical theory, noting how it illuminates and how it comes up short. Students are given a list of U.S. immigration statistics from 1820 to 1977. This is followed by a quotation from *Das Kapital* on exploitation of labor. They are shown *An Andalusian Dog,* and it is followed by a definition of surrealism. They are given the Barthelme short story "Game" (a story of two deranged soldiers locked indefinitely in a missile bunker) and then given a series of quotations on the environmental etiology of insanity. And so on. Thus it is that students are shown the power and limitation of an explanatory framework, the crucial role some such framework plays in analysis (and that no analysis takes place atheoretically), how to assume—and question—an analytic framework, and how to present the result of that analysis in writing. In writing up their analysis they usually have to rely on previously learned strategies: serializing, summarizing, comparing, etc.

---

24. This point is convincingly made by Karl Popper in *Conjectures and Refutations: The Growth of Scientific Knowledge* (New York: Harper and Row, 1968). After working up my analysis exercises, I discovered Marc Belth's *The Process of Thinking* (New York: David McKay, 1977). Belth equates *all* thinking with the bringing to bear of models, metaphors, analogies onto internal and external events.

Our institutions create deplorable conditions for our remedial writing programs and our students—labeled intellectually substandard, placed in the conceptual basements of English departments, if placed in the department at all, ghettoized. No need to polemicize further; we all know the complaints. But what we teachers must remember is that the very nature of many remedial writing courses contributes to institutional insularity, to second-class citizenship and fragmented education, to a limiting of our students' abilities to grow toward intellectual autonomy. Oddly enough, the nature of our programs is nearly synchronized with the narrow reality created for them by our institutions.

Clearly we must work to change our institutions, but we must also question our assumptions about our students' abilities and the pedagogies we have built on these assumptions. All too often these days we hear that remedial writers are "cognitively deficient," locked, for example, at the Piagetian level of concrete (vs. formal) operations. These judgments are unwarranted extrapolations from a misuse (or overuse) of the developmental psychologist's diagnostic instruments, for as Jean Piaget himself reminded us in one of his final articles, if we are not seeing evidence of formal operations in young adults, then we should either better acquaint them with our diagnostics or find more appropriate ones ("Intellectual Evolution from Adolescence to Adulthood," *Human Development,* 15 [1972], 1-12). The problem might well lie with our tools rather than with our students' minds. We must assume, Piaget warns, that in their daily lives our students can generalize and analyze, can operate formally.[25] What they can't do—applying this to the writing teacher's domain—is successfully operate within the unfamiliar web of reasoning/reading/writing conventions that are fundamental to academic inquiry. Our students are not cognitively "deficient" in the clinical sense of the term; if they were, they wouldn't be able to make the progress they do. Our students are not deficient; they are raw. Our job, then, is to create carefully thought-out, appropriate, undemeaning pedagogies that introduce them to the conventions of academic inquiry. Bartholomae presents one such pedagogy. The foundations for another can be found in Shaughnessy's seventh chapter of *Errors and Expectations* (see footnote 21). At various

25. I will quote from Piaget's article:

> In our investigation of formal structures we used rather specific types of experimental situations which were of a physical and logical-mathematical nature because these seemed to be understood by the school children we sampled. However, it is possible to question whether these situations are, fundamentally, very general and therefore applicable to any school or professional environment. . . . It is highly likely that [young adults] will know how to reason in a hypothetical manner in their speciality, that is to say, dissociating the variables involved, relating terms in a combinatorial manner and reasoning with propositions involving negations and reciprocities. They would, therefore, be capable of thinking formally in their particular field, whereas faced with our experimental situations, their lack of knowledge or the fact they have forgotten certain ideas that are particularly familiar to children still in school or college, would hinder them from reasoning in a formal way, and they would give the appearance of being at the concrete level. (p. 10)

For further evidence that many of our young adults are not "deficient"—that is, that once appropriately exposed to Piagetian diagnostic procedures they can evince "formal" operations—see Fred W. Danner and Mary Carol Day, "Eliciting Formal Operations," *Child Development,* 48 (1977), 1600-1606; and Deanna Kuhn, Victoria Ho, and Catherine Adams, "Formal Reasoning Among Pre- and Late-Adolescents," *Child Development,* 50 (1979), 1128-1135. (My thanks to Professor Deborah Stipek, a former post-doctoral fellow at Piaget's Geneva Institute, for her helpful conversations on these issues.)

Since Piagetian tests of formal operations are based on mathematics, formal logic, and physics, I wonder how many of us—being as far removed as we are from our college physics labs—would fumble about with them and be labeled "concrete" thinkers.

points in the present essay, I have suggested a third approach;[26] let me summarize it here.

In my opinion, a remedial writing curriculum must fit into the intellectual context of the university. Topics should have academic substance and, when possible, should require the student to work from text. The expressive, exploratory dimension of writing ought to be exploited here—academic topics as much as personal ones demand a working through, a talking to and making meaning for the self. The richness of the composing process must be revealed. Too many of our students come to us with narrow, ossified conceptions of writing. Our job is to create opportunities so they can alter those conceptions for themselves. We have to allow our writers to be ambitious and to err. Error vigilance creates safe, not meaningful, prose. We need to integrate thinking and reading and writing, and we must pressure our training institutions to give us fuller, richer definitions of writing competence. Finally, we should seriously consider the central role discourse schemata play in discovering, organizing, and presenting information. It is these structures/strategies, rather than sentence-level error, which should be the fundament of our courses.

How flat some of our remedial courses feel. And how distant the eyes of too many of our students. We sometimes take this flatness, this distance as signs of intellectual dullness. They are more likely the signs of boredom, humiliation, even anger. But in my experience anyway the flatness dispels and the distant gazes revitalize when students are challenged, engaged, brought fully into the milieu they bargained for. Yes, we teachers will work slowly, scale carefully, provide as much assistance as we can. But we will still be creating an edge to our "remedial" classroom. Our students will grumble about the strain—grumbling is part of the student's drama—but they will know they are participating in the university. And that is a strain that can make one feel worthwhile.

---

26. This third approach is essentially that embodied in the regular year Freshman Preparatory Program composition courses offered at UCLA. During the summer, however, we employ a variation of this approach in our Freshman Summer Program, an adjunct program in which composition courses are linked to introductory breadth courses. I will be glad to mail a description of the Freshman Summer Program to interested readers. Send requests to Mike Rose, Department of English, 2225 Rolfe Hall, UCLA, 405 Hilgard Avenue, Los Angeles, California 90024.

Elza C. Tiner

Elements of Classical and Medieval Rhetoric in the Teaching of Basic Composition

The concept of teaching basic writing as a process has become increasingly popular in recent years.[1] Teachers of process composition may divide and label the stages of the sequence variously, but most agree on a progression from broad planning to increasingly fine-tuned revisions. Frances Kurilich and Helen Whitaker teach prewriting, drafting, revising, and producing the final copy. Kathleen McWhorter and Candalene McCombs list five stages in the writing of an essay: prewriting, composing, organizing, revising, and proofreading. Joy Reid guides the student through prewriting, organizing, developing supporting details, and correcting grammar and punctuation.

All of these methods are derivative of a process that was developed over two thousand years ago for the composition of orations in political and legal arenas. As Janet Emig points out, the concept of a composing process is not new to the writing classroom, but has its origins in the rhetorical theory of ancient Greece and Rome (15-16). Winifred Bryan Horner and Edward P. J. Corbett have developed textbooks based on the classical process for use in more advanced college writing classes. By the first century A.D., Quintilian's *Institutio Oratoria* recommended the rhetorical process in connection with integrated study of reading, writing, and speaking. Extant medieval and late classical textbooks show that whole language instruction was one of the teaching methods in the grammar schools of the past. These classical and medieval rhetorical handbooks offer a systematic process that can be used today in the teaching of remedial composition, a course often considered too basic for the application of classical theory. Moreover, case studies of students' natural composing processes appear to challenge the classical system: left to their own devices, students write in a variety of sequences, just as professional writers have no one way of working. Emig has observed that high school students prewrite, plan, start, stop, and revise the whole or parts in various ways, particularly when developing material which is "reflexive" or self-motivated, as opposed to school-sponsored writing for which there is often less time for revision (91-93). If the classical process is treated as an open, variable system, it can be adapted to the needs of individ-

From *Virginia English Bulletin* 41 (Spring 1991): 114-20. Reprinted by permission.

1. This paper was originally presented at the Mid-Hudson Modern Language Association Conference, Marist College, November 1988; subsequently revised and published in the *Virginia English Bulletin* 41, 1 (1991): 114-120; and reprinted in the journal of the Lynchburg College Symposium Readings Program, *Agora* 3 (1992): 65-71. The present version has since been revised.

ual remedial writers, giving them clear guidelines with flexibility, steps to assemble for their personal composing processes.

In her article "Classical Rhetoric and Contemporary Basics," Susan Miller argues for a composition program based on classical education. In a more recent study, Andrea Lunsford and Lisa Ede suggest that classical theory can be used to teach reading, speaking, writing, and listening. Moreover, Lynn Troyka shows that classical rhetorical theory can help to increase basic writers' confidence in their learning abilities. The basis for her teaching is Edward P. J. Corbett's *Classical Rhetoric for the Modern Student.* While at New York Institute of Technology and now as director of the Writing Center at Lynchburg College, I have been developing a teaching strategy for remedial writing based on a systematic, but variable, application of Ciceronian and medieval processes of composition.[2]

Ancient rhetorical theory has two main branches: (1) oral and (2) written composition. According to Frederic Wheelock, the term "rhetoric" can refer to both:

> Both "rhetor" and "rhetoric" derive from the same Greek root, which means "to speak." Consequently rhetor is the etymological equivalent of Latin *orator,* "pleader, speaker"; and rhetoric by a little extension comes to mean the art and practice of the effective use of language in writing and speaking. (12)

Oral and written rhetoric have always been interdependent; a similar composition process applies to both, and one helps the other. In the composition of a speech, the orator must first produce a text, whether in note or prose form, or else clearly framed in the mind. Conversely, it helps a writer to read his or her paper aloud, to hear its flow and catch problems of organization, logic, or grammar. In Book 3 of the *Rhetoric,* Aristotle describes a three-part composition process: (1) first consider what facts will make a persuasive speech; (2) find effective language with which to present the facts; and (3) organize the speech. Later Aristotle adds that the orator must also find the proper method of delivery. By the time the *Rhetorica ad Herennium* was written (c. 86-82 B.C.), Roman rhetoric had been divided into five parts which formed a unified sequence: invention, arrangement, style, memory, and delivery. These parts were to be learned in three stages: theory, imitation, and practice, which are henceforth called the "three levels of learning" throughout this paper. Taken together, the parts of rhetoric and the levels of learning form a matrix on which the process of composition can be based in the remedial writing classroom. In fact, this matrix can be used as a *cyclic* process in which students go through the five parts as they produce successive drafts of an essay until they determine the final copy. As well, the sequence can be rearranged so that the parts of rhetoric become each student's most comfortable composing process.

The five parts of rhetoric are defined in Book I, Section 3, of the *Rhetorica ad Herennium.* Here there is no evidence that they are to be treated as a sequence. The anonymous author is simply telling his student, Herennius, what an orator should know:

---

2. During a visiting appointment as a preceptor in Richard Marius's course in Expository Writing at Harvard University in the spring of 1993, I also had the opportunity to adapt the ancient rhetorical system to the variable composing processes of more advanced writers. I thank both Richard Marius and Nancy Sommers for their support and encouragement of this research.

> The speaker, then, should possess the faculties of Invention, Arrangement, Style, Memory, and Delivery. Invention is the devising of matter, true or plausible, that would make the case convincing. Arrangement is the ordering and distribution of the matter, making clear the place to which each thing is to be assigned. Style is the adaptation of suitable words and sentences to the matter devised. Memory is the firm retention in the mind of the matter, words, and arrangement. Delivery is the graceful regulation of voice, countenance, and gesture.

Thirty years later (55 B.C.), Cicero describes the parts as a process in his *De oratore,* Book I, Section 142:

> And, since all the activity and ability of an orator falls into five divisions, I learned that he must first hit upon what to say; then manage and marshal his discoveries, not merely in orderly fashion, but with a discriminating eye for the exact weight as it were of each argument; next go on to array them in the adornments of style; after that keep them guarded in his memory; and in the end deliver them with effect and charm.

Cicero's words "not merely in orderly fashion" suggest that he expects speakers to think critically and adapt material to intended audience, i.e., to go beyond the set format of the six-part oration. Similarly, teachers may rearrange the parts of the composing process to reflect each student's own composing habits. Some students may need to freewrite, then edit out repetition, then organize, then edit some more, and finally add supporting details. Other students may plan carefully, write an outline, and then draft, edit, and proofread.

Medieval authors also describe the parts of rhetoric as steps in a process. However, the later authors introduce a distinction that is specifically intended for producing a written text. They are careful to distinguish between planning in the mind, when invention and arrangement are set up, and the actual writing, when more attention is given to style. Memory and delivery were also very important, because students had to present their written compositions orally. For example, the statutes of Oxford University, written before 1313, specify what kinds of writing assignments grammar masters should give their students:

> Likewise, they are required to assign fifteen verses and prose compositions in appropriate language, not overblown or wordy, with succinct, graceful phrases and clear metaphors, and to the best of their ability, replete with meaning. Those who receive the assignments should write them on parchment on the next holiday or prior, and then, on the next day when they return to school, they should recite them by heart to their master and hand in their writing. (Gibson 20)

As well, they memorized and recited passsages of classical poetry and scripture for development of reading comprehension and skill in writing. The handbooks of this period make clear the interdependence of oral and written composition. Note, for example, the whole language definition of the *ars dictaminis* (business writing), given by the eleventh-century Alberic of Monte Cassino in his textbook, *Rationes dictandi (The Rules of Business Writing):* "*dictamen* is the fitting and appropriate composition of letters about any topic, whether retained in the mind, expressed in speech, or written" (Rockinger 9).

According to the anonymous author of *De rhetorica quae supersunt (What's Most Important in Rhetoric)* the orator first evaluates and defines the type of case and the issues involved, then he figures out how to organize the material. Invention and arrangement are part of the thinking process that precedes the actual composition of the speech. Then the orator produces a written text of the speech, with further attention to arrangement and finally to style. (Joseph Miller 7; Halm 137) In the late twelfth-century *Anticlaudianus,* Alan of Lille describes the art of rhetoric in the figure of a beautiful and eloquent maiden. The description includes the five-step composition sequence. Oral presentation is the end of the whole process (Sheridan 98-99; Bossuat 94). In the opening lines of the *Poetria nova (New Poetry),* written c. 1200-1215, author Geoffrey of Vinsauf emphasizes the distinction between the thinking stage, when invention and arrangement take place, and the writing stage when style is added, and the oral presentation when memory and delivery are required:

> If a man has a house to build, his impetuous hand does not rush into action. The measuring line of his mind first lays out the work, and he mentally outlines the successive steps in a definite order. . . . When due order has arranged the material in the hidden chamber of the mind, let poetic art come forward to clothe the matter with words. (Nims 16-17; Faral 198-199)

Although Geoffrey of Vinsauf does not discuss memory in the introduction, he later devotes a whole chapter (5) to it. The last stage of the process is delivery, covered in chapter 6.

The classical authors *Auctor ad Herennium,* Cicero, and Quintilian all explain how the composition process is mastered through the three levels of learning: theory, imitation, and practice. Theory is the set of rules that provide a system for composition; imitation is the use of writing and speaking models to illustrate theory; and practice is careful exercise of what one has learned. In a modern basic composition course, these three levels of learning can be used at any of the five stages of the writing process. At first the steps of the writing process are introduced singly, and later put together to form a composition sequence that students can modify according to their individual preferences. In remedial composition, because many students lack the skills needed to write at all, the five parts of rhetoric and the three levels of learning give a clear path to follow when developing reading and writing strategies.

For invention, students can develop topics based on their interests. At first many students find it difficult to write about material which has been assigned to them. Therefore, one way to facilitate invention is to let them use material about which they have some prior knowledge. In this way, they first concentrate on mastering the writing process. Later they apply it, no longer a stumbling block, to unfamiliar topics. Progress through the three levels of learning at the invention level is thus: students are given the "theory" in a set of instructions on how to write an essay. Each teacher can design his or her own method, or refer to the textbook models. Then students find reading material in their interest areas; such reading provides models for imitation and ideas for development of topics. Providing models for imitation also motivates students to improve their reading comprehension, because they have a purpose for reading: to get ideas for writing assignments. Once the ideas begin to form, arrangement follows.

In Book III, Section 16, the *Rhetorica ad Herennium* distinguishes between two kinds of arrangement: (1) traditional and (2) creative. The first is based on the rules of rhetoric from the classical textbooks, and the second is accommodated to particular circumstances, according to the judgment of the orator. Under traditional arrangement, the *Rhetorica ad Herennium* gives two subcategories: (1) arrangement of the whole speech and (2) arrangement of individual arguments within the speech. Since students at the basic level work best when given clear directions, the traditional arrangement can be taught first, using the two subcategories for essay format. Later, as they become more comfortable with the idea of organization, they can creatively experiment with different kinds of arrangement.

Using the model of traditional arrangement, students can be taught to look at the organization of their essays on two levels: (1) as a whole, and (2) within each paragraph. The *Rhetorica ad Herennium* offers a logical format which can be used for the arrangement of the whole essay and for individual paragraphs within it. This five-part format, based on the parts of an argument, would help students to practice logical reasoning in an essay: proposition, reason, proof of reason, embellishment, and conclusion. The proposition is the "to prove" or thesis statement of the whole paper, or topic sentence of a paragraph. The reason briefly explains the importance or validity of the proposition. The proof of the reason supports the reason with facts or examples. The embellishment is a section of rhetorical ornament designed to reinforce the main points. At the paragraph level it can be left out, but in an essay it would be an ideal place to let the students practice emotional appeal, or experiment with style, figuring out ways to convince the audience (perhaps other students) of the importance of their main points. The concluding paragraph ties the whole argument together and affirms the proposition, now proved.

Like invention, arrangement can be taught via the three levels of learning. First the class is shown how to set up an essay, then they read examples of logically arranged essays. Next they plan an essay according to the format described previously. Later, after they have mastered the basic parts, in further revisions they can try adding, deleting, and rearranging paragraphs and argument parts in different orders, to see the effects of creative arrangement. Recombinations will also help to free students of dependency on the five-paragraph theme.

After students have learned how to invent and arrange material, they can revise it for style. At the basic level, style is no more than proper word choice, clear word order, and correct grammar. Like invention and arrangement, style can be taught through explanation of principles, models for imitation, and practice. Note that models for imitation can come from a variety of sources, including student writing. Students are often quicker to catch what is unclear in someone else's paper before they see problems in their own papers. Moreover, especially well written papers from the class make inspiring models for imitation and new ideas for invention and arrangement.

Memory is a useful learning technique that is often ignored in teaching composition. Not only does memory help students assemble material in their minds before writing, but it also helps them to assimilate correct phrasing and vocabulary from their models for imitation. In the process of writing, a profitable exercise might be to have students memorize short paragraphs so that they retain word, phrase, and sentence patterns for later use in their own compositions. The use of memory as an aid to invention, arrangement, and style is an area in which further work needs to be done.

Delivery, which once meant the oral presentation of a speech or written text, now becomes the actual writing of a composition on paper or word processor. After a student has assembled freewriting, plan sheets, and notes, and has a clear idea of how an essay is to be set up, the time comes to write it. After the first draft is written, the student carefully checks each of the other areas—invention, arrangement, style—using memory to prepare for each subsequent draft. Students may also read their papers aloud, to hear sentences that flow beautifully or else are awkward. In this way, the students can learn from hearing one another's papers, which again serve as models for imitation and critical analysis.

Most of the current textbooks offer methods of finding material, arranging paragraphs, and revising for style. However, what they do not always make clear is how to put the whole rhetorical strategy together. One way to apply the classical divisions of rhetoric is in cyclic form, where the steps are repeated for each successive draft of a paper. As the student composes, he or she first concentrates on each individual step, revising a paper with respect to each of these divisions. As the paper nears completion, a student checks over the product of each of the steps: content, organization, and style, while keeping in mind (memory) the goal of the paper, and producing successive drafts and ultimately the final copy (delivery).

In a basic class, it is not possible to teach the whole process at once. Thus the three levels of learning help the students to master each aspect of the composition process, and later, when the class knows the whole process, the cycle of revisions suggested earlier can be followed. Note that this is not meant to become a rigid system, but can be adjusted to the level and individual learning styles of the students. Once they learn the steps, some may prefer to organize ideas first, then develop them in the invention stage. Others may work with words and phrases, compose further, and finally organize. Still others may need to repeat steps, such as checking arrangement, after adding and supporting ideas. Nonetheless, no matter which way students use the system, the five-step process and the three levels of learning make a unified plan to follow in teaching basic writing.

Works Cited

Alan de Lille. *Anticlaudianus*. Ed. R. Bossuat. Paris: J. Vrin, 1955.

———. *Anticlaudianus on the Good and Perfect Man*. Trans. James C. Sheridan. Toronto: Pontifical Institute of Mediaeval Studies, 1973.

Aristotle. *Rhetoric. The Complete Works of Aristotle: The Revised Oxford Translation* 2. Bollingen Series 71. Princeton: Princeton UP, 1984.

[Cicero] ad C. Herennium de ratione dicendi (Rhetorica ad Herennium). Trans. Harry Caplan. Loeb Classical Library. Cambridge: Harvard UP, 1954.

Cicero. *De oratore*. Trans. E. W. Sutton and H. Rackham. Loeb Classical Library. Cambridge: Harvard UP, 1948.

Corbett, Edward P. J. *Classical Rhetoric for the Modern Student*. 3rd. edition. New York: Oxford UP, 1990.

Emig, Janet. *The Composing Processes of Twelfth Graders*. NCTE Research Report 13. Urbana: NCTE, 1971.

Faral, Edmond. *Les arts poétiques du XII et du XIII siècle: Recherches et documents sur la technique littéraire du Moyen Age.* Paris: Champion; rep. 1971.

Geoffrey of Vinsauf. *Poetria nova.* Trans. Margaret F. Nims. Toronto: Pontifical Institute of Mediaeval Studies, 1967.

Gibson, Strickland, ed. *Statuta antiqua Vniversitatis oxoniensis.* Oxford: Oxford UP, 1931.

Halm, Karl F. von, ed. *Rhetores Latini Minores.* Leipzig: Teubner, 1853.

Horner, Winifred Bryan. *Rhetoric in the Classical Tradition.* New York: St. Martin's, 1988.

Kurilich, Frances, and Helen Whitaker. *Re:Writing: Strategies for Student Writers.* New York: Holt, 1988.

Lunsford, Andrea A., and Lisa S. Ede. "Classical Rhetoric, Modern Rhetoric, and Contemporary Discourse Studies." *Written Communication* 1 (1984): 78-100.

McWhorter, Kathleen, and Candalene McCombs. *Write to Read: Read to Write.* Boston: Little, 1983.

Miller, Joseph M., Michael H. Prosser, and Thomas W. Benson, eds. *Readings in Medieval Rhetoric.* Bloomington: Indiana UP, 1973.

Miller, Susan. "Classical Rhetoric and Contemporary Basics." *The Rhetorical Tradition and Modern Writing.* Ed. James J. Murphy. New York: Modern Language Association, 1982. 46-57.

Quintilian. *Institutiones oratoriae.* Trans. H. E. Butler. Loeb Classical Library. Cambridge: Harvard UP, 1953.

Reid, Joy. *The Process of Composition.* Englewood Cliffs: Prentice-Hall, 1982.

Rockinger, R. Ludwig von, ed. *Briefsteller und Formelbücher des elften bis vierzehnten Jahrhunderts* 1. Munich, 1863; rep. New York: Burt Franklin, 1961.

Schoeck, Richard J. "The Practical Tradition of Classical Rhetoric." *Rhetoric and Praxis: The Contribution of Classical Rhetoric in Practical Reasoning.* Ed. Jean Dietz Moss. Washington, D.C.: Catholic UP, 1986. 23-41.

Tiner, Elza C. "Evidence for the Study of Rhetoric in the City of York to 1500." M.S.L. Research Report. Toronto: Pontifical Institute of Mediaeval Studies, 1984.

Troyka, Lynn Quitman. "Classical Rhetoric and the Basic Writer." *Essays on Classical Rhetoric and Modern Discourse.* Ed. Robert J. Connors, Lisa S. Ede, and Andrea Lunsford. Carbondale and Edwardsville: Southern Illinois UP, 1984. 193-202.

Wheelock, Frederic. *Quintilian as Educator.* New York: Twayne, 1974.

COMPUTERS

Gail E. Hawisher and Cynthia L. Selfe

The Rhetoric of Technology and the Electronic Writing Class

Since the mass production of the first fully-assembled microcomputer in 1977, technological change has influenced not only the ways in which we write but also, for many of us, the ways in which we teach writing.[1] Increasing numbers of writing instructors now depend on computer-supported classrooms and use on-line conferences that take place over computer networks as teaching environments. Writing instructors who hope to function effectively in these new electronic classrooms must assess ways in which the use of computer technology might shape, for better and worse, their strategies for working with students. Along with becoming acquainted with current composition theory, instructors, for example, must learn to recognize that the use of technology can exacerbate problems characteristic of American classrooms and must continue to seek ways of using technology that equitably support all students in writing classes. All too frequently, however, writing instructors incorporate computers into their classes without the necessary scrutiny and careful planning that the use of any technology requires.

Such scrutiny will become increasingly important with computers, given the considerable corporate and community investment accompanying this technology as its use expands within our educational system. Unfortunately, as writing instructors, we have not always recognized the natural tendency when using such machines, as cultural artifacts embodying society's values, to perpetuate those values currently dominant within our culture and our educational system. This tendency has become evident as we continue to integrate computers into our efforts at writing instruction. In many English composition classes, computer use simply reinforces those traditional notions of education that permeate our culture at its most basic level: teachers talk, students listen; teachers' contributions are privileged; students respond in predictable, teacher-pleasing ways.

With the new technology, these tendencies are played out in classrooms where students labor at isolated workstations on drill-and-practice grammar software or in word-processing facilities where computers are arranged, rank and file, so that teachers can examine each computer screen at a moment's notice to check on what students are writing. What many in our profession have yet to realize is that electronic technology, unless it is considered carefully and used critically, can and will support any one of a number of negative pedagogical approaches that also grow out of our cultural values and our theories of writing.

From *College Composition and Communication* 42 (February 1991): 55-65.

1. We gratefully acknowledge the insightful comments and excellent advice provided by Marilyn Cooper, Michigan Technological University, and Ron Fortune, Illinois State University.

As editors of *Computers and Composition,* a professional journal devoted to the exploration of computer use in English classes, we read primarily of the laudatory influence of computers in promoting a social construction of knowledge. Scant attention is paid, either in the manuscripts we receive or in the articles we read in other journals, to the harmful ways in which computers can be used even by well-meaning teachers who want to create community and social awareness within their classrooms. If electronic technology is to help us bring about positive changes in writing classes, we must identify and confront the potential problems that computers pose and redirect our efforts, if necessary, to make our classes centers of intellectual openness and exchange. We offer our critical perspectives as members of the composition community who strongly support the use of computers and electronic conferences for writing instruction. Our objections lie not in the use of computer technology and on-line conferences but rather in the uncritical enthusiasm that frequently characterizes the reports of those of us who advocate and support electronic writing classes.

In this paper, we examine the enthusiastic discourse that has accompanied the introduction of computers into writing classes and explore how this language may influence both change and the status quo in electronic classrooms. We do this by looking at published reports of computer use that appear in professional journals, by examining data about computer use collected through questionnaires completed by writing instructors at the 1988 Conference on Computers in Writing and Language Instruction (sponsored by the University of Minnesota at Duluth), and by comparing these analyses with a series of on-site classroom observations. After comparing these accounts of computer use, described through what we call the "rhetoric of technology," and our observations of electronic writing classes, we discuss how electronic technology can intensify those inequitable authority structures common to American education. Finally, we argue that computer technology offers us the chance to transform our writing classes into different kinds of centers of learning if we take a critical perspective and remain sensitive to the social and political dangers that the use of computers may pose.

All too often, those who use computers for composition instruction speak and write of "the effects of technology" in overly positive terms as if computers were good in and of themselves. As editors of a journal devoted to studies in computers and composition, we are most often sent glowing reports that fail to reconcile the differences between a visionary image of technology—what we want computers to do—and our own firsthand observations of how computers are being used in many classrooms around the country. Indeed, this distinctive "rhetoric" of technology seems to characterize more conference presentations, as well as many articles on computer use in other journals. This rhetoric—one of hope, vision, and persuasion—is the primary voice present in most of the work we see coming out of computers-and-composition studies,[2] and it is positive in the sense that it reflects the high expectations of instructors committed to positive educational reform in their writing classes. This same rhetoric, however, may also be dangerous if we want to think critically about technology and its uses.

---

2. Exceptions to this optimistic discourse exist, of course, but these critical voices are less pervasive. For an interesting discussion of how an electronically networked writing class 'mutinied" and lost "all sense of decorum about what [was] appropriate to say or write in an English class," see Marshall Kremers's article, "Adams Sherman Hill Meets *ENFI.*"

The Rhetoric of Technology and Electronic Conferences

For an example of what we call the rhetoric of technology and of how it influences our perceptions and use of technology, we can turn to one specific computer application: electronic bulletin boards and conferences (i.e., conversing over networked computers). Among the claims made about using these electronic conference exchanges in writing classes are the following representative examples:

> Networks create an unusual opportunity to shift away from the traditional writing classroom because they create entirely new pedagogical dynamics. One of the most important is the creation of a written social context, an online discourse community, which presents totally new opportunities for effective instruction in writing. (Batson 32)

> Although I thought I might resent students intruding into my own time after school hours, I find instead that I enjoy our correspondences [over the network]—that I get to know students better and they know me better, too, a benefit that transfers to our classroom. (Kinkead 41)

> All the instructors in the pilot project [using an electronic conference for writing instruction] reported never having seen a group of first-year students, thrown randomly together by the registrar's computer, become as close as their students had. Students set up meetings in the library and in campus computer labs, came early to class and stayed late, made plans together for the next semester, and exchanged addresses. The computer, far from making the class more impersonal, fostered a strikingly close community in one of the nation's largest universities. (Shriner and Rice 476)

> Once people have electronic access, their status, power, and prestige are communicated neither contextually . . . nor dynamically. . . . Thus, charismatic and high status people may have less influence, and group members may participate more equally in computer communication. (Kiesler, Siegel, and McGuire 1125)

> On the network, students can work collaboratively to brainstorm, solve problems, experience writing as real communication with real people. (Thompson 92)

> Those people with powerful ideas will have more influence than those with powerful personalities. . . . The democratization fostered by computer conferencing has other consequences as well. Just as nonverbal cues are missing in conferencing, so too are clues about an individual's status and position. (Spitzer 20)

The above comments represent a number of claims about writing instruction and how it can improve in carefully designed electronic settings: students experience different kinds of intellectual "spaces" in which they can learn differently and sometimes more effectively than in more traditional academic forums; instructors can become better acquainted with their students; many of the status cues marking face-to-face discourse are eliminated, thus allowing for more egalitarian discourse, with greater attention to the text at hand. Collaborative activities increase along with a greater sense of community in computer-supported classes.

Although these remarks reflect claims that we have also made and emphasize pedagogical goals that we too are committed to as specialists in computers and composition, they foreground positive benefits of using networked computers without acknowledging possible negative influences as well. The preceding comments suggest what the use of such networks should encourage, and in the best cases *is* encouraging, but they do not necessarily describe the less desirable outcomes that networks are also capable of supporting. More importantly, we have observed that this highly positive rhetoric directly influences the ways in which teachers perceive and talk about computer use in their classes. When we ask computer-using teachers about word processing in their writing classes (with and without networking) more often than not we again hear echoes of these same optimistic reports.

The Rhetoric of Technology and Computer-Supported Writing Classes

At the 1988 Conference on Computers in Writing and Language Instruction, we distributed lengthy open-ended questionnaires to writing instructors in an attempt to learn how the environment of a writing class—its social structures, discourse, and activities—might be shaped by the use of computers. Although we cannot claim that the answers regarding teaching and technology are representative of the profession as a whole, when considered with other commentary from publications and presentations we have seen, they seem typical of the rhetoric of computer-using instructors and are similar to the language that we ourselves use when talking of our electronic classrooms.[3]

Specifically, the instructors responding to our survey were asked the question, "Do you prefer teaching writing with traditional methods or with computers? Why?" As might be expected at a computers-and-writing gathering, all the respondents preferred teaching writing with computers and gave the following as their reasons, listed in order of their frequency:

1. Students spend a great deal of time writing.
2. Lots of peer teaching goes on.
3. Class becomes more student-centered than teacher-centered.
4. One-on-one conferences between instructor and students increase.
5. Opportunities for collaboration increase.
6. Students share more with other students and instructor.
7. Communication features provide more direct access to students, allowing teachers to "get to know" students better.

---

3. The open-ended questionnaires we analyzed were completed by 25 instructors from 10 different states, in addition to Washington, DC. Seventeen of the instructors taught in four-year colleges, four in community colleges, and four in high schools. First-year college writing classes were most frequently given as the course conducted on computers, but instructors also used computers to teach advanced composition, technical writing, business writing, pedagogy courses in composition instruction, and high-school writing courses. Although the majority of the 25 respondents taught in classrooms where stand-alone computers were the rule, several taught in networked environments in which students and instructors shared writing through electronic mail and bulletin boards.

These comments are remarkably similar to the published claims about the use of on-line conferences that we have already examined. Note that these writing teachers, like their colleagues, also concluded that positive changes such as increased student participation and collaboration occurred in classes when they are computer-supported.

These comments illustrate the commitment of the teachers we surveyed to establishing a new kind of cooperative activity in their writing classes, one in which teaching and learning are shared by both instructors and students and through which traditional notions of teaching are altered. These instructors consider themselves not primarily as dispensers of knowledge but rather as collaborators within a group of learners supported by technology. In this sense, we considered the rhetoric of these instructors to be a reflection of their commitment to positive educational change; the survey respondents used the rhetoric of technology to describe a new cooperative electronic classroom shaped by a theory of teaching in which we understand knowledge as socially constructed by both teachers and students rather than as traditionally established. These teachers had come to see and talk about their classrooms in terms of groups of learners-in-progress working with instructors who are also learners (Lunsford and Glenn 186).

As we continued to analyze the open-ended responses to the questionnaires, however, it became clear that when instructors foregrounded the beneficial influences of using computers, they often neglected to mention any negative effects of using the new technology. We recognized, as well, that this perspective was widespread and that the observations the survey teachers made were the same as those we had heard from writing instructors at our own institutions. Moreover, at workshops we have conducted during the past two years, we continue to hear similar, positive reports that correspond to these earlier, more formal analyses.

Teaching Practices and the Computer-Supported Writing Class

Neither the published claims nor the survey responses, however, helped us to explain the less positive, more problematic uses of computers that we encountered during the past five years as we visited many other electronic writing classes around the country and made informal observations. Notes from a sampling of computer-supported classes we observed more formally in 1988 provided us with information about some of the more problematic social and pedagogical changes in electronic classes.[4] Both the formal and informal observations we made supported neither the teachers' responses in the questionnaires nor the published rhetoric of technology that had been our impetus for this study. In other words, we began to see that the language teachers used when they wrote about using computers sometimes provided incomplete stories that omitted other possible interpretations. Let us explain by using examples from those representative classes we formally observed in stand-alone computer classrooms.

First, however, it is important to note that our observations were limited and that we may well have missed day-to-day classroom dynamics. On other days, in some of the classes, the use of computers may indeed have fostered positive changes in the intellec-

4. We observed ten first-year writing classes taught on computers during the summer and fall of 1988. All instructors had taught composition with computers for at least one year, and several had taught composition for five years or more. Some were teaching assistants, and some were full-time composition instructors.

tual climate of the classroom. But we hope that by concentrating on some of the problematic aspects of these electronic classes, we can emphasize that computers do not automatically create ideal learning situations. This is not to say that electronic technology cannot encourage social interaction and cooperative undertakings but rather to stress, in Michel Foucault's words, that a technology cannot "guarantee" any behavior alone "simply by its nature" ("Space" 245); according to Foucault, the "architecture" of such electronic spaces is a highly political act in itself. Like the traditional classroom, the architecture of electronic spaces can put some students at a disadvantage, thwarting rather than encouraging learning.

In each of the ten classes we observed, with a few exceptions, there was a lot of writing going on. In fact, there was so much writing that we wondered sometimes why the time was set aside as class time, rather than as time that students could spend on their writing in a computer lab. We looked for exchanges and talk between instructor and student, and between students—but what we commonly saw were not careful, two-way discussions of the writing problems students were encountering in their papers. Rather the instructors answered a series of one-time queries often having to do with mechanics or coming from the "does-this-sound-right" category. There were exceptions: sometimes an instructor moved from student to student and spent several minutes with each, talking about specific writing problems highlighted on the computer screen. For the most part, though, instructors walked around the room, looking eager, we might add, for someone in the class to need them in some capacity. Although this observation seems to fit with one of the more frequent claims that teachers made for electronic writing classes—students do a lot of writing—the claim does not completely represent the classes we observed. The use of computers in these classes seemed to come between teachers and students, pre-empting valuable exchanges among members of the class, teachers and students alike.

Another kind of computer-supported class we observed reflected traditional practices of writing instruction in American classrooms. The instructor projected a student paper on an overhead projection system, and students critiqued various aspects of the paper. In each instance, classmates seemed to be searching for answers to the instructor's preset questions. And only three or four students were participating in these rather contrived discussions. This sort of class we saw as a variation on George Hillocks's presentational mode. Although the instructors were not lecturing, they had in mind answers that the students were to supply; hence, the discussion, in effect, became the instructor's "presentation." At these times we wondered about the advantage of having computers in the classroom. The use of technology in these classes, far from creating a new forum for learning, simply magnified the power differential between students and the instructor. Ostensibly computers were being used to "share" writing, but the effect of such sharing was to make the class more teacher-centered and teacher-controlled. Hence, describing technology as a mechanism for increasing the sharing of texts or bringing students and teachers together on a more equal basis again told only a part of the story.

Still another typical class we observed was one in which students were meeting in groups, often focusing on something written on the monitor or producing text on the screen. Yet the conversations we overheard only sometimes related to the task at hand—and often, once again, the effort put forth by students seemed to be one aimed at pleasing the instructor rather than one illustrative of active engagement with their classmates or the texts. This type of class seemed to fit with responses from the questionnaires that credited technology with encouraging "lots of peer teaching" or "more opportunities for collaboration." While such claims seemed outwardly to reflect the electronic writing class, they did not take into account the groups that we observed in which neither peer teaching nor collaboration among students occurred.

This realization, then, leads us to believe that it is not enough for teachers to talk about computer use in uncritical terms. We can no longer afford simply, and only, to dwell on the best parts, to tell stories about the best classroom moments, and to feature the more positive findings about computers. Rather, we must begin to identify the ways in which technology can fail us. We need to recognize the high costs of hardware and software, recognize that computers can, and often do, support instruction that is as repressive and lockstep as any that we have seen. We need to be aware of the fact that electronic classrooms can actually be used to dampen creativity, writing, intellectual exchanges, rather than to encourage them. We need to talk about the dangers of instructors who use computers to deliver drill-and-practice exercises to students or of instructors who promote the use of style analyzers to underscore student errors more effectively than they did five years ago with red pens.

How do we proceed then? We do not advocate abandoning the use of technology and relying primarily on script and print for our teaching without the aid of word processing and other computer applications such as communication software; nor do we suggest eliminating our descriptions of the positive learning environments that technology can help us to create. Instead, we must try to use our awareness of the discrepancies we have noted as a basis for constructing a more complete image of how technology can be used positively *and* negatively. We must plan carefully and develop the necessary critical perspectives to help us avoid using computers to advance or promote mediocrity in writing instruction. A balanced and increasingly critical perspective is a starting point: by viewing our classes as sites of both paradox and promise we can construct a mature view of how the use of electronic technology can abet our teaching.

Teaching Practices and Electronic On-Line Conferences

As a more specific example of how a critical perspective can help us to identify, and we hope avoid, the dangers that can accompany computer technology in writing classes, we turn again to the use of electronic conferences and bulletin boards. A critical reexamination of these on-line exchanges suggests that while conferences can help teachers create new and engaging forums for learning, they can also serve in ways that might inhibit open exchanges, reduce active learning, and limit the opportunities for honest intellectual engagement.

In the context of Foucault's description of disciplinary institutions as presented in *Discipline and Punish,* we can speculate as to how such conferences might work to the detriment of students and their learning. The electronic spaces created through networking, we learn by reading Foucault, might also be used as disciplinary technologies, serving to control students and their discourse. Of such technologies, Foucault writes:

> [They are] no longer built simply to be seen . . . , or to observe the external space . . . , but to permit an internal, articulated and detailed control—to render visible

> those who are inside it; in more general terms, an architecture that would operate to transform individuals: to act on those it shelters, to provide a hold on their conduct, to carry the effects of power right to them, to make it possible to know them, to alter them. (172)

This particular theoretical perspective, while it is highly incongruent with existing interpretations of conferences and what goes on in them, may at the same time enrich and problematize those interpretations.

A powerful metaphor to help us critically examine the uses of electronic forums is further elaborated in Foucault's discussion of Bentham's Panopticon, the perfect disciplinary mechanism for the exercise of power.[5] Originally designed as a circular prison building with a guard tower in the middle and the prisoners' cells arranged along the outside, the Panopticon, writes Foucault, is a "mechanism of power reduced to its ideal form" (*Discipline* 205), making it possible for wardens and guards to observe the behavior of inmates without they themselves being observed. Foucault argues that within such a space, because inmates do not know when they are being observed or ignored, prisoners are constantly and unrelentingly self-disciplining. Moreover, because surveillance is "unverifiable," it is all the more effective and oppressive. Although panoptic space differs from electronic bulletin boards and conferences in that students, unlike Bentham's inmates, can converse with one another over networks, those who have conversed over computers will recognize how eavesdropping and watching are made easy through the architecture of an electronic network.

Writing instructors can use networks and electronic bulletin boards as disciplinary mechanisms for observing students' intellectual contributions to written discussions. The institutional requirement of student evaluation contributes to this practice as instructors seek ways "to give students credit" for conference participation. Under certain conditions, without carefully thinking out the theoretical consequences, instructors enter conferences to read and monitor students' conversations without revealing themselves as readers and evaluators. We know after all that electronic conferences are, in some ways, spaces open to public scrutiny, places where individuals with the power of control over technology can observe conversations and participants without being seen and without contributing. When instructors take samples from network discussions into the classroom and use these as positive or negative examples, they are employing electronic conferences to discipline, to shape the conversations and academic discourse of their students.

Such a theoretical perspective reminds us that electronic spaces, like other spaces, are constructed within contextual and political frameworks of cultural values, a point that Shoshana Zuboff makes in her study of computer networking in the corporate environment. As in corporate settings, the architecture of computer networking may encourage "surveillance" of participants. Writing instructors praise on-line communication programs for helping them "get to know" students better, a phrase that survey instructors used in a positive sense but that Foucault includes to describe an architecture of

---

5. We are grateful to Vicki Byard, Purdue University, for bringing Foucault's treatment of Bentham's Panopticon to our attention at the 1989 CCCC in Seattle. In her insightful paper, "Power Play: The Use and Abuse of Power Relationships in Peer Critiquing," she suggested that even those approaches we use with the most liberating intentions may well prove disciplinary in nature.

control. Teachers who have easy access to students through a network can also "keep tabs" on student participation, blurring the thin line between "evaluating" contributions students make to electronic conferences and "inspecting" conversations that occur electronically.

Instructors inspecting electronic spaces and networked conversation have power that exceeds our expectations or those of students. In addition, many students who know a teacher is observing their conversation will self-discipline themselves and their prose in ways they consider socially and educationally appropriate. Constructing such spaces so that they can provide room for positive activities—for learning, for the resistant discourse characteristic of students thinking across the grain of convention, for marginalized students' voices—requires a sophisticated understanding of power and its reflection in architectural terms.

Conclusion

In this paper, we have suggested that the current professional conversation about computer use in writing classes, as evidenced in published accounts, is incomplete in at least one essential and important way. While containing valuable accounts of electronic classes, this conversation fails to provide us with a critical perspective on the problematic aspects of computer use and thus with a full understanding of how the use of technology can affect the social, political, and educational environments within which we teach. In making this point, we are not arguing against the use of computers in general or, more specifically, against the promising use of electronic conferences and bulletin boards. The central assumption underlying our argument is that writing instructors, by thinking critically and carefully about technology, can succeed in using it to improve the educational spaces we inhabit.

Our view of teaching and of how students learn invariably shapes our behavior in the classroom. The metaphors we build to house our professional knowledge exert powerful influence over us. Few of us, we would argue, construe our role as that of "controller," "gatekeeper," or "guard." We are more likely in the context of the writing class to think of ourselves as "teacher," "writer," and perhaps "expert." If we plan carefully and examine our integration of technology critically, computers have the potential for helping us shift traditional authority structures inherent in American education. We can, if we work at it, become learners within a community of other learners, our students. But the change will not happen automatically in the electronic classroom anymore than in a traditional classroom. We have to labor diligently to bring it about.

As teachers we are authority figures. Our culture has imbued us with considerable power within the confines of the classroom: we are the architects of the spaces in which our students learn. Although the use of computer technology may give us greater freedom to construct more effective learning environments, it may also lead us unknowingly to assume positions of power that contradict our notions of good teaching. Unless we remain aware of our electronic writing classes as sites of paradox and promise, transformed by a new writing technology, and unless we plan carefully for intended outcomes, we may unwittingly use computers to maintain rigid authority structures that contribute neither to good teaching nor to good learning.

Works Cited

Batson, Trent. "The ENFI Project: A Networked Classroom Approach to Writing Instruction." *Academic Computing* Feb.-Mar. 1988: 32-33.

Byard, Vicki. "Power Play: The Use and Abuse of Power Relationships in Peer Critiquing." Conference on College Composition and Communication Convention. Seattle, Mar. 1989.

Foucault, Michel. *Discipline and Punish: The Birth of the Prison*. Trans. Alan Sheridan. New York: Vintage, 1979.

———. "Space, Knowledge and Power." *The Foucault Reader*. Ed. Paul Rabinow. New York: Pantheon, 1984. 239-256.

Hillocks, George, Jr. *Research on Written Composition*. Urbana: NCTE, 1986.

Kiesler, Sara, Jane Siegel, and Timothy W. McGuire. "Social Psychological Aspects of Computer-Mediated Communication." *American Psychologist* 39 (1984): 1123-1134.

Kinkead, Joyce. "Wired: Computer Networks in the English Classroom." *English Journal* 77 (1988): 39-41.

Kremers, Marshall. "Adams Sherman Hill Meets *ENFI*." *Computers and Composition* 5 (1988): 69-77.

Lunsford, Andrea A., and Cheryl Glenn. "Rhetorical Theory and the Teaching of Writing." *On Literacy and Its Teaching: Issues in English Education*. Ed. Gail E. Hawisher and Anna O. Soter. Albany: State U of New York, 1990. 174-189.

Shriner, Delores K., and William C. Rice. "Computer Conferencing and Collaborative Learning: A Discourse Community at Work." *College Composition and Communication* 40 (1989): 472-478.

Spitzer, Michael. "Writing Style in Computer Conferences." *IEEE Transactions on Professional Communications* 29 (1986): 19-22.

Thompson, Diane P. "Teaching Writing on a Local Area Network." *T.H.E. Journal* 15 (1987): 92-97.

Zuboff, Shoshana. *In the Age of the Smart Machine: The Future of Work and Power*. New York: Basic, 1988.

Dawn Rodrigues and Raymond Rodrigues

How Word Processing Is Changing Our Teaching: New Technologies, New Approaches, New Challenges

We would like to begin with an anecdote. Last fall, a conscientious graduate student walked into the new computer classroom at Colorado State, thinking it was just another computer lab. He sat down at the only available computer (which happened to be connected to the only available printer) and began working on a project. The teacher, who did not appreciate his rude entry into the middle of her composition class, immediately walked over to him and quietly explained that a class was in session. "It won't bother me," he said. Realizing that he needed to be addressed more directly, the teacher continued, "You'll have to leave the room." Incredulous, the student dutifully complied. Later that day, he complained to the lab director about the rude treatment he had been forced to endure. When the lab director reminded him that he had interrupted a teacher's class, he responded, "But the teacher wasn't teaching." The lab director asked how he knew that no teaching was taking place. The student replied, "All the students were doing was typing."

All the students were doing was typing. Hardly. For when teachers teach writing with computers, their teaching changes—in ways that most casual observers probably cannot appreciate, in ways that teachers themselves can't always predict, and in ways that early researchers (focusing on the effects of word processing on the quality of written products) were unable to imagine. In this paper, we would like to explore a few of the diverse ways teachers have responded with ingenuity and vision to the challenges of technology.

We will begin with some background about teachers' early experiences with computer software. Then, we will move to a presentation of some variations on teaching with a word-processing package and related tools that have made it possible for teachers to develop different computer-writing pedagogies for their distinct contexts: traditional classroom, computer lab, or some combination. The point we want to make is that, regardless of their contexts, as long as students have access to a word-processing program, teachers who re-envision their teaching as a result of the available technology can create dynamic classroom environments for writing.

From *Computers and Composition* 7 (November 1989): 13-25. Reprinted by permission of the publisher.

From Supplemental to Integrative Roles for Computers

Word processing is now regarded as an invaluable tool for teaching writing. It was not always so highly rated. Initially, most teachers sought elaborate software, not word-processing programs, for their computer-assisted classrooms and labs; in the early part of the 1980s, teachers flocked to sessions focusing on computers and composition at the Conference on College Composition and Communication, listening in amazement as the speakers described what computers could do: Pre-writing programs like Hugh Burns' Topoi (1984) and our own Creative Problem-Solving (1984) could engage students in dialogue; style analysis programs like the WRITER'S WORKBENCH (Kiefer and Smith, 1983) could analyze students' text for possible usage errors; and commenting programs like RSVP, developed at Miami-Dade Community College (Anandam, 1980), could help teachers grade papers.

Teachers were fascinated primarily by the instructional features of these early programs—the way the computer software could, purportedly, handle a variety of onerous chores and thus free teachers for other tasks. Reflecting this interest, William Wresch's essay, "Computers in English Class: Finally Beyond Grammar and Spelling Drills" (1982), focuses on only the mentoring features of computer software. Similarly, our "Computer-Assisted Invention: Its Place and Potential" (Rodrigues and Rodrigues, 1984) indicates that a primary benefit of pre-writing and invention programs is their ability to function as supplemental tutoring systems. We, too, were dazzled by the software and—in envisioning the role of pre-writing software as an out-of-class teaching supplement—failed to consider ways students might use these programs in a computer classroom. With teachers present, pre-writing programs can become writing tools rather than teaching machines. With a teacher present, students are more likely to judiciously use pre-writing programs, selecting programs appropriate for their individual writing needs and suitable to their varying writing tasks.

Teachers' first attempts to integrate word processing into their classes were similarly shortsighted. Teachers viewed word processing as a possible solution to current pedagogical problems instead of as a teaching tool. For instance, in their hope that students would revise and edit more extensively if they typed their drafts at a computer terminal, teachers willingly and eagerly sent students to the new computer labs being established in their schools. Sending students to a computer for help was, they soon discovered, unproductive. Word-processing packages themselves do not teach students how to revise. The results of early research are disappointing: Students revised no more using a word-processing program on a computer than they did using pencil and paper (Kurland, 1982). Subsequent research has confirmed these results: Other than two reports of slight improvement in the quality of basic writers' texts (King, Birnhaum, and Wageman, 1984; Pivarnik, 1985), there has not been any evidence to suggest that the word processing alone makes a significant difference in writing quality (Hawisher, 1988). And why should it? Word-processing packages themselves do not teach students how to write and revise—but teachers can.

By focusing on the idea that computer programs can replace teachers in their ability to handle certain subtasks of teaching—what Fred Kemp has called the "replacement fallacy" (1987)—teachers had temporarily failed to grasp a more powerful aspect of computers: their capacity to expand human understanding "much like telescopes can extend human vision'' (Kemp, 1987). With experience, however, teachers recognized the limitations of their early use of computers and software, and they began to view their teaching through technological lenses. As a result, teachers have learned how to create exciting computer environments in their classes, environments that are beginning to have a powerful impact on their teaching.

Different Contexts, Different Pedagogies

Anyone who has the opportunity to look closely at a variety of classrooms across the country will see that teachers have developed a variety of ways to respond to the availability of word processing. Their responses vary depending on their individual contexts: the number of computers teachers have available to them, the placement of their computers, and the frequency with which their students can gain access to the computers. And, of course, their teaching strategies vary according to the ability of the teacher to continue learning and adjusting to both the changing students and the changing technology.

As teachers experience the effect word processing has on their students' writing behaviors, they tend to make significant changes in their teaching strategies. For example, instead of having students complete pre-writing activities individually, teachers frequently encourage students to generate ideas in pairs or in small groups. Instead of recommending that all students use the word processing in the same way, teachers now are inclined to encourage students to develop their own combinations of pencil and paper and computer-writing strategies. Teachers are finding Bridwell's early observations (1984-1985) to be accurate: Writers do indeed react to word processing in vastly different ways. Not only do students who are accustomed to doing a lot of pre-planning on paper have difficulty when they attempt to compose at the monitor, students with limited access to computers rarely learn how to compose freely and comfortably at the computer monitor. Even the notion of what constitutes a computer-composer has changed. As Hawisher (1988) notes:

> Whereas at one time computer composing tended to demand that every task or subprocess entailed in the activity of writing be performed at a computer, there seems to be growing agreement that writers who use some combination of strategies involving both hard copy and on-line composing are indeed computer composers. (p. 9)

Paul LeBlanc's recent essay in *Computers and Composition* (1988) emphasizes the need for teachers to acknowledge students' different computer-writing processes in their teaching. LeBlanc explains that writers with different writing processes have distinct kinds of problems when they become computer composers; thus, teachers need to develop an assortment of pedagogies to ''maximize the computer's helpfulness'' (p. 30). For example, if a teacher notices that a student revises excessively, the teacher needs to know when to intervene and help the student writer overcome the writing problem caused by the ease of revising with a word-processing program. If another student does not revise enough, the teacher needs to be certain that the student can indeed use the commands for moving and copying text.

Because the contexts in which teachers teach with computers vary and because those contexts determine much of how and what they teach, the strategies teachers use and the innovations they develop differ. We have grouped those primary contexts for teaching with computers in three categories: teaching in a traditional classroom while students have some access to computers in other contexts; teaching in a traditional classroom with regular or occasional use of a computer lab; and teaching in a computerized classroom.

The Traditional Classroom Context. This is probably the most common context that composition instructors find themselves in, and, given the financial constraints of most institutions, it may be the model for the near future. In this context, the instructor teaches in a room without computers, but the students have access to computers outside the classroom. In some schools, they can go to a campus-wide or department computer laboratory, or they can write on their personal computers in their rooms or in residence hall computer labs.

Many instructors in these contexts continue teaching in traditional ways. But instructors can—and many do—modify their traditional classrooms. They bring computers into the room to demonstrate composing techniques, hooking the computer to a large-screen monitor or video-data display, or they bring in a videotape of different writers' emerging text. (The instructor easily can create the latter by hooking a computer to a videotape machine and recording while someone is writing.) Then, the instructor plays the tape in class, demonstrating how the writer on each segment of the tape actually writes—with false starts, deletions, insertions, paragraph moves, or any other writing techniques that the students should be aware of. An excellent use for this technique is in demonstrating revision strategies: As a segment of text appears on the screen, the instructor can stop the videotape and ask the students to suggest ways the text might be improved. Then, the instructor can start the tape and let the students see how the text was moved and reshaped by the writer in the process of revising.

Teachers make other changes in their traditional classroom approaches. Because they know that students have access to word processing outside of class, they frequently assume that all drafts used in class will be printed drafts. The ease of reading these means that students will be able to work collaboratively—and more rapidly—as they offer revision suggestions. Realizing that their peers all have entered their drafts at a word processor encourages students to suggest substantive changes, knowing that their advice will be received more willingly by their partners, who will not have to recopy their texts to revise.

As students work on a given assignment, the instructor can ask different students to bring in printed copies of their drafts in a variety of stages. Drafts now take on a new meaning for writers, who can save each successive "draft" with a new name and thus chart the movement of their text through a cycle of revisions. Teachers either can photocopy the computer printouts, make overhead projector transparencies of them, or both. The ease by which these multiple drafts can be produced and duplicated means that writing becomes more immediate and, therefore, more meaningful for students.

In discussing writing with the entire class, many instructors relate specific writing strategies to the word-processing capabilities. The language of the instructor changes accordingly. Instead of speaking only about "revising," "adding details," or "reorganizing," the instructor also speaks of "deleting," "inserting," or "moving." Brainstorming moves from being a classroom exercise that many students do not actually do when they work independently with pencil and paper to an activity teachers can choreograph with an assignment such as the following:

> List all the things you want to write about at the bottom of your file. Then start writing and move back and forth from your list of ideas to your draft. Bring along your printed list and your draft to class next time.

Technical advice, especially necessary when some students are new to word processing, also enters the discussions: "Be sure to save your files as often as possible"; "If you write in the computer lab, be sure to print a hard copy before you leave"; or even, "Make sure you protect your floppy disk when you carry it out of the lab."

In short, when students have access to word processing, teaching in a traditional classroom need not be traditional teaching. Instructors can adjust their teaching if they are aware of the possibilities offered by word processing. When writing tasks are completed at the computer, expectations for what students will bring to class change, and the nature of classroom interactions changes, too.

The Traditional Classroom Supplemented by Computer Laboratories. Even though most composition classrooms may not have computers within them, many instructors now have the option of reserving the computer laboratory—either regularly or occasionally—as an extension of the classroom. The laboratory can be used by students either individually or as a class. If an entire class is to use the computer laboratory, and if other instructors also use the laboratory, scheduling must be correlated with the pedagogical intent of the laboratory sessions. Some instructors schedule the class into the laboratory for one session a week. In such a case, instructors often use the laboratory for a specific phase of writing, such as drafting the initial draft or revising after peer critiques. The original laboratory at Colorado State, for example, was used for style analysis but now has multiple uses. Some English Departments reschedule laboratory times every two or three weeks, allowing instructors to vary their purposes for taking classes to the laboratories. And other laboratories are scheduled on a first-come, first-serve basis, with appropriate controls to accommodate any instructor who desires to use the laboratory.

If the lab that students have available to them outside of class is "networked" with an electronic mail system, students may be able to collaborate on common writing projects without needing to be present at the same time. Asynchronous systems do not require that the intended recipient of the message be present at the time the message is sent, but the systems allow students to check their computer mailboxes whenever their schedules permit. And if students can communicate with the laboratory through a modem, either by reaching the laboratory through a campus-wide network from anywhere on campus or by phoning from off-campus, the use of the laboratory is multiplied. Teachers at New Jersey Institute of Technology have conducted entire courses in what they describe as a "virtual classroom"—a classroom that "meets" only when students access their mail and respond to teacher or peer questions.

At Colorado State, instructors in the English Department have used electronic mail capabilities for a variety of purposes. After students have read a common reading, they respond to the writing in free-response entries or in specifically cued responses, posting their responses in a mail file that others can access only in "read-only" mode. In their next electronic assignment, teachers might ask students to respond to or comment on the responses of the other writers. In one instructor's class, students have used mail in a less-structured way, carrying on extensive conversations debating issues through the mail system. In all of these examples, electronic mail functions as an electronic double-entry notebook, a computer arena for collaborative reflection.

Although each class can converse in written messages with all members in the class or even with members of other classes, some instructors prefer, at times, to divide their classes into small groups so that these students collaborate more efficiently with one another on specialized projects. In short, the computer laboratory has allowed instructors to continue their instruction beyond the traditional classroom, even when the instructor cannot be present.

The Computerized Classroom. We first envisioned the computerized classroom as the ideal context for teaching writing, but economics have not allowed that to materialize for many English departments, and some instructors have decided that they prefer to have at least one class period per week in a context away from the computers. Nevertheless, the ability to teach writing in a classroom in which writers can compose and revise at their own computers provides new variations. As Ronald Sudol (1985) has remarked, teaching writing in a computer classroom "offers an opportunity to reinvent the workshop classroom model in the context of the new technology" (p. 331). We will discuss computer laboratories that are not networked and those that are networked.

Both networked and non-networked classrooms allow the instructor to observe, coach, and monitor students as they write. In a non-networked lab, an instructor may be able to use a large monitor or a data view projector (a device that is used in conjunction with a computer and an overhead projector) to allow everyone to view the writing on one computer. Students can be asked to come up to that computer and demonstrate specific writing techniques that others might want to emulate. By being able to monitor students as they write, teachers can be alert for "model ways" that students use to generate or revise text. They can then ask students to re-play that strategy for all to see.

Guiding the work of students in such a lab requires a flexible instructor capable of shifting from a student who is still generating ideas to a student who is revising a draft. The workshop environment that evolves when writers work on essays in a computer classroom has little resemblance to a traditional classroom. A computer classroom is noisy, seemingly chaotic, and demands new teaching strategies. Students in various stages of writing challenge the instructor to find or schedule specific times to work with students on common writing experiences. In this particular computer-classroom context, teachers at the University of Massachusetts at Amherst found a need for in-service training on how to read and respond quickly to student writing on-line rather than on paper (Moran and LeBlanc, 1989).

How the room is physically arranged also influences the way that an instructor teaches. For example, in one writing laboratory at Colorado State with computers arranged around the four walls, inexperienced computer-writing instructors expressed dismay at not having a "front of the room." We are still learning how well and in how many ways they have adjusted to this perceived problem. But it does serve to symbolize the influence of the computer upon instructors. Should the monitors all be visible from the center of the room (and students, therefore, be facing the walls) so that the instructor can easily read what each person is writing? Or should the monitors face toward the walls (with students facing the center of the room) so that the instructor can observe the faces of the students and respond to facial cues? Personal preferences may have to yield to the will of the majority within a department.

If the computers are networked, a new set of possibilities influences the instructor. If the laboratory is capable of video data switching (the capability of displaying the text on any screen or on any other screen or screens), for example, the instructor can enable all students in the class to observe another student's writing by switching that student's input to all terminals in the room. The other students can observe the writing as it evolves or can critique it at a particular moment. If students are grouped on different "channels," the instructor can effectively promote collaborative activities within those groups, can enable the groups to work on common writing tasks, and can develop a wide variety of peer writing tasks.

With video-switching and CB-chat utilities (a feature that allows students to send messages to one another, to small groups, or to the entire class in "real-time"), students can carry out actual dialogues in writing. By using such written dialogues, students can learn how written "talk" can be used for inventing ideas (Batson, 1989). This experience may help students develop the ability to sustain a line of discussion for amazingly long stretches of time, not only in casual written conversations, but, eventually, in their own drafts.

A potential problem that instructors may experience when they teach with this "chat mode" occurs when student responses to one another's questions do not appear in sequence on the screen but are interspersed by other students' responses to still other students' questions. For example, a student may respond to one particular comment only to find eight other responses interspersed before his or hers and to discover that the topic shifted. Thus,a continuous, meaningful interchange may be destroyed. As with other networked approaches, having small groups or pairs of students on separate channels will overcome this problem. The Annenberg Corporation has funded the ENFI project (Electronic Networks for Interaction), a consortium of several universities who are doing research in the benefits, methods, and potential problems of this use of synchronous networked conversations.

As these varied situations demonstrate, computer classrooms allow teachers to create new kinds of writing environments, and, as S. Bailey Shurbutt (1987) has noted, when students learn how to use the computer in such settings, they develop control over their writing. By creating powerful environments for writers, teachers discover that word processing presents opportunities for them to help students in ways that traditional techniques simply did not allow.

Conclusion: New Approaches, New Challenges

As we have prepared to teach with computers, our concept of our teaching roles has changed. And as our view of computers and computer software has broadened in scope, we have learned how to orchestrate our students' learning experiences. Instead of searching for software to solve our teaching problems, we have begun actively to reinvent classroom teaching. Thus, neither research findings, nor computer contexts, nor our limited understanding of computers and software stand in the way of our teaching students to write.

Teachers now realize that the activities of students as they write both on and off the computer have to flow smoothly, or else the computer-writing experiences can lead to improved motivation without improved writing. Researchers now realize that they are unlikely to document actual improvement in the quality of student writing as a result of students' using the computer if the teacher and students have not tapped the full potential of the computer. Accordingly, researchers have begun to re-examine their research designs.

In short, teachers who integrate computers into their writing classes really are teaching more than just what they would teach if they did not use computers, or if they relied only on fancy software. They are teaching a new way of thinking about and working with writing—a way of thinking of text as fluid and movable, a way of thinking about communication as dynamic and purposeful.

When visitors observe what is happening in computer classrooms across the country, what do they see? Are the students typing, revising, or writing? Are the teachers teaching? When we teach writing with computers, our teaching and our students' writing behaviors change—in ways that a casual observer might not appreciate.

Works Cited

Anandam, K., Eisel, E., and Kotler, N. "Effectiveness of a Computer-based Feedback System for Writing." *Journal of Computer-Based Instruction* 6 (1980): 125-133.

Batson, T. "Teaching in Networked Classrooms." *Empowering Teachers: Teacher-Education for Computers*. Ed. C. Selfe, D. Rodrigues, and W. Oates. Urbana, IL: NCTE, 1989. 247-256.

Bridwell, L. S., P. Johnson, and S. Brehe. "Composing and Computers: Case Studies of Experienced Writers." *Writing in Real Time: Modeling Production Processes*. Ed. A. Matsuhashi. Norwood, NJ: Ablex, 1986.

Bridwell, L. S., G. Sire, and R. Brooke. "Revising and Computing: Case Studies of Student Writers." *The Acquisition of Written Language: Revision and Response*. Ed. S. Freedman. Norwood, NJ: Ablex, 1985. 172-194.

Burns, H. "Recollections of First-Generation Computer-Assisted Prewriting." *The Computer in Composition Instruction: A Writer's Tool*. Ed. W. Wresch. Urbana, IL: NCTE, 1984. 15-33.

Hawisher, G. "Research Update: Writing and Word Processing." *Computers and Composition* 5 (1988): 7-27.

Kemp, F. "Getting Smart with Computers: Computer-Aided Heuristics for Student Writers." *The Writing Center Journal* 8 (1987): 3-10.

Kiefer, K., and C. Smith. "Textual Analysis with Computers: Tests of Bell Laboratories' Computer Software." *Research in the Teaching of English* 17 (1983): 201-214.

King, B., J. Birnbaum, and J. Wageman. "Word Processing and the Basic College Writer." *The Written Word and the Word Processor*. Ed. T. Martinez. Philadelphia: Delaware Valley Writing Council, 1984.

Kurland, M. "The Development of the Bank Street Writer." Second annual Conference on Microcomputers and Education. Phoenix, Arizona, 1982.

LeBlanc, P. "Interactive Networking: Creating Bridges Between Speech, Writing, and Composition." *Computers and Composition* 5.2 (1988): 29-42.

Moran, C., and P. LeBlanc. "Adapting to a New Environment: Word Processing and the Training of Writing Teachers at the University of Massachusetts at Amherst." *Computers in English and the Language Arts: The Challenge of Teacher Education.* Ed. C. Selfe, D. Rodrigues, and W. Oates. Urbana, IL: NCTE, 1989. 111-130.

Pivarnik, B. "The Effect of Training in Word Processing on the Writing Quality of Eleventh Grade Students." (Doctoral dissertation, University of Connecticut, 1985) *Dissertation Abstracts International* 46 (1985): 1827-A.

Rodrigues, D., and R. Rodrigues. "Computer-Assisted Creative-Problem Solving." *The Computer in Composition Instruction: A Writer's Tool.* Ed. W. Wresch. Urbana, IL: NCTE, 1984. 34-46.

Rodrigues, R., and D. Rodrigues. "Computer-Assisted Invention: Its Place and Potential." *College Composition and Communication* 35 (1984): 78-87.

Selfe, C., D. Rodrigues, and W. Oates, eds. *Computers in English and the Language Arts: The Challenge of Teacher Education.* Urbana, IL: NCTE, 1989.

Shurbutt, S. B. "Integration of Classroom Computer Use and the Peer Evaluation Process." *The Writing Center Journal* 8 (1987): 35-42.

Sudol, R. "Applied Word Processing: Notes on Authority, Responsibility, and Revision in a Workshop Model." *College Composition and Communication* 36 (1985): 331-335.

Wresch, W. "Computers in English Class." *College English* 44 (1982): 483-490.

Wayne M. Butler and James L. Kinneavy

The Electronic Discourse Community: god, Meet Donald Duck

Introduction: Chapter 6

In an early issue of *Focuses,* Joseph Trimmer chronicled the history of writing centers in an anecdotal fashion, drawing upon his years of experience in various settings. His history included six chapters: A Writing Workshop (circa 1960), A Writing Laboratory (circa 1965), A Writing Clinic (circa 1970), A Writing Center (circa 1975), A Learning Center (circa 1985), and A Culture Club (circa 1990). As he describes the characteristics of each setting and era, Trimmer emphasizes that each variety of the writing center reflects the dominant rhetorical, composition, and pedagogical theories of the time. For instance, the activities of the 1975 writing center Trimmer set up in "a small side street across from the English Department" seemed "suspicious if not downright revolutionary." Trimmer writes, "Our password was 'process." In the beginning we saw it alluded to in xeroxed copies of manuscripts written by anonymous colleagues. Then we heard it debated at conferences, matched with the names of the inner circle—Emig, Elbow, Murray—and magnified by new words—*freewriting, drafting, revising*" (30-31). His forecast for the future, Chapter 6: A Culture Club (circa 1990), is clearly speculative: "Perhaps it would be best if you imagined this story yourself, because one story will not please everyone" (34). After our experiences in a networked microcomputer classroom, our version of Chapter 6 would be called The Electronic Discourse Community (circa 1990).

Such a community thrives on the harmonious combination of various contemporary technologies, rhetorical and composition theories, and pedgogical practices. Just as writing teachers' emphasis on product shaped the practices of the writing workshop of the early 1960s, an emphasis which privileged error detection and correctness, thus driving students to writing workshops like the ill seeking out emergency rooms, the electronic discourse community reflects the writing profession's recent concerns with social constructivist epistemology, the knowledge making functions of discourse communities, the power of collaborative learning practices, and the role of computers in the English classroom and curriculum. In this essay we will describe an environment which we believe effectively redefines the traditional writing center, and writing classrooms in general.

From *Focuses* 4 (Winter 1991): 91-108. Reprinted by permission of the publisher.

From Writing Center to Electronic Discourse Community. Historically, writing centers have been places where students receive individualized help with writing skills and concepts and particular writing assignments and tasks, usually from experienced peer tutors, graduate students, or instructors, and generally outside and in addition to the activities of the writing classroom. The role of the writing center has been to help students get what they are not getting in class for the purpose of becoming better students in their writing classes. The electronic discourse community, as we have designed it, tends to extend the activities of the writing classroom rather than supplement them. While the traditional writing center remains a place where students come in to get help, to be remediated, to be "fixed," the electronic discourse community becomes a new kind of center, a place where writers write to and for other writers.

The introduction of the personal computer to the writing lab, of course, is not new or revolutionary. In fact, because of sundry constraints that limited English departments to only a few computers, the writing lab was a natural place to install the new technology. By 1987, microcomputers in the writing lab were so prevalent that *The Writing Center Journal* dedicated a special-topic issue to "explore some of the questions, concerns, and benefits that have resulted from our use of computers in the writing center" (Harris and Kinkead 1). The articles focused on the technology of the time—stand-alone microcomputers running wordprocessing, invention heuristic, and style-checking programs. The issue's contributing authors could not have anticipated how Local Area Network (LAN) technology might affect the places where writing is done. Nevertheless, the editors discovered a phenomenon that would anticipate the major strength of the networked writing classroom when they noted that "computers can supplement the dialogue between tutor and writer" (Harris and Kinkead 1). In the networked classroom, it is this dialogue, though not so much between tutor and writer as between writer and writer, that serves as the significant difference between both traditional writing centers and writing classrooms and the electronic discourse community. If the traditional writing center is akin to individualized music lessons in which the music teacher demonstrates and then evaluates the individual performance of the individual player, then the electronic discourse community is the site of the jam session where the musicians learn music by playing music for and with others, learning from one another, becoming musicians by being musicians.

The Site of the Community

The setting for our electronic community is a windowless former storage room in the basement of the undergraduate library at the University of Texas. This medium-sized room and its adjacent offices at the end of a long corridor deviates from the types of learning spaces usually found in schools of any size. It is a hybrid space. At times it is a classroom where a variety of English classes meet, some on a full-time and others on a part-time basis, including Freshman English, lower-division writing electives, upper-division literature courses, and graduate rhetoric courses (not to mention the occasional foreign language class). But since classes are not scheduled during every hour of the day, there are significant open lab times when students enrolled in the scheduled classes can drop in to use the equipment, complete class activities, or meet with other students to collaborate on projects. Most of the instructors using the facility also hold their office hours in an adjoining room where they meet with students, prepare future classes, read students' drafts on-line, or catch up on class e-mail. On a normal day, the hallway outside these rooms is a bazaar of mingling freshmen, upper-division students, and graduate students. As they mill around waiting for a class to end so they can work on the equipment, they engage in social banter and discussions about the technology, a writing project, or a new idea to be struggled with.

The topography of the community has changed over the last several years as our knowledge about the potential of the center has grown. In its earliest form, the computers sat two to a table. The tables were arranged to approximate a traditional classroom, one complete with chalk board in front. More recently, however, the tables have been pushed around the perimeter of the room, and a large hollow-square conference table dominates the room. Today our networked writing classroom feels more like a writing center than it did in the past because students have the flexibility to turn their chairs to gather around the conference table for large group or face-to-face discussions. Such an arrangement combines the advantages of the traditional classroom (face-to-face interaction) and the networked classroom (electronic, textual interaction).

The physical setting of our electronic discourse community consists of microcomputer hardware, Local Area Network (LAN) hardware and software, and LAN-based instructional software *(The Daedalus Instructional System)*. Each of the 24 workstations includes an IBM PC with a 20 megabyte hard disk; 2 floppy drives; an EGA color monitor; and an IBM Proprinter. All the micros are networked using IBM Token Ring network hardware and Novell Netware network software, with an IBM AT as the network server. A LAN links individual microcomputers together through a network server and permits several options. One is program and file sharing—all users on the network can share a word processor or invention program, for example, and with proper security clearance, can access files from the network server's hard disk. In the writing center, the files consist of reading selections, students' drafts, and final versions of the papers.

Another use is electronic conferencing, of which there are two forms: synchronous and asynchronous. The former, also known as "real-time" communication, permits live, text-based discussions. Such discussions share certain features with oral classroom communication. For instance, participants log onto the program during the same time period (a class meeting) and are given a topic by the teacher (e.g., "Who are the dead of the title of the short story 'The Dead'?") or initiate their own topic ("Hi. My name is Alma, and I think the story stunk."). When participants have something to contribute, they compose a message in a private text editor box (the core of a word processor which permits simple text entry, deleting, and backspacing), edit it, and then send the message to a central transcript that appears on all the other participants' monitors. As in an oral discussion, all contributions are public. Unlike a public discussion, however, turn-taking is not necessary. All participants can be reading and writing at their own pace and contributing simultaneously—the computer system merely appends the contributions to an ever growing public transcript as the messages are received.

The other type of conferencing, asynchronous, is commonly referred to as electronic mail (e-mail), a name that is as metaphorical as it is descriptive. In the particular version of e-mail used in our classes *(Contact)*, a participant composes a message, addresses it to a particular member of the electronic community or to "all," and sends it to a central bulletin board, a place from which other users can access and read the mail. Where in synchronous communication all participants have to be logged onto the network simultaneously, participants in asynchronous conferencing are not bound by spatial or temporal constraints. When students become members of a writing community mediated by the network and electronic mail, they can have access to one another's drafts and critical feedback when it is convenient for them, not just when the class meets. For example, Kenny can write his draft at home on his personal computer and have it done by 4:00 pm on Monday. When he next comes to campus, he can drop by the networked center, log on, copy his draft to the network server, and leave a message on the e-mail system to his writing partner, Alma, informing her of the draft's file name and requesting her to read it and give him some feedback on the draft's structure, argument, and development. (If Kenny has a modem, and the writing center has a bulletin board on the network as ours does, he can do all this from home.) When Alma drops by the center to work on a draft of her paper, she checks her mail messages, sees Kenny's request, reads his draft (a ''read-only'' version she can view but not alter), and uses the e-mail to fire him back a reply. All this drafting, reading, and critiquing, the events of the collaborative classroom, can (and have) occurred outside of class time without the participants meeting face to face.

From Local Area Network to Electronic Discourse Community

The mere presence of a LAN in the writing center, however, does not necessarily nurture the growth of an electronic discourse community. Networks can be used to support nearly any type of pedagogy, even some that have come under criticism recently (Hawisher and Selfe 63). When used primarily for program sharing, LANs can be the backbone of elaborate instructional delivery devices that allow students access to drill and practice tutorials. When used primarily as file sharing systems, LANs can be used as electronic chalk boards with which teachers display samples of prose, errors, style sheets, or nearly any type of text that could otherwise be distributed on ditto or displayed on an overhead transparency. The electronic conferencing capabilities, moreover, can be used to support a product-oriented, teacher-centered classroom in which the teacher controls the electronic discussions by requiring tightly patterned responses. Some network programs allow teachers to monitor student writing processes without the writer's knowledge while others permit the instructor to wrest control of the student's text-in-progress and interject comments as the student composes. Such uses of a LAN, however, may undermine the potential of the technology to transform the classroom from a place of knowledge transmission to a working model of how communities construct knowledge through writing. What is necessary to encourage a truly interactive electronic discourse community is a theoretical base for the use of the LAN.

Contemporary Rhetorical Theory and the Networked Classroom. Recent discussions in rhetoric have focused on the social aspects of language and communities, and these rhetorical theories buttress the electronic discourse community. James Berlin's ''social-epistemic,'' Lester Faigley's ''social view'' and Kenneth Bruffee's ''social contructivism'' are, as Berlin notes, related ''because they share a notion of rhetoric as a political act involving a dialectical interaction engaging the material, the social, and the individual writer, with language as the agency of mediation'' (488). In Berlin's social epistemic rhetoric, ''the real is located in a relationship that involves the dialectical interaction of the observer, the discourse community (social group) in which the observer is functioning, and the material conditions of existence. . . . Most important, this dialectic is grounded in language: the observer, the discourse community, and the material conditions of existence are all verbal constructs'' (488). Faigley writes that social views of language, rhetoric, and literary theory all develop out of the central assumption that ''human language (including writing) can be understood only from the perspective of a society rather than a single individual. . . . The focus of a social view of writing, therefore, is not on how the social situation influences the individual, but on how the individual is a constituent of culture'' (''Competing Theories'' 535). Bruffee, using the term Social Constructionism to describe the theory that rejects Cartesian, foundational notions of knowledge, claims,

> A social constructionist position . . . assumes that entities we normally call reality, knowledge, thought, facts, texts, selves . . . are constructs generated by communities of like-minded peers. Social construction understands reality, knowledge . . . as community-generated and community-maintained linguistic entities—or, more broadly speaking, symbolic entities—that define or ''constitute'' the communities that generate them. (''Social Construction'' 774)

Collaborative Learning Research. Collaborative learning, the classroom practice which attempts to turn the classroom into a meaning-making discourse community, has gained popularity in traditional paper and pencil classrooms, and has had an important influence on the development of electronic discourse communities in the English classroom. The effects of collaborative learning have been studied since the 1920s, with many of the reports concluding that cooperative methods have a greater effect than competitive or individualistic learning on academic achievement and non-cognitive outcomes (see reviews in Johnson, et al; Johnson and Johnson; Slavin). In English and composition research, various experimental, ethnographic, and case studies have concluded that group techniques, when compared with traditional techniques, have greater effect on overall quality of essays, students' understanding of the writing process and the function of audience, and writing apprehension (Bouton and Tutty; Clifford; Hillocks; Snipes, Thie). In his meta-analysis of composition research, George Hillocks defines three dominant teaching modes: presentational, naturalistic, and environmental. The environmental, in which ''the instructor plans and uses activities which result in high levels of student interaction concerning particular problems parallel to those they encounter in certain kinds of writing'' (247), subsumes collaborative learning techniques. When comparing the three modes, Hillocks writes,

> In contrast to the presentational mode, [the environmental] mode places priority on high levels of student involvement. In contrast to natural process, the environmental mode places priority on structured problem-solving activities, with clear objectives, planned to enable students to deal with similar problems in composing. On pre-to-post measures, the environmental mode is over four times more effective than the traditional presentational mode and three times more effective than the natural process mode. (247)

Thomas Barker and Fred Kemp argue that the networked writing classroom, one that capitalizes on the file-sharing and electronic conferencing capabilities of a LAN, places social constructivist theory at the very center of pedagogy, thereby undermining the proscenium style of teaching that places the teacher on the stage and the students in the audience. The proscenium classroom, claim Barker and Kemp, enforces notions of the teacher as sole authority and arbiter of knowledge, a model that is in direct opposition to contemporary learning, pedagogical, and critical theories that view knowledge as a social construct. Writing in the computer networked setting becomes a tool for communication, not simply a product for an evaluator, and therefore, students' texts rather than instructors' evaluations become the focus of class activity. Barker and Kemp's network theory "empowers the student writer and de-neutralizes his sense of text . . . [and] rejects elitist epistemologies which support an 'information transfusion' or teacher-centered model of instruction in favor of social constructivist models which privilege a communal process of knowledge making" (26).

Computers and Writing. Since the electronic discourse community combines various microcomputer technologies to create and transmit knowledge (via word processing, synchronous and asynchronous electronic conferencing, electronic file transfer), the characteristics and effects of each component need to be looked at separately before they can be examined as an integrated environment. Studies examining computers in the English classroom are dominated by investigations of word processing and its effects on writing processes and writing quality. With the advent of networked microcomputers, however, researchers are turning their attention away from the effects of word processing to the dynamics of text-sharing pedagogies and synchronous and asynchronous conferencing.

Previous research on word processing and writing has described students' writing behaviors, has analyzed the content and revisions made on word-processed essays, and has reported the effects of word processing on the overall quality of essays. Although users and instructors continue to intuit the benefits of using word processors, a review of experimental research indicates that there does not seem to be a consensus on the effect of word processors on writing (Daiute; Harris; Hawisher, "Effects"). Gail Hawisher, in a review and critique of nearly 40 published articles and entries on computers and writing research appearing in *Dissertation Abstracts International,* concludes, "If we examine the two variables of revision and quality, contradictory results continue to emerge" ("Research" 7).

Even though the results of word processing studies led (and continue to lead) to contradictory conclusions, instructors have continued using microcomputers in their classrooms, even if they have not had enough for every student to use. The shortage of computers, in fact, has led researchers to examine collaboration on computers. Those who have theorized about and described the use of computers for collaborative learning and writing have concluded that the marriage of word processors and collaborative learning techniques can facilitate the types of social interaction that affect student learning (Dickinson; Skubikowski and Elder; Sudol). After her review of the literature on cooperation in computer settings, Hawisher concludes "one might hypothesize that a spirit of cooperation rather than competitiveness prevails in a classroom with computers" ("Research" 16).

Electronic Conferencing. The early history of electronic conferencing was dominated by wide-area networks (WANs), large main frame supported systems that link long distance participants. Although WANs continue to proliferate, LAN technology has permitted a much less expensive way of linking computers (and users) together. Despite differences in cost and capabilities, many functions (program and file sharing and electronic conferencing) are similar, and the early research on WANs informs LAN applications.

During the early 1970s, WANs were used primarily by research scientists and business people to share information with large networks of peers. Hiltz and Turoff, using the generic phrase *computer conferencing,* examined some of the early uses, practices, and behaviors of adults on such systems. When discussing the unique capabilities of computer conferencing, they note that time and distance barriers are removed, allowing participants to enter ''conversations'' and compose their contributions when it is convenient for them. They also note that group sizes can be expanded without decreasing actual participation, since each person can ''talk'' whenever she wishes without the constraints of turn-taking protocol. Because all participants can contribute simultaneously, no one person can dominate, no one can be interrupted, and multiple leaders are more likely to emerge (106). Furthermore, since it is possible to read much faster than to listen, more total information can be exchanged in a given amount of time (9). Computer conferencing is also less intimate and self-exposing than face-to-face verbal interaction because only participants' words (which can be carefully composed and edited) are transmitted. Hence, other nonverbal cues characteristic of face-to-face communication, such as personal mannerisms and physical handicaps, do not influence the message. ''Thus,'' Hiltz and Turoff theorize, participants ''can feel more free to express disagreements or suggest potentially unpopular ideas. In addition, statements may be considered on their merit rather than by the status of their proponents'' (27). Of synchronous conferencing, they hypothesize that there is a strong tendency toward more equal participation as compared to face-to-face groups, more opinions tend to be asked and offered, there is less explicit agreement or disagreement with the opinions and suggestions of others, and there is a great deal of informal, explicit sociability (123-125).

Hiltz and Turoff's work focused on adult professionals using WANs, but recently numerous researchers and scholars have focused on the perceived benefits and effects of electronic conferencing in the English classroom. Some of these researchers have drawn their conclusions based on observations of their students' efforts on LAN synchronous applications (Barker and Kemp; Batson; Bump; Faigley ''Subverting''; Peyton; Peyton and Batson; Sirc; Thompson, ''Teaching,'' ''Conversational,'' ''Interactive''). In her review of the network literature, Selfe summarizes many of the claims that have been made about networks in writing classrooms: network interaction can encourage students to become active learners, enhance collaborative activity, change the focus of social interaction from who is speaking to what is being said, allow for more egalitarian discourse by eliminating status marking cues of face-to-face interaction, and make reading and writing inherently social.

The Practice: One Freshman and One Advanced Comp Course

To illustrate a good deal of this theory and research, we will take examples from several specific courses exemplifying the use of the network. One class was a course in advanced composition, usually taken by juniors and seniors. Of the twenty students who took the course, five majored in English, five in journalism, three in history, two in business, and one each in economics, art, linguistics, radio-television-film, and biology.

There were five papers: an analysis of the university's newspaper, an analysis of the argumentation techniques of a Nobel prize winner writing a popular article on the origin of life, an evaluation of an object or process in the student's major field, a research paper exploring different positions on a controversial issue in the student's major field, and a persuasive paper on the same controversial subject. All of these kinds of papers had been taught for several years in a traditional classroom.

Two of the papers for the class illustrate the use of the network quite differently and dramatically. The second paper was an analysis of the argumentation techniques of Nobel prize winner George Wald in "The Origin of Life," published originally in 1954 in *Science*. The students were assigned the piece over the weekend and told that they would "discuss" it on the network on Monday. Since the issue was clearly controversial, it was decided to use the network's synchronous discussion program because one of the features of the synchronous discussion program used in this class *(InterChange)* is its ability to accept pseudonyms, a mode which has the tendency to make students more frank in their comments. Students were advised to choose historical or fictional names and to give their reactions to the piece and then to each other. This technique resulted in several interesting differences in the discussions, as compared to discussions in the traditional classroom.

Discussion Differences. The names chosen were interesting; in the order of appearance they were: god, Donald Duck, Gandhi, Darwin, Cleopatra, John Locke, Sally Roe, Big Bird, Daphne du Maurier, B. C., RZ, Clarence Darrow, Cinderella, T. Jefferson, Ptolemaic, Othello, Medusa, Joan of Arc, Yosemite Sam, Phaedrus (the instructor was Darwin but the students did not know this). All the names above as well as the transcripts below are reproduced here exactly as they were typed by the participants themselves. Behind these pseudonyms the students presented a much more candid discussion of the essay than those held by students in the previous seven years of advanced expository writing classes. A few typical initial comments follow:

> *John Locke:* In general, I was overwhelmed with the Wald article. I found nothing (yet) to contend. I guess I am a creationist, but also an evolutionist of sorts. I've always had a problem with evolution because it left the question of how life began unanswered. Wald's theory is the first I've ever read that attempted to answer that problem. I would question some of his scientific analysis—I don't follow his calculations very well (hey, I am a liberal arts major—not math)—and is it true that molecules are moving in a table, and that if they all moved together the table would rise? What about gravity. Anyway, I thought it was a pretty good paper.
>
> *Ptolemaic:* My first impression of Wald was arrogance. "The more rational elements of society, however, tended to make a more naturalistic view of the matter" [creationism versus a scientific view of the origins of life]. Many theologians who have devoted most of their lives to the ministry, received their doctorate degrees, but are committed to creationism I'm sure consider themselves as 'rational'.
>
> *Joan of Arc:* I'm sorry to say I didn't give this paper a chance. I got to paragraph 10

and quit. I guess I'm a creationalist and I guess I have a lot of questions in my own mind, with all the evidence that has been presented so that reading this kind of stuff makes me nervous and uncomfortable.

Othello: I agree with Wald that it is very possible that god in fact created life, but not necessarily in man. We have evidence of some creatures evolving through time, such as the horse. Horses started out as animals no bigger than an average dog today, and now have grown to a size of about 5 times the size of a dog. Based on this idea, man could have begun as a small living organism, growing over the centuries to the size he is today. . . . people are larger today than they were 100 years ago due to scientific research and differences in diet.

Medusa: I agree . . . the earth is not the center of everything. I believe in god and all, but I question the ''Seven Day Creation'' bit. I suppose I believe in the creation scoop because I am comfortable believing in this way. But, Darwin made Ugh Ugh . . . never mind.

What differentiates these responses from those we have had in the traditional classroom is their candor, their depth, and their length. Students are not afraid to reveal their religious backgrounds. They also have more measured reactions to the piece, given the written medium. And the responses are considerably longer than oral responses typical of students in the traditional classroom. The caliber of the discussion continues in the next stage, the replies of students to one another. Here are a few exchanges on a single theme (it should be pointed out that these responses were actually followed by other remarks, but which related to other topics; consequently there is not a strict sense of transitions):

god: Wald has no clue as to the definitions of probability. According to his theory, if we wait around long enough, not just spontaneous generation, but everything, and I mean everything will occur, no matter how slight the probability.

god: The word impossible means ''not possible.'' It does not mean unlikely, nor does it mean improbable. I think this is a breach of all sorts of laws for Wald to have something impossible to occur.

Darwin: Dear god, Not exactly accurate. Only those things which can happen will happen—things following the rules of chemistry and physics.

Donald Duck: He criticizes ''probability,'' yet he makes probability statements on lawyers.

Donald Duck: His probability argument violates the law of large numbers: as the number of events are repeated, the deviation from the probability minimizes, and therefore, flipping a coin 1,000 or one time has a probability of 1/2. This is from an intro to bus stats class.

Clarence Darrow: I thought it a very good argument for the most part. The only problem I had was his notion of time. He presents the proverbial: if a perpetual typewriter is left under an apple tree, falling apples will eventually type the Lord's Prayer. Even given the extremities of geological time I'm sure that's not enough. Wald attempts to go from probabilities of six-sided dice to amino acid chains in the same breath. That doesn't quite wash with me. How about y'all?

John Locke: I agree with Darwin—things will "probably" happen given time and given the laws of physics. Yes, I agree there is probably life out there somewhere—maybe not quite like us—but given time (and what else does the universe really have?) surely life has emerged somewhere else. Or are we so self-centered that life on earth is the only life in the universe? Sounds like Aristotle to me—everything revolves around us.

Sally Roe: Clarence, I agree. I think Wald needs a good shot in the arm. I also couldn't accept his notion of probability because it started to get ridiculous.

John Locke: I think that we are being too self-centered again by saying that his argument of time and probability are "irrational." Are they? Or are we just looking at it from a too-human perspective?

Yosemite Sam: I think people have difficulty relating time as a continuous concept to the brief instant known as their life. I think Wald was trying to overwhelm people with this concept when he spoke of probability. I also think he was trying to speak in a way he thought would be understood, this might have necessarily meant speaking down in 1954.

RZ: I have not read all of Wald's work, but I did find some very interesting things in it. In paragraph four I think he defines the creationists very well, or at least points to them interestingly. His argument on probability reads like the odds given by a pyramid scam artist. He also over-complicates simple organisms and the evolutionary process.

All of these people stayed in this discussion for some time, and Big Bird, Gandhi, Cinderella, T. Jefferson, Daphne du Maurier, and Clarence Darrow also got into the act. There were 26 separate comments on the probability issue (out of 85 comments for the entire fifty minute discussion). This was entirely appropriate since the notion of probability was at the core of Wald's argument.

This strand of the discussion illustrates several additional advantages of the synchronous conferencing: students can narrow in on a specific issue and not be side-tracked by other issues. And there is much more active participation than in many regular classroom discussions. There were 20 participants and 85 responses to the reading. There were 22 printed pages for the interchange, averaging about 250 words a page, so that each participant averaged about 275 words; that represents somewhat more than a page of text. The teacher had about the same average number of responses as the students and certainly did not dominate the discussion. No oral class discussion can come anywhere close to this extent of participation.

These discussions continued in the succeeding classes. This was a long project (with the rough draft—sometimes several drafts—and the final draft); it took slightly under a month, so it would be inappropriate to suggest that the large amount of time spent in the sequel would be representative of typical composition themes.

Invention, Planning, Rough Drafts, and Revision. Let us take a shorter theme, a freshman composition that took about two and a half weeks, to exemplify the methodology after the initial reading exercise. The assignment was to evaluate something, and the assigned chapter in the course rhetoric gave three examples of evaluations and suggested a full range of campus, town, and national objects or procedures which could be evaluated. After an electronic conference session in which they reacted to the chapter, the students were expected to have a tentative topic ready for the second class. In this class period, everyone made a quick oral presentation of the projected topic and students were divided into electronic sub-conference groups of four members each. (Another feature of *InterChange,* the conferencing software used in our classes, is that it permits instructors or participants to create sub-conferences off the main class conference in which participants can break into smaller electronic discussion groups.) They were told to critique the topics of the members of their group for feasibility given the time framework, for breadth and depth, and for audience reception (each paper had to be understood by all of the members of the group). The discussions were lively and useful: each student received four reactions (the teacher shuttled back and forth among the groups). The topics ranged from the evaluation of a surgical technique on brains in mice to learn about brain operations in humans, to the evaluation of a day center, several restaurants, a computer, a specific dormitory, etc. The same technique was applied to the criteria for evaluation proposed for the paper; the groups spoke to their understandibility, to their group acceptance (the audience had to accept the norms), and to the ability of the writer to gather objective evidence for the evaluation. Then a tentative plan (not quite an outline) was sketched on the computer by each member of the cluster groups and reactions to it were given by the three other members and the teacher.

Finally, the first full draft, composed with a word processor, was put on the network and hard copy was presented to each of the group members and to the teacher on a Friday. On Monday, the hard copies were returned to each writer (again with four critiques). The drafts and the accompanying critiques were then discussed in an oral group session. On Wednesday, the groups worked on their themes on the network and critiqued the improvements brought to the paper. Then on Friday, the final draft was copied to the network server. This final step is all important because it is akin to publishing—once the final version is on the network, all members of the community can retrieve and read it at will, much like an electronic library except that the holdings are the writings of the members of the class, as well as members of other classes held in networked labs.

At each stage of this process, therefore, from reading through choice of topic, invention of criteria for evaluation, plan, rough drafts, revision, and final draft there was extensive feedback, occasionally from the whole class, but always from three peers and the teacher. Thus, beginning with the choice of a topic, each student received twenty-one different reader reactions to his or her piece. In addition, whenever relevant, the textbook was brought in to illustrate materials. In effect, the entire invention process was radically shifted to a student-centered and collaborative classroom. This includes the reading and discussion stage, which as outlined above, became more candid, more measured and thoughtful, longer, more directed to single important strands, and involved many more students than in the usual classroom situation. And, perhaps most significantly, the central activity of the writing class is writing.

For both the freshman and the advanced class, the results were superior to anything produced in previous years by similar students—the results, as compared to our previous teaching experiences, were astonishing. In effect, the network gave several significant new dimensions to our whole approach to teaching writing.

The Computer and the Issue of Content vs. Skill. Possibly the most important theoretical issue raised by the use of networked computers in the English classroom has to do with the problem of what the students in writing classes should write about. Granted that we are trying to get students to be able to define, to describe, to prove a point by evidence, to explore an issue, to evaluate, etc. Granted also that we are trying to get students to write mechanically correct expository themes. These are the skills which composition classes traditionally have been concerned with.

But what content should we interface with these skills to produce the best results? Are these skills best taught in writing themes about literature, as some members of the English profession maintain? Others believe that students should choose their own topics—even if they are trivial in our eyes. Still others believe that students are more knowledgeable in their major fields and that writing about such subjects elicits more mature writing.

Discussing this issue in a recent committee meeting about our controversial freshman English course, one colleague argued that one of the major merits of the new syllabus was that it gave students something serious to write about. Given this initial impetus, the concern about the various writing skills (defining, summarizing, exploring, proving, etc.) came naturally in the investigation of the topic. This is the merit of courses in writing across the curriculum, many argue.

Computer programs for the major skills (defining, dividing, describing, evaluating, narrating, proving, exploring, informing, persuading, self-expression) can be written, though they do not yet all exist. If students can be explained the general range of these options and when to use each, then they could start with a subject matter and weave in the proper rhetorical and grammatical skills when necessary. Such an approach, one that is indeed the pedagogical cornerstone of the electronic discourse community, gives priority to subject matter and yet considers the rhetorical and grammatical issues when necessary. It could solve the eternal dilemma of content and form in teaching writing.

Conclusion

We are not alone or the first in our evaluation of this new site of writing instruction, this place for the 1990s. Various writers have attested to the possible and real benefits of such classrooms, including the notion that electronic conferencing serves to bridge some of the gaps between speaking and writing, bringing reading, writing, and speaking closer together than previously believed possible (Faigley, "Subverting"; Peyton and Batson; Peyton; Sirc; Thompson). When discussing the possible advantages of synchronous communication with basic writers, Thompson notes that "the ENFI (Electronic Networks for Interaction) method is . . . promising for basic writers because it offers them a link between their generally adequate oral language skills and their weak, reluctant writing" ("Interactive" 17). Joy Kreeft Peyton and Trent Batson, pioneers in using synchronous communications with deaf students, argue convincingly that "the network provides the conditions now considered crucial for writing development: a close relationship between reading and writing activities; writing with a strong sense of context with a known and present audience; and experiencing peer pressure as well as pressure from the teacher to write well" (6).

Faigley's analysis of real-time conferencing reveals positive alterations in classroom interaction, behavior, and attitudes. Discourse structures change because students participate whenever and however they decide to, and the teacher cannot control classroom discourse in traditional ways. Real-time interaction, furthermore, increases participation by all students, particularly those who normally would not participate in the oral classroom. Because of increased level of participation and changed roles for the teachers and students alike, the traditional classroom pattern of teacher initiation/student response/teacher evaluation (IRE) disintegrates. Faigley writes, "When we create space for students, we allow them to discover that meaning is not fixed but socially reconstituted each time language is used. . . . The play of language in [synchronous communication] subverts authoritarian discourse by showing human discourse is composed of many voices" ("Subverting" 310).

The electronic discourse community is the place where many voices of human discourse can meet, mix, and mingle. It may be a room (with or without windows) housing networked computers in, near, or on the other side of campus from the English department, and whether it is used as a classroom or a writing center (though the distinction will undoubtedly decrease as the distinction between where one learns writing and where one does writing narrows) is not all that crucial. What is important, however, is that the site is used in such a way so that writers and their teachers can, in what Marilyn Cooper and Cynthia Selfe see as the truly revolutionary potential of computers in the teaching and learning of writing, "learn from the clash of discourses, to learn through engaging in discourse" (867).

Works Cited

Barker, Thomas T., and Fred O. Kemp. "Network Theory: A Post-Modern Pedagogy for the Writing Classroom." *Computers in Society: Teaching Composition in the Twenty-First Century*. Ed. Carolyn Handa. Upper Montclair, NJ: Boynton/Cook, 1990. 1-28.

Batson, Trent. "The ENFI Project: A Networked Classroom Approach to Writing Instruction." *Academic Computing* 2.5 (1988): 32-33.

Berlin, James A. "Rhetoric and Ideology in the Writing Class." *College English* 50 (1988): 477-494.

Bouton, Kathleen, and Gary Tutty. "The Effects of Peer-Evaluation of Compositions on Writing Improvement." *The English Record* 26 (1975): 64-67.

Bruffee, Kenneth A. *A Short Course in Writing*. 3rd ed. Boston: Little, 1985.

———. "Social Construction, Language, and the Authority of Knowledge: A Bibliographical Essay." *College English* 48 (1986): 733-790.

Bump, Jerome. "Radical Changes in Class Discussion Using Networked Computers." *Computers and the Humanities* 24 (1990): 49-65.

Clifford, John. "Composing in Stages: The Effects of a Collaborative Pedagogy." *Research in the Teaching of English* 15 (1981): 37-55.

Cooper, Marilyn M., and Cynthia L. Selfe. "Computer Conferences and Learning: Authority, Resistance, and Internally Persuasive Discourse." *College English* 52 (1990): 847-869.

The Daedalus Instructional System. (Including *InterChange* and *Contact*). Computer software. The Daedalus Group, 1988.

Daiute, Colleen. "Physical and Cognitive Factors in Revising: Insights from Studies with Computers." *Research in the Teaching of English* 20 (1986): 141-159.

Dickinson, D. K. "Cooperation, Collaboration, and a Computer: Integrating a Computer into a First-Second Grade Writing Program." *Research in the Teaching of English* 20 (1986): 357-378.

Faigley, Lester. "Competing Theories of Process: A Critique and a Proposal." *College English* 48 (1986): 527-543.

———. "Subverting the Electronic Workbook: Teaching Writing Using Networked Computers." *The Writing Teacher as Researcher: Essays in the Theory of Class-Based Research.* Ed. D. Daiker and M. Morenberg. Portsmouth, NH: Heinemann, 1990. 290-311.

Harris, Jeanette. "Student Writers and Word Processing: A Preliminary Evaluation." *College Composition and Communication* 36 (1985): 323-330.

Harris, Jeanette, and Joyce Kinkead. "From the Editors." *The Writing Center Journal* 8.1 (1987): 1-2.

Hawisher, Gail. E. "The Effects of Word Processing on the Revision Strategies of College Freshmen." *Research in the Teaching of English* 21 (1987): 145-160.

———. "Research Update: Writing and Word Processing." *Computers and Composition* 5.2 (1988): 7-27.

Hawisher, Gale E., and Cynthia Selfe. "The Rhetoric of Technology and the Electronic Writing Class." *College Composition and Communication* 42 (1991): 55-65.

Hillocks, George, Jr. *Research on Written Composition: New Directions for Teaching.* Urbana, IL: NCTE, 1986.

Hiltz, Susan R., and Murray Turoff. *The Network Nation: Human Communication via Computer.* Reading, MA: Addison-Wesley, 1978.

Johnson, David W., and Roger T. Johnson. "Internal Dynamics of Cooperative Learning Groups." *Learning to Cooperate, Cooperating to Learn.* Ed. R. E. Slavin, S. Sharan, and S. Kagen. New York: Plenum, 1985. 103-124.

Johnson, David W., et al. "Effects of Cooperative and Individualistic Goal Structures on Achievement: A Meta-Analysis." *Psychological Bulletin* 89 (1981): 47-62.

Peyton, Joyce K. "Computer Networks for Real-Time Written Interaction in the Writing Classroom: An Annotated Bibliography." *Computers and Composition* 6.3 (1989): 105-122.

Peyton, Joyce K., and Trent Batson. "Computer Networking: Making Connections Between Speech and Writing." *ERIC/CLL News Bulletin* 10.1 (1986): 1-7.

———. "Computer Networking: Providing a Context for Deaf Students to Write Collaboratively." *Teaching English to Deaf and Second Language Students* 6.2 (1988): 19-24.

Selfe, Cynthia. "An Open Letter to Computer Colleagues: Notes from the Margin." Fifth Annual Computers and Writing Conference. Minneapolis, May 1989.

Sirc, Geoffrey M. "Learning to Write on a L.A.N." *T.H.E. Journal* 15.8 (1988): 100-104.

Skubikowski, Kathleen, and Janet Elder. "Word Processing in a Community of Writers." *College Composition and Communication* 38 (1987): 198-201.

Slavin, Robert E. *Cooperative Learning.* New York: Longman, 1983.

Snipes, Wilson. "An Inquiry: Peer Group Teaching in Freshman Writing." *College Composition and Communication* 22 (1971): 169-174.

Sudol, Robert A. "Applied Word Processing: Notes on Authority, Responsibility and Revision in a Workshop Model." *College Composition and Communication* 36 (1985): 331-335.

Thie, Theodora. ''Testing the Efficiency of the Group Method.'' *English Journal* 14 (1925): 134-137.

Thompson, Diane. ''Conversational Networking: Why the Teacher Gets Most of the Lines.'' *Collegiate Microcomputer* 6.3 (1988): 193-201.

———. ''Interactive Networking: Creating Bridges Between Speech, Writing and Composition.'' *Computers and Composition* 5.3 (1988): 2-27.

———. ''Teaching Writing on a Local Area Network.'' *T.H.E. Journal* 15.2 (1987): 92-97.

Trimmer, Joseph F. ''Story Time: All About Writing Centers.'' *Focuses* 1.2 (1988): 27-35.

Additional Sources

JOURNALS

The following journals publish articles of interest to college writing teachers:

College Composition and Communication
College English
Composition Studies (formerly *Freshman English News*)
Computers and Composition
Focuses
Journal of Advanced Composition
Journal of Basic Writing
Journal of Teaching Writing
Rhetoric Review
Teaching English in the Two-Year College
Writing Center Journal
Writing on the Edge
Written Communication

SUGGESTED READINGS

Because the articles in this edition of *The Writing Teacher's Sourcebook* focus on pedagogical theory and practice to a greater extent than did those in previous editions, we have continued that focus in the following list of suggested readings. Not too many years ago, we could have listed all the important books on the college teaching of writing on a page or two. Because that is no longer possible, we have been content to provide a representative sampling of books that we believe will interest teachers, ranging from two classics from the 1970s to three or four interesting titles that have just been announced as we prepare this list.

Bartholomae, David, and Anthony Petrosky. *Facts, Artifacts, and Counterfacts: Theory and Method for a Reading and Writing Course*. Portsmouth, NH: Boynton/Cook, 1986.

Belanoff, Pat, and Marcia Dickson, eds. *Portfolios: Process and Product*. Portsmouth, NH: Boynton/Cook, 1991.

Berlin, James A., and Michael Vivion, eds. *Cultural Studies in the English Classroom*. Portsmouth, NH: Boynton/Cook, 1993.

Berthoff, Ann E. *The Making of Meaning*. Upper Montclair, NJ: Boynton/Cook, 1981.

Bullock, Richard, and John Trimbur, eds. *The Politics of Writing Instruction: Postsecondary*. Portsmouth, NH: Boynton/Cook, 1991.

Caywood, Cynthia L., and Gillian R. Oreering, eds. *Teaching Writing: Pedagogy, Gender, and Equity*. Albany: State U of New York P, 1987.

Note: For a more extensive list of journals, see Chris M. Anson and Hildy Miller, "Journals in Composition: An Update," *College Composition and Communication* 39 (May 1988): 198-216.

Corbett, Edward P. J. *Classical Rhetoric for the Modern Student.* 3rd ed. New York: Oxford UP, 1990.

Donahue, Patricia, and Ellen Quandahl, eds. *Reclaiming Pedagogy: The Rhetoric of the Classroom.* Carbondale: Southern Illinois UP, 1989.

Donovan, Timothy R., and Ben W. McClelland, eds. *Eight Approaches to Teaching Composition.* Urbana, IL: National Council of Teachers of English, 1980.

Elbow, Peter. *Embracing Contraries: Explorations in Learning and Teaching.* New York: Oxford UP, 1986.

———. *Writing Without Teachers.* New York: Oxford UP, 1973.

Faigley, Lester. *Fragments of Rationality: Postmodernity and the Subject of Composition.* Pittsburgh: U of Pittsburgh P, 1992.

Fontaine, Sheryl I., and Susan Hunter, eds. *Writing Ourselves into the Story: Unheard Voices from Composition Studies.* Carbondale: Southern Illinois UP, 1993.

Foster, David. *A Primer for Writing Teachers.* 2nd ed. Portsmouth, NH: Boynton/Cook, 1992.

Gere, Anne Ruggles. *Writing Groups: History, Theory, and Implications.* Studies in Writing and Rhetoric. Carbondale: Southern Illinois UP, 1987.

Graves, Richard L., ed. *Rhetoric and Composition: A Sourcebook for Teachers and Writers.* 3rd ed. Portsmouth, NH: Boynton/Cook, 1990.

Harkin, Patricia, and John Schilb, eds. *Contending with Words: Composition and Rhetoric in a Postmodern Age.* New York: Modern Language Association, 1991.

Harris, Jeanette. *Expressive Discourse.* Dallas: Southern Methodist UP, 1990.

Haswell, Richard H. *Gaining Ground in College Writing: Tales of Development and Interpretation.* Dallas: Southern Methodist UP, 1991.

Hawisher, Gail E., and Paul LeBlanc, eds. *Re-imagining Computers and Composition: Teaching and Research in the Virtual Age.* Portsmouth, NH: Boynton/Cook, 1992.

Hawisher, Gail E., and Cynthia L. Selfe, eds. *Evolving Perspectives on Computers and Composition Studies.* Urbana, IL: National Council of Teachers of English, 1991.

Hill, Carolyn Ericksen. *Writing from the Margins: Power and Pedagogy for Teachers of Composition.* New York: Oxford UP, 1990.

Horner, Winifred Bryan. *Rhetoric in the Classical Tradition.* New York: St. Martin's, 1988.

Hurlbert, C. Mark, and Michael Blitz, eds. *Composition and Resistance.* Portsmouth, NH: Boynton/Cook, 1991.

Knoblauch, C. H., and Lil Brannon. *Rhetorical Traditions and the Teaching of Writing.* Upper Montclair, NJ: Boynton/Cook, 1984.

Lawson, Bruce, Susan Sterr Ryan, and W. Ross Winterowd, eds. *Encountering Student Texts: Interpretive Issues in Reading Student Writing.* Urbana, IL: National Council of Teachers of English, 1990.

Lindemann, Erika. *A Rhetoric for Writing Teachers.* 2nd ed. New York: Oxford UP, 1987.

Lindemann, Erika, and Gary Tate. *An Introduction to Composition Studies.* New York: Oxford UP, 1991.

Lunsford, Andrea, and Lisa Ede. *Singular Texts/Plural Authors: Perspectives on Collaborative Writing.* Carbondale: Southern Illinois UP, 1990.

Lunsford, Andrea, Helene Moglen, and James Slevin, eds. *The Right to Literacy.* New York: Modern Language Association, 1990.

Macrorie, Ken. *Searching Writing.* Upper Montclair, NJ: Boynton/Cook, 1984.

Murray, Donald M. *Learning by Teaching.* Upper Montclair, NJ: Boynton/Cook, 1982.

Newkirk, Thomas, ed. *Nuts and Bolts: A Practical Guide to Teaching College Composition.* Portsmouth, NH: Boynton/Cook, 1993.

Ponsot, Marie, and Rosemary Deen. *Beat Not the Poor Desk.* Upper Montclair, NJ: Boynton/Cook, 1982.

Rose, Mike. *Lives on the Boundary.* New York: Penguin, 1989.
Shaughnessy, Mina. *Errors and Expectations.* New York: Oxford UP, 1977.
Shor, Ira. *Critical Teaching and Everyday Life.* Chicago: U of Chicago P, 1987.
Tobin, Lad. *Writing Relationships: What Really Happens in the Composition Class.* Portsmouth, NH: Boynton/Cook, 1993.

BIBLIOGRAPHIC RESOURCES

Serial Bibliographies

CCCC Bibliography of Composition and Rhetoric. [1987-]. Carbondale: Southern Illinois UP, 1990-. Formerly *Longman Bibliography of Composition and Rhetoric.* [1984-86]. New York: Longman, 1987-88.
Educational Resources Information Center (ERIC). *Resources in Education (RIE)* and *Current Index to Journals in Education (CIJE).* [Bound Volumes: *RIE* 1966-, *CIJE* 1969-]. [Silverplatter CD-ROM Disc 1982-]. [Online Computer Library Center (OCLC) 1966-]. Washington, D.C.: U.S. Department of Health, Education, and Welfare.
MLA International Bibliography of Books and Articles on the Modern Languages and Literatures. [Bound Volumes 1922-]. [Silverplatter CD-ROM Disc 1981-]. [Online Computer Library Center (OCLC) 1963-]. New York: Modern Language Association.

Specialized Bibliographies

Beach, Richard, and Lillian S. Bridwell, eds. *New Directions in Composition Research.* New York: Guilford, 1984.
Bizzell, Patricia, and Bruce Herzberg. *The Bedford Bibliography for Teachers of Writing.* 3rd ed. Boston: Bedford, 1991.
Braddock, Richard, Richard Lloyd-Jones, and Lowell Schoer. *Research in Written Composition.* Champaign, IL: National Council of Teachers of English, 1963.
Cleary, James W., and Frederick W. Haberman, eds. *Rhetoric and Public Address: A Bibliography, 1947-61.* Madison: U of Wisconsin P, 1964.
Cooper, Charles R., and Lee Odell, eds. *Research on Composing: Points of Departure.* Urbana, IL: National Council of Teachers of English, 1978.
Hillocks, George, Jr. *Research on Written Composition: New Directions for Teaching.* Urbana, IL: National Conference on Research in English and ERIC, 1986.
Horner, Winifred Bryan, ed. *Historical Rhetoric: An Annotated Bibliography of Selected Sources in English.* Boston: G. K. Hall, 1980.
———. *The Present State of Scholarship in Historical and Contemporary Rhetoric.* Rev. ed. Columbia: U of Missouri P, 1990.
McClelland, Ben W., and Timothy R. Donovan, eds. *Perspectives on Research and Scholarship in Composition.* New York: Modern Language Association, 1985.
Malton, Ronald J. *Index to Journals in Communication Studies Through 1985.* Annandale, VA: Speech Communication Association, 1987.
Moran, Michael, and Martin J. Jacobi, eds. *Research in Basic Writing: A Bibliographic Sourcebook.* Westport, CT: Greenwood, 1990.
Moran, Michael, and Debra G. Journet, eds. *Research in Technical Communication: A Bibliographic Sourcebook.* Westport, CT: Greenwood, 1985.
Moran, Michael G., and Ronald F. Lunsford, eds. *Research in Composition and Rhetoric: A Bibliographic Sourcebook.* Westport, CT: Greenwood, 1984.
Tate, Gary, ed. *Teaching Composition: Ten Bibliographical Essays.* 1976. Rev. and enlarged. *Teaching Composition: Twelve Bibliographical Essays.* Fort Worth: Texas Christian UP, 1987.